Oberwolfach Seminars

Volume 53

The workshops organized by the *Mathematisches Forschungsinstitut Oberwolfach* are intended to introduce students and young mathematicians to current fields of research. By means of these well-organized seminars, also scientists from other fields will be introduced to new mathematical ideas. The publication of these workshops in the series *Oberwolfach Seminars* (formerly *DMV seminar*) makes the material available to an even larger audience.

Paul Breiding • Kathlén Kohn • Bernd Sturmfels

Metric Algebraic Geometry

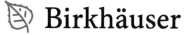 Birkhäuser

Paul Breiding
Universität Osnabrück
Osnabrück, Germany

Kathlén Kohn
KTH Royal Institute of Technology
Stockholm, Sweden

Bernd Sturmfels
Max Planck Institute
for Mathematics in the Sciences
Leipzig, Germany

ISSN 1661-237X   ISSN 2296-5041   (electronic)
Oberwolfach Seminars
ISBN 978-3-031-51461-6   ISBN 978-3-031-51462-3   (eBook)
https://doi.org/10.1007/978-3-031-51462-3

Mathematics Subject Classification (2020): 13P25, 14Q30, 53Z50, 62R01, 90C23

This work was supported by Max-Planck-Institut für Mathematik in den Naturwissenschaften and MPDL.

This book is published under the imprint Birkhäuser, www.birkhauser-science.com by the registered company Springer Nature Switzerland AG
The registered company address is: Gewerbestrasse 11, 6330 Cham, Switzerland

Paper in this product is recyclable.

To Andrea, Jonas, Hyungsook, and our families.

♡

Preface

This book grew out of the lecture notes we developed for the Oberwolfach Seminar on Metric Algebraic Geometry. An Oberwolfach Seminar is essentially a summer school for PhD students and postdocs. Ours was held in the week of May 29 to June 2, 2023. Each of the three lecturers presented five of the 15 chapters. The lectures were supplemented by intense working sessions and inspiring evening discussions.

While developing our material for the school, we were motivated by the following thought. In the early 19th century, there was no difference between algebraic geometry and differential geometry. The two were part of the same subject. Geometers studied natural properties of curves and surfaces in 3-space, such as curvature, singularities, and defining equations. In the 20th century, the threads diverged. The standard curriculum now offers algebraic geometry and differential geometry in rather disconnected courses.

In the present days, geometry plays an important role for data science, and this requires us to rethink the schism bemoaned above. Many applied problems center around metric questions, such as optimization with respect to distances. These require tools from different areas in geometry, algebra and analysis.

To offer a path towards integration, we propose a circle of ideas which we call metric algebraic geometry. This term is a neologism which joins the names metric geometry and algebraic geometry. It first appeared in the title of Madeleine Weinstein's PhD dissertation (UC Berkeley 2021). Building on classical foundations, the field embarks towards a new paradigm that combines concepts from algebraic geometry and differential geometry, with the goal of developing practical tools for the 21st century.

Many problems in the sciences lead to polynomial equations over the real numbers. The solution sets are real algebraic varieties. Understanding distances, volumes and angles – in short, understanding metric properties – of those varieties is important for modeling and analyzing data. Other metric problems in optimization and statistics involve, for instance, minimizing the Euclidean distance from a variety to a given data point. Furthermore, in topological data analysis, computing the homology of a submanifold depends on curvature and on bottlenecks. Related metric questions arise in machine learning, in the geometry of computer vision, in learning varieties from data, and in the study of Voronoi cells.

This book addresses a wide audience of researchers and students, who will find it useful for seminars or self-study. It can serve as the text for a one-semester course at the graduate level. The key prerequisite is a solid foundation in undergraduate mathematics, especially in algebra and geometry. Course work in statistics, computer science, and numerical analysis, as well as experience with mathematical software, are helpful as well.

We hope that you enjoy this book. It will invite you to develop your own perspective, and to share it.

Osnabrück   *Paul Breiding*
Stockholm   *Kathlén Kohn*
Leipzig   *Bernd Sturmfels*
November 2023

About the Authors

Paul Breiding is a professor for mathematical methods in data science at the University of Osnabrück. He is also an Emmy-Noether Research Group Leader and a member of the Academy of Sciences and Literature Mainz. In 2021 he received the Early Career Prize of the SIAM Activity Group on Algebraic Geometry. Paul's interests lie in numerical and random algebraic geometry.

Kathlén Kohn is a tenure-track assistant professor at KTH in Stockholm. Her research investigates the underlying geometry in computer-vision, machine-learning and statistical problems, using algebraic methods. For her research in computer vision, Kathlén received the Best Student Paper Award at the International Conference on Computer Vision (ICCV) in 2019 and the Swedish L'Oréal-Unesco For Women in Science prize in 2023.

After many years at UC Berkeley, **Bernd Sturmfels** now serves as a director at the Max-Planck Institute for Mathematics in the Sciences in Leipzig, Germany, where he leads the Nonlinear Algebra group. He has published 10 books and 300 articles, and he has mentored over 60 doctoral students and countless postdocs. Bernd's interests range from algebraic geometry and combinatorics to statistics, optimization and theoretical physics.

Acknowledgements

The authors of this book wholeheartedly thank the following 21 enthusiastic and hard-working participants of the Oberwolfach Seminar on Metric Algebraic Geometry, held in the week of May 29 to June 2, 2023. Their feedback, both during the Seminar and thereafter, was tremendously important for the development of this book. Without their input this work would not have been possible:

Patience Ablett, Yueqi Cao, Nick Dewaele, Mirte van der Eyden, Luca Fiorindo, Sofia Garzon Mora, Sarah-Tanja Hess, Emil Horobet, Nidhi Kaihnsa, Elzbieta Polak, Hamid Rahkooy, Bernhard Reinke, Andrea Rosana, Felix Rydell, Pierpaola Santarsiero, Victoria Schleis, Svala Sverrisdóttir, Ettore Teixeira Turatti, Máté László Telek, Angelica Marcela Torres Bustos, and Beihui Yuan.

In addition to the participants from Oberwolfach, we obtained feedback from many others. We are very grateful to the following 26 mathematicians for sending us "bug reports" on draft versions of this book:

Viktoriia Borovik, Shelby Cox, Karel Devriendt, Sarah Eggleston, Hannah Friedman, Lukas Gustafsson, Marvin Hahn, Leonie Kayser, Yelena Mandelshtam, Orlando Marigliano, Stefano Mereta, Dmitrii Pavlov, Rosa Preiß, Kemal Rose, Anna-Laura Sattelberger, Elima Shehu, Luca Sodomaco, Monroe Stephenson, Francesca Tombari, Raluca Vlad, Ada Wang, Madeleine Weinstein, Maximilian Wiesmann, Daniel Windisch, Elias Wirth, and Maksym Zubkov.

We thank Chiara Meroni for her important contribution to Chapter 14. The material on SDP hierarchies in Section 14.3 is due to Chiara.

Finally, we are extraordinarily grateful to Thomas Endler for creating the figures in this book. Many thanks also to Madeline Brandt for providing us a draft of Figure 8.1.

Contents

Chapter 1
Historical Snapshot

Throughout this book, we will encounter the interplay of metric concepts with algebraic objects. In classical texts such as Salmon's book [156], metric properties of algebraic varieties were essential. This includes the curvature of algebraic curves and computing their arc lengths and circumscribed areas using integral calculus. Conversely, many curves of interest were defined in terms of distances or angular conditions. This chapter provides an introduction to this classical approach to algebraic curves.

The chapters that follow develop a modern view on *metric algebraic geometry*. This involves algebraic structures in distance minimization (in Chapter 2), optimal transport (in Chapter 5), machine learning (in Chapter 10), and computer vision (in Chapter 13). We will often use the word *model* when talking about an algebraic variety because in these settings algebraic varieties serve as abstract mathematical models for data. In particular, we pose the *variety hypothesis*: information in data can be described by polynomial equations.

Our starting point is typically a *real algebraic variety* X in real affine space $\mathbb{R}^n$. To utilize methods from algebraic geometry, we pass to the Zariski closure of X in complex affine space $\mathbb{C}^n$, or in complex projective space $\mathbb{P}^n$. In most cases, to avoid heavy notation, we use the same symbol for both X and its (projective or affine) complexification. The metric in metric algebraic geometry enters in the form of a notion of distance. The default is the distance induced by the standard Euclidean structure on $\mathbb{R}^n$. Over the real numbers, we use the Euclidean inner product. For $\mathbf{p}, \mathbf{q} \in \mathbb{R}^n$, this inner product is defined as

$$\langle \mathbf{p}, \mathbf{q} \rangle := \mathbf{p}^\top \mathbf{q} = \sum_{i=1}^{n} p_i q_i .$$

This induces the Euclidean norm, which is defined by $\|\mathbf{p}\| := \sqrt{\langle \mathbf{p}, \mathbf{p} \rangle}$. The Euclidean distance function $d(\mathbf{p}, \mathbf{q}) := \|\mathbf{p} - \mathbf{q}\|$ turns $\mathbb{R}^n$ into a metric space. The usual metric in the complex space $\mathbb{C}^n$ is given by the *Hermitian inner product* $\langle \mathbf{p}, \mathbf{q} \rangle_{\mathbb{C}} = \mathbf{p}^* \mathbf{q}$.

However, complex conjugation is not an algebraic operation, and we avoid using it. Instead, throughout this book we consider the *algebraic extension* of the Euclidean inner product and write $\langle \mathbf{p}, \mathbf{q} \rangle = \mathbf{p}^\top \mathbf{q}$ also when $\mathbf{p}, \mathbf{q} \in \mathbb{C}^n$ are complex vectors. This yields a non-degenerate bilinear form on $\mathbb{C}^n$, which we use to identify $\mathbb{C}^n$ with its dual space. In

P. Breiding et al., *Metric Algebraic Geometry*, Oberwolfach Seminars 53,
https://doi.org/10.1007/978-3-031-51462-3_1

this manner, we can consider an algebraic variety and its dual within the same ambient space. We emphasize that, over the complex numbers, the bilinear form $\langle \cdot, \cdot \rangle$ is *not* positive definite, and hence does not induce a metric. For instance, we have $\langle (1, i), (1, i) \rangle = 0$. Nevertheless, using the algebraic extension of the Euclidean inner product, we will be able to study metric problems in $\mathbb{R}^n$ using methods from complex algebraic geometry. In some chapters, we will use other metrics, but this will be spelled out explicitly. For instance, in Chapter 5 we care about distance functions induced by norms whose unit balls are polytopes. However, the standard choice of metric is the Euclidean metric above.

With these 20th-century basics out of the way, we shall now step into the 19th century.

1.1 Polars

Our first historical snapshot concerns polars of algebraic curves in the plane. This will lead us to metric definitions of special families of curves.

For a fixed line L in the real plane $\mathbb{R}^2$ with a distinguished point $\mathbf{o} \in L$, we can choose a positive and a negative direction and define a signed Euclidean distance on L. More concretely, we fix a unit vector $\mathbf{v}$ through $\mathbf{o}$, such that $\mathbf{v}$ spans L. We define

$$\overline{\mathbf{op}} := \lambda \in \mathbb{R}, \quad \text{where } \mathbf{p} = \mathbf{o} + \lambda \mathbf{v} \in L.$$

In particular, this definition depends on the line L, the reference point $\mathbf{o}$, and the chosen direction given by the vector $\mathbf{v}$. The unsigned distance is the Euclidean distance

$$|\overline{\mathbf{op}}| = \|\mathbf{o} - \mathbf{p}\|.$$

According to Salmon [156, §56], the following theorem was first proven by Roger Cotes (1682–1716) in his Harmonia Mensurarum. It is valid both over $\mathbb{R}$ and over $\mathbb{C}$.

Theorem 1.1 (Cotes [51]) *Fix a point $\mathbf{o}$ in the plane and an algebraic curve C of degree d. Consider any line L through the point $\mathbf{o}$ that intersects the curve C in d distinct points $\mathbf{r}_1, \ldots, \mathbf{r}_d$. We denote by $\mathbf{p}_L$ the point on L whose signed distance to $\mathbf{o}$ satisfies*

$$\frac{d}{\overline{\mathbf{op}_L}} = \frac{1}{\overline{\mathbf{or}_1}} + \frac{1}{\overline{\mathbf{or}_2}} + \cdots + \frac{1}{\overline{\mathbf{or}_d}}. \tag{1.1}$$

Then, the following subset of the plane is a straight line:

$$K := \bigcup_{L \text{ is a line through } \mathbf{o}} \{ \mathbf{p}_L \mid \mathbf{p}_L \text{ satisfies Equation (1.1)} \}.$$

Salmon called K the *polar line* of the curve C and the point $\mathbf{o}$. Note that our definition of the polar line made use of the Euclidean metric.

Example 1.2 ($d = 2$) Consider a conic C and a point $\mathbf{o}$ outside of C. Then the polar line is spanned by the two points on C whose tangent line contains $\mathbf{o}$; see Figure 1.1.   ◇

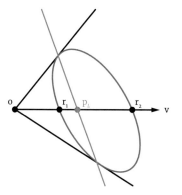

Fig. 1.1: A conic and its polar line (blue) with respect to the point **o**.

The relation (1.1) means that the signed distance from **o** to $\mathbf{p}_L$ is the harmonic mean of the signed distances from **o** to the intersection points $\mathbf{r}_i$. We rewrite this as

$$\sum_{i=1}^{d} \left(\frac{1}{\overline{\mathbf{op}_L}} - \frac{1}{\overline{\mathbf{or}_i}} \right) = 0. \tag{1.2}$$

Suppose now that the point **o** lies on the line at infinity. Then the point $\mathbf{p}_L$ becomes the average of the points $\mathbf{r}_1, \mathbf{r}_2, \ldots, \mathbf{r}_d$. This is proved in the following corollary.

Corollary 1.3 *Let C be a plane curve of degree d and $\mathcal{L}$ a pencil of parallel lines. Let $\mathbf{p}_L$ be the average of all (complex) intersection points of the curve C with some line L in $\mathcal{L}$:*

$$\mathbf{p}_L := \frac{1}{d} \sum_{i=1}^{d} \mathbf{r}_i, \quad \text{where } L \cap C = \{\mathbf{r}_1, \mathbf{r}_2, \ldots, \mathbf{r}_d\}. \tag{1.3}$$

Then the set of all points $\mathbf{p}_L$, as L ranges over $\mathcal{L}$, forms a straight line in the plane.

Proof We start by fixing a line $L \in \mathcal{L}$. We bring the two fractions in Equation (1.2) to a common denominator, we use that $\overline{\mathbf{or}_i} - \overline{\mathbf{op}_L} = \overline{\mathbf{or}_i} + \overline{\mathbf{p}_L\mathbf{o}} = \overline{\mathbf{p}_L\mathbf{r}_i}$, and we multiply the result by $\overline{\mathbf{op}_L}$. In this way, we see that (1.2) is equivalent to

$$\sum_{i=1}^{d} \frac{\overline{\mathbf{p}_L\mathbf{r}_i}}{\overline{\mathbf{or}_i}} = 0. \tag{1.4}$$

In the latter calculation, we kept the same choice of direction on the line L, but we changed the reference point from **o** to $\mathbf{p}_L$. To investigate what happens in the limit when **o** goes to infinity, we fix a point $\mathbf{o}_0$ on the line L such that $\lambda_i := \overline{\mathbf{o}_0\mathbf{r}_i} > 0$. Let **v** be the vector defining the line L and its direction. We can express the points on L using $t \in \mathbb{R}$ as

$$\mathbf{o}_t := \mathbf{o}_0 - t\mathbf{v}.$$

For every fixed $t \geq 0$, we have $\overline{\mathbf{o}_t \mathbf{r}_i} = t + \lambda_i$ and we write $\mathbf{p}_{L,t}$ for the point $\mathbf{p}_L$ that satisfies (1.4) for $\mathbf{o}_t$. Multiplying (1.4) with t, we obtain

$$\sum_{i=1}^{d} \frac{t}{t + \lambda_i} \cdot \overline{\mathbf{p}_{L,t} \mathbf{r}_i} = 0.$$

In the limit $t \to \infty$, the point $\mathbf{p}_{L,t}$ converges to a point $\mathbf{p}_L$. We conclude that $\sum_{i=1}^{d} \overline{\mathbf{p}_L \mathbf{r}_i} = 0$. In words, the point $\mathbf{p}_L$ is the average of the points $\mathbf{r}_1, \mathbf{r}_2, \ldots, \mathbf{r}_d$. For $\mathbf{o}$ at infinity, (1.1) describes the average point $\mathbf{p}_L$ in (1.3), and Theorem 1.1 specializes to Corollary 1.3.   □

Salmon attributes Corollary 1.3 to Newton, who called the resulting straight line the *diameter* of the curve C corresponding to the parallel-lines pencil $\mathcal{L}$ [156, §51]. We point out that Cotes and Newton knew each other. In fact, Cotes edited the second edition of Newton's Principia before its publication.

Corollary 1.3 is important in contemporary applied mathematics. It has been extended to higher-dimensional varieties and is known as the *trace test* in numerical algebraic geometry [161, Chapter 15.5].

Remark 1.4 Here is another delightful theorem on distances of a point to intersection points between a curve and lines. It was first given by Newton in his Enumeratio Linearum Tertii Ordinis: For a plane curve C of degree d, a point $\mathbf{o}$ in the plane, and two distinct lines passing through $\mathbf{o}$, consider the ratio

$$\frac{\overline{\mathbf{or}_1} \cdot \overline{\mathbf{or}_2} \cdots \overline{\mathbf{or}_d}}{\overline{\mathbf{os}_1} \cdot \overline{\mathbf{os}_2} \cdots \overline{\mathbf{os}_d}}, \tag{1.5}$$

where $\mathbf{r}_1, \mathbf{r}_2, \ldots, \mathbf{r}_d$ and $\mathbf{s}_1, \mathbf{s}_2, \ldots, \mathbf{s}_d$ are the intersection points of the two lines with C. The ratio (1.5) is invariant under translating the point $\mathbf{o}$ and the two intersecting lines [156, §46].

In [156, §57], Salmon extends the construction of polar lines to polar curves of higher order. Using notation as in Theorem 1.1, he shows that the locus of points $\mathbf{p}_L$ satisfying

$$\sum_{1 \leq i < j \leq d} \left(\frac{1}{\overline{\mathbf{op}_L}} - \frac{1}{\overline{\mathbf{or}_i}} \right) \left(\frac{1}{\overline{\mathbf{op}_L}} - \frac{1}{\overline{\mathbf{or}_j}} \right) = 0$$

instead of (1.1) is a conic, called the *polar conic* of the curve C and the point $\mathbf{o}$. If $\mathbf{o}$ is at infinity, the polar conic is also referred to as the *diametral conic*. More generally, the locus of points $\mathbf{p}_L$ satisfying

$$\sum_{i_1 < \ldots < i_k} \left(\frac{1}{\overline{\mathbf{op}_L}} - \frac{1}{\overline{\mathbf{or}_{i_1}}} \right) \left(\frac{1}{\overline{\mathbf{op}_L}} - \frac{1}{\overline{\mathbf{or}_{i_2}}} \right) \cdots \left(\frac{1}{\overline{\mathbf{op}_L}} - \frac{1}{\overline{\mathbf{or}_{i_k}}} \right) = 0$$

is the *polar curve* of order k associated with the curve C and the point $\mathbf{o}$ [156, §58]. If $\mathbf{o}$ is at infinity, that polar curve is called the *curvilinear diameter* of order k. In conclusion, the 17th-century definition of polar curves has a metric origin. Measuring distances plays a crucial role in what we have seen so far.

In the 20th-century algebraic geometry literature, on which our Chapter 4 on polar varieties will rest, the definition of polar curves makes no reference to any metric. Namely, polar curves are characterized without involving distances, as follows.

The defining equation of the curve C in homogeneous coordinates equals $f(x, y, z) = 0$, where f has degree $d := \deg C$. For $\mathbf{o} = (a : b : c)$, we define the differential operator

$$\Delta_{\mathbf{o}} := a\frac{\partial}{\partial x} + b\frac{\partial}{\partial y} + c\frac{\partial}{\partial z}. \tag{1.6}$$

Then $\Delta_{\mathbf{o}}^{d-k} f$ is a homogeneous polynomial of degree k in x, y, z. This is the defining equation of the polar curve to C of order k. This was proved by Salmon in [156, §63].

By exploiting symmetry in Taylor series, Salmon also shows the following beautiful duality between polar curves that "may be written at pleasure" [156, §63]. Denoting by $\mathbf{p}$ the point with homogeneous coordinates $\mathbf{p} = (x : y : z)$, Salmon's result states:

$$\frac{1}{(d-k)!}\Delta_{\mathbf{o}}^{d-k} f(x, y, z) = \frac{1}{k!}\Delta_{\mathbf{p}}^{k} f(a, b, c).$$

In particular, the polar curve of order k that is associated with the degree d curve C and the point $\mathbf{o}$ is the locus of all points $\mathbf{p}$ such that the polar curve of order $d - k$ associated with C and $\mathbf{p}$ passes through $\mathbf{o}$.

1.2 Foci

"... we believe that it will be found that every point which has any special relation to any curve will be found either to be a singular point of the curve, or a focus of it"
(George Salmon [156, §125])

An ellipse is the set of points in $\mathbb{R}^2$ whose sum of distances to two fixed points is constant. The two points are the *foci* of the ellipse. In 1832, Plücker generalized the definition of foci to arbitrary plane curves, in the manner described below. Consider a circle in the plane. If we embed the affine plane into the projective plane $\mathbb{P}^2$ by sending $(x, y) \in \mathbb{C}^2$ to $(x : y : 1)$, then every circle passes through the two *circular points at infinity* $(1 : i : 0)$ and $(1 : -i : 0)$. In fact, passing through both circular points characterizes circles.

Lemma 1.5 *An irreducible quadratic curve is a circle if and only if it passes through the two circular points at infinity.*

Proof Let $f(x, y, z) = (x - az)^2 + (y - bz)^2 - r^2 z^2$ be the polynomial defining a circle. Then, $f(1 : \pm i : 0) = 1^2 + (\pm i)^2 = 0$. This shows one direction. For the other direction, let $f = a_1 x^2 + a_2 xy + a_3 y^2 + a_4 xz + a_5 yz + a_6 z^2$ be an arbitrary real quadric. If f passes through the circular points, then $0 = f(1 : i : 0) = a_1 + a_2 i - a_3$ and $0 = f(1 : -i : 0) = a_1 - a_2 i - a_3$. This implies $a_1 = a_3$ and $a_2 = 0$. Hence, f is a circle in the affine plane $\{z = 1\}$. □

The following definition goes back to Plücker [147].

Definition 1.6 Consider a plane curve C. A point $\mathbf{f}$ in the plane is a *focus* of the curve C if both lines spanned by the point $\mathbf{f}$ and the circular points at infinity are tangent to C.

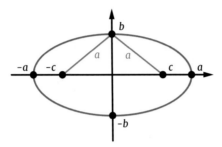

Fig. 1.2: The real foci of an ellipse with width $2a$ and height $2b$ ($a \geq b$) and its axes aligned with the x-axis and y-axis of the real affine plane are $(\pm\sqrt{a^2 - b^2}, 0)$.

Example 1.7 We determine the foci (in the sense of Plücker's Definition 1.6) of an ellipse. Since the condition that a line in $\mathbb{P}^2$ passes through the point $(1 : \pm i : 0)$ is invariant under translations and rotations of the real affine plane $\{z \neq 0\} \subseteq \mathbb{P}^2$, we may translate and rotate the ellipse such that its two defining foci (whose sum of distances is constant along the ellipse) become $(c, 0)$ and $(-c, 0)$ in the real affine plane (with $c \geq 0$). Now the ellipse is the locus of points (x, y) that satisfy, for some constant $a > 0$, the equation

$$\sqrt{(x - c)^2 + y^2} + \sqrt{(x + c)^2 + y^2} = 2a.$$

From this, we see that the two points $(\pm a, 0)$ lie on the ellipse. Moreover, the ellipse intersects the y-axis at $(0, \pm b)$ such that $b^2 = a^2 - c^2$; see Figure 1.2. Hence, the width and height of the ellipse are $2a$ and $2b$, respectively, (with $a \geq b > 0$) and its defining equation can be written as

$$\frac{x^2}{a^2} + \frac{y^2}{b^2} = 1.$$

If the ellipse is not a circle (i.e., $a > b$), then it has two tangent lines that pass through the circular point at infinity $(1 : i : 0)$, namely the tangent lines at $\mathbf{p}_+ := (a^2 : ib^2 : c)$ and $\mathbf{p}_- := (-a^2 : -ib^2 : c)$. The tangent lines at the complex conjugate points $\bar{\mathbf{p}}_+$ and $\bar{\mathbf{p}}_-$ contain the other circular point $(1 : -i : 0)$. The foci à la Plücker are the four points of intersection of the two tangent lines through $(1 : i : 0)$ with the two tangent lines through $(1 : -i : 0)$. Since the tangent lines at $\mathbf{p}_\pm$ and $\bar{\mathbf{p}}_\pm$ are complex conjugates, they meet at a real point, namely $(\pm c : 0 : 1)$. This shows that the real foci we used to define the ellipse are indeed foci in the sense of Plücker's Definition 1.6. The other two foci are obtained by intersecting the tangent line at $\mathbf{p}_\pm$ with the tangent line at $\bar{\mathbf{p}}_\mp$. They are the imaginary points $(0 : \mp ic : 1)$.

If the ellipse is a circle (i.e., $a = b$), then it passes through the two circular points at infinity. The tangent lines at those points are the only ones that pass through the circular points at infinity. They are given by $x \pm iy = 0$ and intersect at the origin $(0 : 0 : 1)$. Thus, in this case, the four foci coincide, and they all lie at the center of the circle.   ◇

In general, the number of foci of an algebraic plane curve depends on its *class*, that is the degree of its dual curve. The *dual projective plane* is the set of lines in the original projective plane $\mathbb{P}^2$. The *dual curve* $C^\vee$ of a plane curve $C \subset \mathbb{P}^2$ is the Zariski closure in the dual projective plane of the set of tangent lines $T_\mathbf{x}C$ at regular points $\mathbf{x}$ of C. Hence, the degree of the dual curve $C^\vee$ (equivalently, the class of C) is the number of tangent lines to C that pass through a generic point in the plane $\mathbb{P}^2$. Using the usual Euclidean inner product $\langle \cdot, \cdot \rangle$, we can view the dual curve as a curve in $\mathbb{P}^2$. More specifically, the dual is

$$C^\vee = \overline{\{\mathbf{y} \mid \text{there is a regular point } \mathbf{x} \in C \text{ with } \langle \mathbf{y}, \mathbf{v} \rangle = 0 \text{ for all } \mathbf{v} \in T_\mathbf{x}C\}} \subset \mathbb{P}^2. \quad (1.7)$$

This means that $C^\vee$ is the Zariski closure of the set of points representing normal lines to C. Here, orthogonality is measured by $\langle \cdot, \cdot \rangle$. If C is defined by the polynomial f and we have a point $\mathbf{x} \in C$, then the normal line of C at $\mathbf{x}$ is spanned by the gradient

$$\nabla f(\mathbf{x}) = \left(\partial f / \partial x_1(\mathbf{x}), \ \partial f / \partial x_2(\mathbf{x}) \right)^\top.$$

Consequently,

$$C^\vee = \overline{\{\nabla f(\mathbf{x}) \mid \mathbf{x} \text{ is a regular point of } C\}}.$$

Suppose now that the curve C has class m. Then, through each of the two circular points at infinity, there are m tangent lines to the curve C. There are m^2 intersection points of these two sets of m lines. Furthermore, if the curve C is real then the intersection points of conjugate pairs of tangent lines are also real. This leads to the following result; see [156, §125]. The case $m = 2$ was featured in Example 1.7.

Proposition 1.8 *A curve C of class m has m^2 complex foci (counted with multiplicity). When the curve C is real, exactly m foci are real.*

In the remainder of this section, we shall examine the foci for additional classes of curves. There are several constructions that generalize ellipses in an obvious way. For instance, an *n-ellipse* is the locus of points in a plane whose sum of distances to n fixed points in the plane is constant. The class of n-ellipses was studied by Tschirnhaus in 1686 [168] and Maxwell in 1846 [129]. As we defined them, n-ellipses are not algebraic, but semialgebraic. In fact, they are boundaries of planar spectrahedra [137].

In what follows, we focus on an alternative generalization. Instead of considering the sum of distances to foci, we consider a *weighted sum* or *product*. This leads us to Cartesian ovals and Cassini ovals, respectively. The *Cartesian oval* is named after Descartes who first studied them in his 1637 *La Géométrie* for their application to optics. This curve is the locus of points in a plane whose weighted sum of distances to two fixed points is constant, i.e., it is the locus of points $\mathbf{p}$ that satisfy the metric condition

$$\|\mathbf{p} - \mathbf{f}_1\| + s\|\mathbf{p} - \mathbf{f}_2\| = r,$$

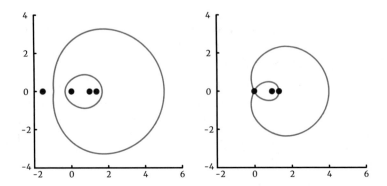

Fig. 1.3: Cartesian ovals with defining foci $\mathbf{f}_1 = (0,0), \mathbf{f}_2 = (1,0)$ and weight $s = 2$ are described by the equation $(-3(x^2 + y^2) + 8x + r^2 - 4)^2 - 4r^2(x^2 + y^2) = 0$. The left picture shows the Cartesian oval for $r = 3$ and the right picture shows it for $r = 2$. The foci are the red points. In both cases, the focus at $(4/3, 0)$ has order three. In the right picture, the focus at $(0, 0)$ has order two. This gives six real foci in total.

where $\mathbf{f}_1$ and $\mathbf{f}_2$ are fixed points in the plane and s, r are fixed constants. The Cartesian oval satisfies an equation of degree four. Hence the Zariski closure of the Cartesian oval is a quartic curve. The real part of this quartic curve consists of two nested ovals; see Figure 1.3. Of the four equations

$$\|\mathbf{p} - \mathbf{f}_1\| \pm s\|\mathbf{p} - \mathbf{f}_2\| = \pm r, \tag{1.8}$$

exactly two have real solutions and those describe the ovals. Salmon shows that a quartic curve is a Cartesian oval if and only if it has cusps at the two circular points at infinity; see [156, §129]. By Plücker's formula [156, §72], it follows that the class of such a quartic is six (except in degenerate cases). Salmon determines the six real foci of Cartesian oval quartic curves [156, §129] (see also Basset [13, §273]): Three of them form a triple focus which is located at the intersection of the cusps' tangent lines. The remaining three foci lie on a straight line. Two of them are the points $\mathbf{f}_1$ and $\mathbf{f}_2$ that define the curve in (1.8). It was already observed by Chasles in [45, Note XXI] that any two of the three single foci can be used to define the Cartesian oval by an equation of the form (1.8). If two of the three single foci come together, the Cartesian oval degenerates to a *Limaçon of Pascal*; see Figure 1.3. Salmon further observes another interesting metric property of foci: whenever a line meets a Cartesian oval in four points, the sum of the four distances from any of the three single foci is constant [156, §218].

A *Cassini oval* (named after Cassini who studied them in 1693 [40]) is the locus of points in a plane whose product of distances to two fixed points $\mathbf{f}_1$ and $\mathbf{f}_2$ is a fixed constant r. Hence the Cassini oval is an algebraic curve defined by the real quartic polynomial

$$\|\mathbf{p} - \mathbf{f}_1\|^2 \cdot \|\mathbf{p} - \mathbf{f}_2\|^2 = r^2.$$

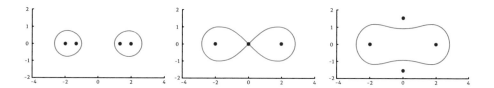

Fig. 1.4: The Cassini ovals in Equation (1.9) with defining foci $f_1 = (-2, 0)$ and $f_2 = (2, 0)$. All foci are shown in red. The pictures display the cases $r = 3$ (left), $r = 4$ (middle) and $r = 5$ (right). The outer foci on the x-axis all have multiplicity 3. The middle focus in the second picture has multiplicity 2.

The circular points at infinity are double points on each Cassini oval. Since these are typically the only singularities, the class of a Cassini oval is eight in general [156, §219].

Basset [13, §247] explains that the two pairs of complex conjugate tangent lines at the two nodes intersect at f_1 and f_2, respectively. Hence, those points are foci in the sense of Plücker's Definition 1.6. Moreover, Basset shows that each of them is a triple focus (the reason being that the nodal tangents are stationary). To describe the remaining two real foci, we translate and rotate (as in Example 1.7) such that $f_1 = (c, 0)$ and $f_2 = (-c, 0)$. Then, the Cassini oval is defined by the quartic equation

$$\big((x - c)^2 + y^2\big)\big((x + c)^2 + y^2\big) = r^2. \tag{1.9}$$

If $r < c^2$, then the real locus consists of two ovals. Otherwise, the real locus is connected, where the degenerate case $r = c^2$ is the *lemniscate of Bernoulli*; see Figure 1.4. In the case of two ovals, the remaining two real foci also lie on the x-axis; they are $(\pm\frac{1}{c}\sqrt{c^4 - r^2}, 0)$. If $r > c^2$, the two foci are $(0, \pm\frac{1}{c}\sqrt{r^2 - c^4})$ and lie on the y-axis. In the degenerate case $r = c^2$, they become a double focus at the origin.

In conclusion, the foci of a plane curve are landmark points of a metric origin. Starting from these foci, one obtains interesting classical curves, like Cassini ovals and Cartesian ovals. These curves are gems in the repertoire of metric algebraic geometry.

1.3 Envelopes

Given a one-dimensional algebraic family of lines in the plane, its *envelope* is a curve such that each of the given lines is tangent to the curve. We can view the family of lines as an algebraic curve $\mathcal{L}$ in the dual projective plane. The envelope is the dual curve $\mathcal{L}^\vee$, which now lives in the primal projective plane. This is a first instance of the biduality, which we shall see more generally in Equation (2.15) and Theorem 4.11.

Example 1.9 The diameters of a plane curve C form a one-dimensional family of lines. Indeed, there is one such diameter for each point on the line at infinity; cf. Corollary 1.3. For a cubic curve C, the envelope of that family is the locus consisting of the centers

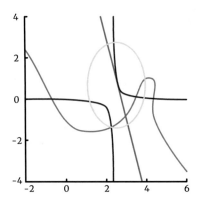

Fig. 1.5: The cubic curve $x^3 + y^3 + 5x^2 - 2xy + x - 1 = 0$ (blue) and the envelope $9xy + 15y - 1 = 0$ (red) of its diameters. The diameter (green) and diametral conic (yellow) are associated with the point $(2 : 1 : 0)$ at infinity. Their equations are given in Example 1.9.

of the diametral conics of C [156, §160]. For instance, for the cubic curve defined by $x^3 + y^3 + 5x^2 - 2xy + x - 1 = 0$, the diameter associated with the point $(a : b : 0)$ is the line

$$3a^2x + 3b^2y + 5a^2 - 2ab = 0.$$

The family of all diameters is the curve $\mathcal{L}$ defined by $25x^2 - 4xy - 30x + 9 = 0$. Its dual curve $\mathcal{L}^\vee$, that is the envelope of the diameters, is $9xy + 15y - 1 = 0$. The diametral conic associated with $(a : b : 0)$ is given by the equation

$$3a\left(x + \frac{5a - b}{3a}\right)^2 + 3b\left(y - \frac{a}{3b}\right)^2 + a - \frac{(5a - b)^2}{3a} - \frac{a^2}{3b} = 0.$$

Its center $(\frac{-5a+b}{3a}, \frac{a}{3b})$ lies on the envelope. In fact, the tangent line of the envelope at that point is the diameter associated with $(a : b : 0)$; see Figure 1.5.   ◇

Evolutes and caustics are instances of envelopes defined by metric properties. The *evolute* of a plane curve C is the envelope of its normals (i.e., the lines orthogonal to its tangent lines). Equivalently, the evolute is the locus of the centers of curvature (see Proposition 6.2). The study of evolutes goes back to Apollonius (ca. 200 BC); see [172].

A recent study of evolutes can be found in [146]. The degree of the evolute of a general smooth curve of degree d is $3d(d - 1)$ (see Corollary 6.9). The class of this evolute equals d^2, by [156, §116]. Moreover, the evolute of the general degree d curve has $\frac{d}{2}(3d - 5)(3d^2 - d - 6)$ double points, $3d(2d - 3)$ cusps, and it has no other singularities.

The evolute of the Cartesian oval in Figure 1.3 is illustrated in Figure 1.6. Its equation is

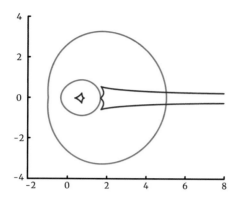

Fig. 1.6: The Cartesian oval from Figure 1.3 (left) together with its evolute.

$$102036672x^{10}y^2 + 433655856x^8y^4 + 833407380x^6y^6 + 917059401x^4y^8$$
$$+ 558336726x^2y^{10} + 143065521y^{12} - 884317824x^9y^2 - 3106029888x^7y^4$$
$$- 4898885832x^5y^6 - 4008450240x^3y^8 - 1331276472xy^{10} + 3251316672x^8y^2$$
$$+ 9515584512x^6y^4 + 12088352844x^4y^6 + 6432939486x^2y^8 + 620191890y^{10}$$
$$- 40310784x^9 - 6758774784x^7y^2 - 16647933888x^5y^4 - 15962551632x^3y^6$$
$$- 4237194240xy^8 + 342641664x^8 + 9145229184x^6y^2 + 18728830368x^4y^4$$
$$+ 11743648812x^2y^6 + 961612425y^8 - 1239556608x^7 - 9234062208x^5y^2$$
$$- 14497919136x^3y^4 - 4640798304xy^6 + 2495722752x^6 + 8064660672x^4y^2$$
$$+ 8003654064x^2y^4 + 835700656y^6 - 3071831040x^5 - 6288399360x^3y^2$$
$$- 2974296960xy^4 + 2390342400x^4 + 3772699200x^2y^2 + 540271200y^4$$
$$- 1173312000x^3 - 1396800000xy^2 + 349920000x^2 + 228000000y^2$$
$$- 57600000x + 4000000 = 0.$$

We will see in Proposition 6.5 that the finite cusps of the evolute correspond to the points of critical curvature of the curve C. Salmon computes the length of an arc of the evolute as "the difference of the radii of curvature at its extremities" [156, §115]. The converse operation to computing the evolute is finding an *involute*; that is, for a given plane curve C, find a curve whose evolute is C. Involutes of an algebraic curve are typically not unique and they might not be algebraic. For instance, the involute of a circle is a transcendental curve [156, §235]. The nonuniqueness of involutes can be seen from the offset curves in Section 7.2. Any two offset curves of a given curve have the same evolute. Finally, we note that the foci of a plane curve are also foci of its evolute and its involutes [156, §127].

We conclude with caustics of plane curves. These come in two flavors. Let us imagine that a fixed point **r** in the plane emits light. The light rays are reflected at each point of a given plane curve C. The *caustic by reflection* is the envelope of the family of reflected rays. Similarly, the *caustic by refraction* is the envelope of the family of refracted rays. At a point **p** of the given curve C, the refracted ray is defined as follows: If $\sphericalangle_1$ is the angle

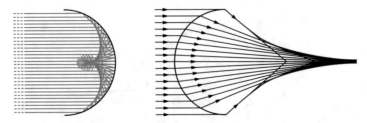

Fig. 1.7: Caustics by reflection (left) and refraction (right) of a circle with light source at infinity.

that the light ray from the radiant point $\mathbf{r}$ to $\mathbf{p}$ makes with the normal of C at $\mathbf{p}$ and $\sphericalangle_2$ is the angle between the refracted ray and the same normal, then the ratio $\frac{\sin \sphericalangle_1}{\sin \sphericalangle_2}$ is a constant independent of $\mathbf{p}$, called *refraction constant*. Figure 1.7 shows both caustics for a circle where the radiant point is at infinity. Those curves can be commonly observed in real life, e.g., when the sun shines on a round glass.

Example 1.10 Caustics by refraction of circles are evolutes of Cartesian ovals. This result is due to Quetelet [156, §124]. For instance, the evolute of the Cartesian oval in Figures 1.3 (left) and 1.6 is the caustic by refraction of the circle with radius $\frac{2}{3}$ which is centered at the triple focus $(\frac{4}{3}, 0)$. Here, the radiant point is $\mathbf{f}_1 = (0, 0)$ and the refraction constant is $\frac{2}{3}$. ◇

Chapter 2
Critical Equations

We consider a model X that is given as the zero set in $\mathbb{R}^n$ of a collection $\{f_1, \ldots, f_k\}$ of polynomials in n unknowns $x_1, \ldots, x_n$. Thus, X is a real algebraic variety. In algebraic geometry, it is preferable to work with the complex algebraic variety defined by the same polynomials. This is the Zariski closure, and we will use the same symbol: $X \subset \mathbb{C}^n$. We assume that X is irreducible, that $I_X = \langle f_1, \ldots, f_k \rangle$ is its prime ideal, and that the set of nonsingular real points is Zariski dense in X. The $k \times n$ Jacobian matrix $\mathcal{J} = (\partial f_i / \partial x_j)$ has rank at most c at any point $\mathbf{x} \in X$, where $c = \mathrm{codim}(X)$. The point $\mathbf{x}$ is *nonsingular* on X if the rank is exactly c. The variety X is called *smooth* if all its points are nonsingular. Elaborations on these hypotheses are found in many textbooks, including [133, Chapter 2].

The following optimization problem arises in many applications, and we shall revisit it again and again throughout this book. Given a data point $\mathbf{u} \in \mathbb{R}^n \backslash X$, compute the distance to the model X. Thus, we seek a point $\mathbf{x}^*$ in X that is closest to $\mathbf{u}$. The answer depends on the chosen metric. We focus on *least-squares problems*, where the metric is a Euclidean metric. Furthermore, we assume that the optimal point $\mathbf{x}^*$ is smooth on X. One can compute $\mathbf{x}^*$ by solving the *critical equations*. In optimization, these are known as first-order conditions or KKT equations, and they arise from introducing Lagrange multipliers. We seek to compute all complex solutions to the critical equations. The set of these *critical points* is typically finite, and it includes all local maxima, all local minima, and all saddle points.

2.1 Euclidean Distance Degree

We begin by discussing the *Euclidean distance (ED) problem*, which is as follows:

$$\text{minimize } \|\mathbf{x} - \mathbf{u}\|^2 = \sum_{i=1}^{n} (x_i - u_i)^2 \text{ subject to } \mathbf{x} = (x_1, \ldots, x_n) \in X. \qquad (2.1)$$

This optimization problem is defined over the real locus of X. We study it by the strategy that we announced above: we first derive the critical equations for (2.1), and then study

© The Author(s) 2024
P. Breiding et al., *Metric Algebraic Geometry*, Oberwolfach Seminars 53,
https://doi.org/10.1007/978-3-031-51462-3_2

them over the complex variety X. Thus, we interpret the optimization problem (2.1) as a problem of solving polynomial equations over $\mathbb{C}$.

The *augmented Jacobian matrix* $\mathcal{AJ}$ is the $(k+1) \times n$ matrix which is obtained by placing the row vector $(x_1 - u_1, \ldots, x_n - u_n)$ atop the Jacobian matrix $\mathcal{J}$. We form the ideal generated by the $(c+1) \times (c+1)$ minors of $\mathcal{AJ}$, we add the ideal of the model I_X, and we then saturate that sum by the ideal of $c \times c$ minors of $\mathcal{J}$. See [60, Equation (2.1)]. The result is the *critical ideal* $C_{X,\mathbf{u}}$ of the model X with respect to the data point $\mathbf{u}$.

Example 2.1 (Plane curves) Let X be the plane curve defined by a polynomial $f(x_1, x_2)$. We wish to compute the Euclidean distance from X to a given point $\mathbf{u} = (u_1, u_2) \in \mathbb{R}^2$. To this end, we form the augmented Jacobian matrix. This matrix is square of size 2×2:

$$\mathcal{AJ} \;=\; \begin{bmatrix} x_1 - u_1 & x_2 - u_2 \\ \partial f / \partial x_1 & \partial f / \partial x_2 \end{bmatrix}. \tag{2.2}$$

We get the critical ideal from f and the determinant of $\mathcal{AJ}$ by performing a saturation step:

$$C_{X,\mathbf{u}} \;=\; \langle\, f, \det(\mathcal{AJ}) \,\rangle \;:\; \langle\, \partial f / \partial x_1, \partial f / \partial x_2 \,\rangle^\infty. \tag{2.3}$$

The ideal $C_{X,\mathbf{u}}$ lives in $\mathbb{R}[x_1, x_2]$. See [52, Section 4.4] for saturation and other ideal operations. Frequently, the coefficients of f and the coordinates of $\mathbf{u}$ are rational numbers, and we can perform the computation purely symbolically in $\mathbb{Q}[x_1, x_2]$. The saturation step in (2.3) removes points that are singular on the curve $X = V(f)$. If X is smooth then saturation is unnecessary, and we simply have $C_{X,\mathbf{u}} = \langle\, f, \det(\mathcal{AJ}) \,\rangle$. On the other hand, if X has singular points, then we must saturate. For a concrete example take the cardioid

$$f \;=\; (x_1^2 + x_2^2 + x_2)^2 - (x_1^2 + x_2^2), \tag{2.4}$$

and fix a random point $\mathbf{u} = (u_1, u_2)$. The ideal $\langle\, f, \det(\mathcal{AJ}) \,\rangle$ is the intersection of $C_{X,\mathbf{u}}$ and an $\langle x_1, x_2 \rangle$-primary ideal of multiplicity 3. The critical ideal $C_{X,\mathbf{u}}$ has three distinct complex zeros. We can express their coordinates in radicals in the given numbers u_1, u_2.

The following code computes the degree of the critical ideal $C_{X,\mathbf{u}}$ of the cardioid X for the data point $\mathbf{u} = (2, 1)$. The code is written for the software Macaulay2 [73].

```
R = QQ[x1, x2];
u1 = 2; u2 = 1;
f = (x1^2 + x2^2 + x2)^2 - (x1^2 + x2^2);
AJ = matrix {{x1- u1, x2 - u2}, {diff(x1, f), diff(x2, f)}};
I = ideal {f, det AJ};
C = saturate(I, ideal {diff(x1, f), diff(x2, f)});
degree C
```

Saturation of ideals is discussed in Chapter 3. See (3.1) for the definition of saturation. $\diamond$

The variety $V(C_{X,\mathbf{u}})$ is the set of complex critical points for the ED problem (2.1). For a random data point $\mathbf{u}$, this variety is a finite subset of $\mathbb{C}^n$, and it contains the optimal solution $\mathbf{x}^*$, provided the latter is attained at a smooth point of X. It was proved in [60] that the number of critical points, i.e. the cardinality of the variety $V(C_{X,\mathbf{u}})$, is independent of $\mathbf{u}$, provided we assume that the data point $\mathbf{u}$ is sufficiently general. See also Example 3.20.

Definition 2.2 The *Euclidean distance degree* (ED degree) of the variety X is the number of complex points in the variety $V(C_{X,\mathbf{u}})$, for generic data points $\mathbf{u}$. We write this as

$$\mathrm{EDdegree}(X) := \#V(C_{X,\mathbf{u}}).$$

In Example 2.1 we examined a quartic curve whose ED degree equals 3. The ED degree of a variety X measures the difficulty of solving the ED problem (2.1) using exact algebraic methods. The ED degree is an important complexity measure in metric algebraic geometry.

Example 2.3 (Space curves) Fix $n = 3$ and let X be the curve in $\mathbb{R}^3$ defined by two general polynomials f_1 and f_2 of degrees d_1 and d_2 in three unknowns x_1, x_2, x_3. The augmented Jacobian matrix has format 3×3, and we compute it as follows:

$$\mathcal{AJ} = \begin{bmatrix} x_1 - u_1 & x_2 - u_2 & x_3 - u_3 \\ \partial f_1/\partial x_1 & \partial f_1/\partial x_2 & \partial f_1/\partial x_3 \\ \partial f_2/\partial x_1 & \partial f_2/\partial x_2 & \partial f_2/\partial x_3 \end{bmatrix}. \tag{2.5}$$

Fix a general data vector $\mathbf{u} \in \mathbb{R}^3$. Then the critical ideal equals $C_{X,\mathbf{u}} = \langle f_1, f_2, \det(\mathcal{AJ}) \rangle$. Hence, the set of critical points is the intersection of three surfaces. These surfaces have degrees d_1, d_2 and $d_1 + d_2 - 1$. By Bézout's Theorem [133, Theorem 2.16], the expected number of complex solutions to the critical equations is the product of these degrees. Hence, the ED degree of the curve X equals $d_1 d_2 (d_1 + d_2 - 1)$.

The same formula can be derived from a formula for general curves in terms of algebraic geometry data. Let X be a general smooth curve of degree d and genus g in any ambient space $\mathbb{R}^n$. By [60, Corollary 5.9], we have $\mathrm{EDdegree}(X) = 3d + 2g - 2$. Our curve in 3-space has degree $d = d_1 d_2$ and genus $g = d_1^2 d_2/2 + d_1 d_2^2/2 - 2d_1 d_2 + 1$. We conclude that

$$\mathrm{EDdegree}(X) = 3d + 2g - 2 = d_1 d_2 (d_1 + d_2 - 1).$$

This formula also covers the case of plane curves (cf. Example 2.1). Namely, if we set $d_1 = d$ and $d_2 = 1$ then we see that a general plane curve X of degree d has $\mathrm{EDdegree}(X) = d^2$. In particular, a general plane quartic has ED degree 16. However, that number can drop a lot for curves that are special. For the cardioid in (2.4) the ED degree drops from 16 to 3.◇

Here is a general upper bound on the ED degree in terms of the given polynomials.

Proposition 2.4 *Let X be a variety of codimension c in $\mathbb{R}^n$ whose ideal I_X is generated by polynomials $f_1, f_2, \ldots, f_c, \ldots, f_k$ of degrees $d_1 \geq d_2 \geq \cdots \geq d_c \geq \cdots \geq d_k$. Then*

$$\mathrm{EDdegree}(X) \leq d_1 d_2 \cdots d_c \cdot \sum_{i_1 + i_2 + \cdots + i_c \leq n-c} (d_1 - 1)^{i_1} (d_2 - 1)^{i_2} \cdots (d_c - 1)^{i_c}. \tag{2.6}$$

Equality holds when X is a generic complete intersection of codimension c (hence $c = k$).

Proof This appears in [60, Proposition 2.6]. We can derive it as follows. Bézout's Theorem ensures that the degree of the variety X is at most $d_1 d_2 \cdots d_c$. The entries in the ith row of the matrix $\mathcal{AJ}$ are polynomials of degrees $d_{i-1} - 1$. The degree of the variety of $(c + 1) \times (c + 1)$ minors of $\mathcal{AJ}$ is at most the sum in (2.6). This follows from the

Giambelli–Thom–Porteous formula, which expresses the degree of a determinantal variety in terms of symmetric functions. The intersection of that determinantal variety with X is our set of critical points. The cardinality of that set is at most the product of the two degrees. Generically, that intersection is a complete intersection and equality holds in (2.6).   □

Formulas and bounds for the ED degree are important when studying exact solutions to the optimization problem (2.1). The paradigm is to compute all complex critical points, by either symbolic or numerical methods (cf. Chapter 3), and to then extract one's favorite real solutions among these. This reveals all local minima in (2.1). The ED degree is an upper bound on the number of real critical points, but this bound is generally not tight.

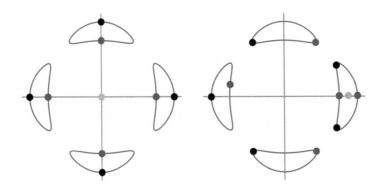

Fig. 2.1: ED problems on the Trott curve: configurations of eight (left) or ten (right) critical real points. Data points are yellow, local minimal are green, and local maxima are red. The coordinates of the critical points are computed by solving the critical equations in (2.8).

Example 2.5 Consider the case $n = 2$, $c = 1$, $d_1 = 4$ in Proposition 2.4, where X is a generic quartic curve in the plane $\mathbb{R}^2$. The number of complex critical points is EDdegree$(X) = 16$. But, they cannot be all real. For an illustration, consider the *Trott curve* $X = V(f)$, given by

$$f = 144(x_1^4 + x_2^4) - 225(x_1^2 + x_2^2) + 350x_1^2x_2^2 + 81. \tag{2.7}$$

This curve is shown in Figure 2.1. Given data $\mathbf{u} = (u_1, u_2)$ in $\mathbb{R}^2$, the critical equations are

$$(x_1 - u_1)\frac{\partial f}{\partial x_2} - (x_2 - u_2)\frac{\partial f}{\partial x_1} = 0 = f. \tag{2.8}$$

Assuming $\mathbf{u}$ to be a general point, these two quartic equations have 16 distinct complex solutions, and these are all critical points in X. Since the Trott curve is smooth, the saturation step in (2.3) is not needed when computing the ideal $C_{X,\mathbf{u}}$.

The ED degree 16 is an upper bound for the number of real critical points of the optimization problem (2.1) on the Trott curve X for any data point $\mathbf{u}$. The actual number of real critical points depends heavily on the specific location of $\mathbf{u}$. For data $\mathbf{u}$ near the origin, eight of the 16 points in $V(C_{X,\mathbf{u}})$ are real. For $\mathbf{u} = (\frac{7}{8}, \frac{1}{100})$, which is inside the rightmost oval, there are 10 real critical points. The two scenarios are shown in Figure 2.1. Local minima are green, while local maxima are red. Finally, consider $\mathbf{u} = (2, \frac{1}{100})$, which lies to the right of the rightmost oval. Here, the number of real critical points is 12. ◇

In general, our task is to compute the complex zeros of the critical ideal $C_{X,\mathbf{u}}$. Algorithms for this computation can be either symbolic or numerical. Symbolic methods usually rest on the construction of a Gröbner basis, to be followed by a floating point computation to extract the solutions. In recent years, numerical methods have become popular. These are based on homotopy continuation. We explain how homotopy continuation works in Chapter 3. That chapter will be devoted to *computations*.

Example 2.6 Let us revisit the cardioid from Example 2.1. For $\mathbf{u} = (2, 1)$ we compute the three critical points using the Julia [20] software HomotopyContinuation.jl [31].

```
using HomotopyContinuation, LinearAlgebra
@var x1 x2
u1 = 2; u2 = 1;
f = (x1^2 + x2^2 + x2)^2 - (x1^2 + x2^2);
AJ = [[x1 - u1; x2 - u2] differentiate(f, [x1; x2])];
solve([f; det(AJ)])
```
◇

We next illustrate current capabilities.

Example 2.7 Suppose X is defined by $c = k = 3$ random polynomials in $n = 7$ variables, of degrees d_1, d_2, d_3. The table below lists the ED degree in each case, and the times used by HomotopyContinuation.jl to compute and certify all critical points in $\mathbb{C}^7$.

$d_1\,d_2\,d_3$	3 2 2	3 3 2	3 3 3	4 2 2	4 3 2	4 3 3	4 4 2	4 4 3
EDdegree	1188	3618	9477	4176	10152	23220	23392	49872
Solve (sec)	3.849	21.06	61.51	31.51	103.5	280.0	351.5	859.3
Certify (sec)	0.390	1.549	4.653	2.762	7.591	17.16	21.65	50.07

These ED degrees can be checked with Proposition 2.4. In our computation, the critical ideal $C_{X,\mathbf{u}}$ represents a system of 10 equations in 10 variables. In addition to the three equations $f_1 = f_2 = f_3 = 0$ in the 7 variables $x_1, \ldots, x_7$, we take the seven equations

$$(1, y_1, y_2, y_3) \cdot \mathcal{AJ} = 0.$$

In these equations, y_1, y_2, y_3 are new variables. The three additional equations ensure that the 4×7 matrix $\mathcal{AJ}$ has rank ≤ 3. This formulation avoids the listing of all $\binom{7}{4} = 35$ maximal minors. It is the preferred representation of determinantal varieties in the setting of numerical algebraic geometry.

The timings above refer to computing all complex solutions to the system of 10 equations in 10 variables. They include the certification step, as described in [30], which proves correctness and completeness. These computations were performed with the software HomotopyContinuation.jl [31] on a 16 GB MacBook Pro with an Intel Core i7 processor working at 2.6 GHz. They suggest that our critical equations can be solved fast and reliably, with proof of correctness, when the ED degree is less than 50000. ◇

We now introduce an alternative version of the ED degree by using a different metric. We fix a weight vector $\Lambda = (\lambda_1, \ldots, \lambda_n) \in \mathbb{R}^n$ whose entries are positive real numbers. The *weighted Euclidean metric* is

$$\|\mathbf{x}\|_\Lambda^2 := \sum_{i=1}^n \lambda_i \, \mathbf{x}_i^2.$$

If Λ is the all-one vector, then we obtain the usual Euclidean metric. For arbitrary Λ, we consider the analogue of the optimization problem (2.1):

$$\text{minimize } \|\mathbf{x} - \mathbf{u}\|_\Lambda^2 = \sum_{i=1}^n \lambda_i (x_i - u_i)^2 \text{ subject to } \mathbf{x} = (x_1, \ldots, x_n) \in X. \qquad (2.9)$$

Let $\mathbf{u} \in \mathbb{R}^n$. We obtain the Λ-augmented Jacobian matrix $\mathcal{AJ}_\Lambda \in \mathbb{R}^{(k+1) \times n}$ by placing the row vector $(\lambda_1(x_1 - u_1), \ldots, \lambda_n(x_n - u_n))$ atop the Jacobian matrix $\mathcal{J}$. The critical ideal of the optimization problem (2.9) is then $C_{X,\mathbf{u}}^\Lambda = I : J^\infty$, where I is generated by the $(c+1) \times (c+1)$ minors of $\mathcal{AJ}_\Lambda$ together with the generators of I_X, and J is the ideal of $c \times c$ minors of $\mathcal{J}$. This leads to the following definition.

Definition 2.8 Let $X \subset \mathbb{R}^n$ be a real variety, let $\Lambda = (\lambda_1, \ldots, \lambda_n)$ have positive entries, and let $\mathbf{u} \in \mathbb{R}^n$ be a general point. The Λ-*weighted Euclidean distance degree* of the variety X is the number of points in the variety $V(C_{X,\mathbf{u}}^\Lambda)$. For general weights Λ, this number is independent of Λ, and we call it the *generic Euclidean distance degree* of X, denoted

$$\text{EDdegree}_{\text{gen}}(X) := \#V(C_{X,\mathbf{u}}^\Lambda).$$

We will discuss the generic ED degree for a class of matrix problems in the next section.

2.2 Low-Rank Matrix Approximation

A specific instance of the ED problem (2.1) arises when considering varieties of matrices of low rank that are constrained to have a special structure. Sometimes these matrices are flattenings of tensors. This version of the problem was studied in the article [142], which focuses on Hankel matrices, Sylvester matrices and generic subspaces of matrices, and which uses a weighted version of the Euclidean metric. In this section, we offer a brief introduction to this special case of our general ED problem. Our point of departure is the following low-rank approximation problem for rectangular matrices:

$$\text{minimize } \|A - U\|^2 = \sum_{i=1}^m \sum_{j=1}^n (a_{ij} - u_{ij})^2 \text{ subject to } \text{rank}(A) \leq r. \qquad (2.10)$$

In this problem, we are given a real *data matrix* $U = (u_{ij})$ of format $m \times n$, where $m \leq n$, and we wish to find a real matrix $A = (a_{ij})$ of rank at most r that is closest to U. The

Euclidean norm on the space of matrices is sometimes called the *Frobenius norm*. We can write this as $\|A\|^2 = \text{Trace}(A^\top A)$. Indeed, if $\mathbf{a}_1, \ldots, \mathbf{a}_n$ denote the columns of A, then the diagonal entries of $A^\top A$ are $\langle \mathbf{a}_i, \mathbf{a}_i \rangle$ for $i = 1, \ldots, n$. This implies

$$\text{Trace}(A^\top A) = \sum_{i=1}^{n} \langle \mathbf{a}_i, \mathbf{a}_i \rangle = \sum_{i=1}^{n} \sum_{j=1}^{m} a_{ij}^2 = \|A\|^2.$$

The solution to (2.10) is given by the *singular value decomposition*

$$U = T_1 \cdot \text{diag}(\sigma_1, \sigma_2, \ldots, \sigma_m) \cdot T_2.$$

Here T_1 and T_2 are orthogonal matrices, and the real numbers in the diagonal matrix satisfy $\sigma_1 \geq \sigma_2 \geq \cdots \geq \sigma_m \geq 0$. These are the singular values of U. The following well-known theorem from numerical linear algebra concerns the variety of $m \times n$ matrices of rank $\leq r$.

Theorem 2.9 (Eckart–Young) *The closest matrix of rank $\leq r$ to the given matrix U equals*

$$U^* = T_1 \cdot \text{diag}(\sigma_1, \ldots, \sigma_r, 0, \ldots, 0) \cdot T_2. \tag{2.11}$$

It is the unique local minimum for generic U. All complex critical points are real. They are found by substituting zeros for $m-r$ entries of $\text{diag}(\sigma_1, \ldots, \sigma_m)$. Thus, $\text{EDdegree}(X) = \binom{m}{r}$.

We present a complete proof of the Eckart–Young Theorem in Section 9.2.

In *structured low-rank approximation*, we are given a linear subspace $\mathcal{L} \subset \mathbb{R}^{m \times n}$, and a data matrix $U \in \mathcal{L}$, and we wish to solve the following restricted optimization problem:

$$\text{minimize } \|A - U\|^2 = \sum_{i=1}^{m} \sum_{j=1}^{n} (a_{ij} - u_{ij})^2 \text{ subject to } A \in \mathcal{L} \text{ and } \text{rank}(A) \leq r. \tag{2.12}$$

A best-case scenario would be the following: if U lies in $\mathcal{L}$ then so does the SVD solution U^* in (2.11). This happens for some linear subspaces $\mathcal{L}$, including symmetric and circulant matrices. However, most subspaces $\mathcal{L}$ of $\mathbb{R}^{m \times n}$ do not enjoy this property, and finding the global optimum in (2.12) can be quite difficult. Our discussion follows the article [142], which studies this optimization problem for both generic and special subspaces $\mathcal{L}$.

We consider structured low-rank approximation using a weighted Euclidean metric. Our primary task is to compute the number of complex critical points of (2.12). Thus, we seek to find the Euclidean distance degree (ED degree) of the determinantal variety

$$\mathcal{L}_{\leq r} := \{A \in \mathcal{L} \mid \text{rank}(A) \leq r\} \subset \mathbb{R}^{m \times n}.$$

We use a Λ-weighted Euclidean distance coming from the ambient matrix space $\mathbb{R}^{m \times n}$. We write $\text{EDdegree}_\Lambda(\mathcal{L}_{\leq r})$ for the Λ-weighted Euclidean distance degree of the variety $\mathcal{L}_{\leq r}$ (see Definition 2.8). The importance of the weights Λ is highlighted in [60, Example 3.2], for the seemingly harmless situation when $\mathcal{L}$ is the space of all symmetric matrices in $\mathbb{R}^{n \times n}$.

Of special interest are the usual Euclidean distance degree, denoted by $\text{EDdegree}(\mathcal{L}_{\leq r})$, when $\Lambda = \mathbf{1}$ is the all-one matrix, and the generic ED degree $\text{EDdegree}_{\text{gen}}(\mathcal{L}_{\leq r})$, when the weight matrix Λ is generic. The generic ED degree is given by a formula that rests on intersection theory. See [60, Theorem 7.7] and Theorem 2.13 below. Indeed, choosing the weights λ_{ij} to be generic ensures that the projective closure of $\mathcal{L}_{\leq r}$ has transversal intersection with the isotropic quadric $\{ A \in \mathbb{P}^{mn-1} \mid \sum_{i=1}^{m} \sum_{j=1}^{n} \lambda_{ij} a_{ij}^2 = 0 \}$.

We next present illustrations for the concepts above. These can also serve as examples for Theorem 2.13 below, as seen by the Macaulay2 [73] calculation in Example 2.22.

Example 2.10 Let $m = n = 3$ and $\mathcal{L} \subset \mathbb{R}^{3 \times 3}$ the 5-dimensional space of Hankel matrices:

$$A = \begin{bmatrix} a_0 & a_1 & a_2 \\ a_1 & a_2 & a_3 \\ a_2 & a_3 & a_4 \end{bmatrix}, \quad U = \begin{bmatrix} u_0 & u_1 & u_2 \\ u_1 & u_2 & u_3 \\ u_2 & u_3 & u_4 \end{bmatrix}, \quad \text{and} \quad \Lambda = \begin{bmatrix} \lambda_0 & \lambda_1 & \lambda_2 \\ \lambda_1 & \lambda_2 & \lambda_3 \\ \lambda_2 & \lambda_3 & \lambda_4 \end{bmatrix}.$$

Our goal is to solve the following constrained optimization problem for $r = 1$ and $r = 2$:

$$\text{minimize } \lambda_0(a_0 - u_0)^2 + 2\lambda_1(a_1 - u_1)^2 + 3\lambda_2(a_2 - u_2)^2 + 2\lambda_3(a_3 - u_3)^2 + \lambda_4(a_4 - u_4)^2$$
$$\text{subject to rank}(A) \leq r.$$

This can be rephrased as an unconstrained optimization problem. For instance, for rank $r = 1$, we get a one-to-one parametrization of $\mathcal{L}_{\leq 1}$ by setting $a_i = st^i$ for $i = 0, 1, 2, 3, 4$. Our optimization problem is then as follows:

$$\text{minimize } \lambda_0(s - u_0)^2 + 2\lambda_1(st - u_1)^2 + 3\lambda_2(st^2 - u_2)^2 + 2\lambda_3(st^3 - u_3)^2 + \lambda_4(st^4 - u_4)^2.$$

The ED degree is the number of critical points in this unconstrained minimization problem, where we require $t \neq 0$. We consider this problem for three weight matrices:

$$\mathbf{1} = \begin{bmatrix} 1 & 1 & 1 \\ 1 & 1 & 1 \\ 1 & 1 & 1 \end{bmatrix}, \quad \Omega = \begin{bmatrix} 1 & 1/2 & 1/3 \\ 1/2 & 1/3 & 1/2 \\ 1/3 & 1/2 & 1 \end{bmatrix}, \quad \Theta = \begin{bmatrix} 1 & 2 & 2 \\ 2 & 2 & 2 \\ 2 & 2 & 1 \end{bmatrix}.$$

Here Ω gives the usual Euclidean metric when $\mathcal{L}$ is identified with $\mathbb{R}^5$. The last weight matrix Θ arises from identifying $\mathcal{L}$ with the space of symmetric $2 \times 2 \times 2 \times 2$-tensors; see Chapter 12 for a study of metrics in the space of tensors. We compute

$$\text{EDdegree}_{\mathbf{1}}(\mathcal{L}_{\leq 1}) = 6, \quad \text{EDdegree}_{\Omega}(\mathcal{L}_{\leq 1}) = 10, \quad \text{EDdegree}_{\Theta}(\mathcal{L}_{\leq 1}) = 4,$$
$$\text{EDdegree}_{\mathbf{1}}(\mathcal{L}_{\leq 2}) = 9, \quad \text{EDdegree}_{\Omega}(\mathcal{L}_{\leq 2}) = 13, \quad \text{EDdegree}_{\Theta}(\mathcal{L}_{\leq 2}) = 7.$$

In both cases, Ω exhibits the generic behavior, so we have

$$\text{EDdegree}_{\text{gen}}(\mathcal{L}_{\leq r}) = \text{EDdegree}_{\Omega}(\mathcal{L}_{\leq r}).$$

See [142, Sections 3 and 4] for larger Hankel matrices and formulas for their ED degrees.◇

Remark 2.11 The distance given by the above weight matrix Θ yields the *Bombieri–Weyl distance*. This plays an important role in our study of condition numbers in Chapter 9.

Example 2.12 Let $m = n = 3, r = 1$ and $\mathcal{L} = \mathbb{R}^{3 \times 3}$. Thus, we consider the weighted rank-one approximation problem for 3×3-matrices. We know that $\text{EDdegree}_{\text{gen}}(\mathcal{L}_{\leq 1}) = 39$; see [60, Example 7.10]. We take a circulant data matrix and a circulant weight matrix:

$$U = \begin{bmatrix} -59 & 11 & 59 \\ 11 & 59 & -59 \\ 59 & -59 & 11 \end{bmatrix} \quad \text{and} \quad \Lambda = \begin{bmatrix} 9 & 6 & 1 \\ 6 & 1 & 9 \\ 1 & 9 & 6 \end{bmatrix}.$$

This instance has 39 complex critical points. Of these, 19 are real, and 7 are local minima:

$$\begin{bmatrix} 0.0826 & 2.7921 & -1.5452 \\ 2.7921 & 94.3235 & -52.2007 \\ -1.5452 & -52.2007 & 28.8890 \end{bmatrix}, \begin{bmatrix} -52.2007 & 28.8890 & -1.5452 \\ 2.7921 & -1.5452 & 0.0826 \\ 94.3235 & -52.2007 & 2.7921 \end{bmatrix}, \begin{bmatrix} -52.2007 & 2.7921 & 94.3235 \\ 28.8890 & -1.5452 & -52.2007 \\ -1.5452 & 0.0826 & 2.7921 \end{bmatrix},$$

$$\begin{bmatrix} -29.8794 & 36.2165 & -27.2599 \\ -32.7508 & 39.6968 & -29.8794 \\ 39.6968 & -48.1160 & 36.2165 \end{bmatrix}, \begin{bmatrix} -48.1160 & 36.2165 & 39.6968 \\ 36.2165 & -27.2599 & -29.8794 \\ 39.6968 & -29.8794 & -32.7508 \end{bmatrix}, \begin{bmatrix} -29.8794 & -32.7508 & 39.6968 \\ 36.2165 & 39.6968 & -48.1160 \\ -27.2599 & -29.8794 & 36.2165 \end{bmatrix},$$

$$\begin{bmatrix} -25.375 & -25.375 & -25.375 \\ -25.375 & -25.375 & -25.375 \\ -25.375 & -25.375 & -25.375 \end{bmatrix}.$$

The first three are the global minima in our ED distance problem. The last matrix is the local minimum where the objective function has the largest value: note that each entry equals $-203/8$. The entries of the first six matrices are algebraic numbers of degree 10 over $\mathbb{Q}$. For instance, the two upper left entries 0.0826 and -48.1160 are among the four real roots of the irreducible polynomial

$$164466028468224x^{10} + 27858648335954688x^9 + 1602205386689376672x^8$$
$$+ 7285836260028875412x^7 - 2198728936046680414272x^6$$
$$- 14854532690380098143152x^5 + 2688673091228371095762316x^4$$
$$+ 44612094455115888622678587x^3 - 41350080445712457319337106x^2$$
$$+ 270391294999043116889674775x - 1977632463563766878765625.$$

Here, the critical ideal in $\mathbb{Q}[x_{11}, x_{12}, \ldots, x_{33}]$ is not prime. A computation in Macaulay2 reveals that it is the intersection of six maximal ideals. The degrees of these maximal ideals over $\mathbb{Q}$ are 1, 2, 6, 10, 10, 10. The sum of these numbers equals $39 = \text{EDdegree}_{\text{gen}}(\mathcal{L}_{\leq 1})$. ◇

Explicit formulas are derived in [142, Section 3] for $\text{EDdegree}_{\text{gen}}(\mathcal{L}_{\leq r})$ when $\mathcal{L}$ is a generic subspace of $\mathbb{R}^{m \times n}$. This covers the four cases that arise by pairing affine subspaces or linear subspaces with either unit weights or generic weights. One important feature of determinantal varieties is that they are not complete intersections. This fact implies that their ED degrees are much smaller than suggested by the upper bound in Proposition 2.4.

2.3 Invitation to Polar Degrees

We have introduced the ED degree of an algebraic variety X as a complexity measure for the ED problem in (2.1). The number 39 in the previous example served as an illustration on how the ED degree controls the number of critical points. But a deeper understanding is needed. In this section, we develop the algebro-geometric roots of the ED degree, which will then yield more advanced algorithms for finding it. Here is a key result for this:

Theorem 2.13 *If the given variety X meets both the hyperplane at infinity and the isotropic quadric transversally, then EDdegree(X) equals the sum of the polar degrees of the projective closure of X.*

This theorem appears in [60, Proposition 6.10]. The hypothesis of the theorem is satisfied for all varieties X in $\mathbb{R}^n$ after a general linear change of coordinates. This implies the following result for EDdegree$_{\text{gen}}(X)$, which is the quantity introduced in Definition 2.8.

Corollary 2.14 *The generic ED degree of a variety is the sum of the polar degrees of its projective closure.*

We shall explain all the terms used in Theorem 2.13. First of all, the *projective closure* of our affine real variety $X \subset \mathbb{C}^n$ is its Zariski closure in complex projective space $\mathbb{P}^n$, which we also denote by X. Algebraically, $\mathbb{P}^n$ is obtained from $\mathbb{C}^n$ by adding one homogenizing coordinate x_0. We identify the affine space $\mathbb{C}^n$ with the open subset $\{\mathbf{x} \in \mathbb{P}^n \mid x_0 \neq 0\}$. Its set complement $\{\mathbf{x} \in \mathbb{P}^n \mid x_0 = 0\} \simeq \mathbb{P}^{n-1}$ is the *hyperplane at infinity* inside $\mathbb{P}^n$. The hypersurface $\{\mathbf{x} \in \mathbb{P}^{n-1} \mid \sum_{i=1}^n x_i^2 = 0\}$ is called the *isotropic quadric*. It lives in the hyperplane at infinity and it has no real points. The hypothesis in Theorem 2.13 means that the intersection of X with these two loci is reduced and has the expected dimension.

If we are given a real projective variety X in $\mathbb{P}^n$ from the start, then we also consider the ED problem for its affine cone in $\mathbb{R}^{n+1}$. For this problem, the data vector has $n + 1$ coordinates, say $\mathbf{u} = (u_0, u_1, \ldots, u_n)$. The augmented Jacobian $\mathcal{AJ}$ is now redefined so as to respect the fact that all polynomials are homogeneous. The general formula for this matrix and the homogeneous critical ideal appear in [60, Equation (2.7)].

For a curve $X \subset \mathbb{P}^2$ with defining polynomial $f(x_0, x_1, x_2)$, we use the 3×3 matrix

$$
\mathcal{AJ} = \begin{bmatrix} u_0 & u_1 & u_2 \\ x_0 & x_1 & x_2 \\ \partial f/\partial x_0 & \partial f/\partial x_1 & \partial f/\partial x_2 \end{bmatrix}.
$$

The homogeneous critical ideal in $\mathbb{R}[x_0, x_1, x_2]$ is computed as follows:

$$
C_{X,\mathbf{u}} = \big\langle f, \det(\mathcal{AJ}) \big\rangle : \big(\langle \partial f/\partial x_0, \partial f/\partial x_1, \partial f/\partial x_2 \rangle \cdot (x_1^2 + x_2^2)\big)^\infty. \tag{2.13}
$$

The critical points are given by the variety $V(C_{X,\mathbf{u}})$ in the complex projective plane $\mathbb{P}^2$. The cardinality of this variety equals EDdegree(X). The factor $(x_1^2 + x_2^2)$ in the saturation step (2.13) is the isotropic quadric. It is needed whenever the hypothesis of Theorem 2.13 is not satisfied. Namely, it removes any extraneous component that may arise from non-transversal intersection of the projective curve X with the isotropic quadric.

Example 2.15 (Cardioid) We consider the homogeneous version of the cardioid that was studied in Example 2.1. This is the curve $X = V(f)$ in $\mathbb{P}^2$ defined by

$$f = (x_1^2 + x_2^2 + x_0 x_2)^2 - x_0^2(x_1^2 + x_2^2). \tag{2.14}$$

This curve has three singular points, namely that at the origin $V(x_1, x_2)$ in the affine plane $\mathbb{C}^2 = \{x_0 \neq 0\}$ and the two points in the isotropic quadric $V(x_1^2 + x_2^2)$ in $\mathbb{P}^1 = \{x_0 = 0\}$.

The homogeneous critical ideal $C_{X,\mathbf{u}}$ is generated by three cubics, and it defines seven points in $\mathbb{P}^2$. Hence the projective cardioid X has EDdegree$(X) = 7$. This is also the ED degree of the affine cardioid in (2.4) but only after a linear change of coordinates. We note that even a modest change of coordinates can have a dramatic impact on the ED degree. For instance, if we replace x_1 by $2x_1$ in (2.4) then the ED degree jumps from 3 to 7. ◇

We next define the polar degrees of a projective variety $X \subset \mathbb{P}^n$. Other definitions and many more details are given in Chapter 4. Recall that points in the dual projective space $(\mathbb{P}^n)^*$ represent hyperplanes in the primal space $\mathbb{P}^n$. The Euclidean bilinear form $\langle \cdot, \cdot \rangle$ is non-degenerate. Therefore, we can use it to identify $\mathbb{P}^n$ and its dual space. To be precise, the point $\mathbf{h} \in \mathbb{P}^n$ represents the hyperplane $\{\mathbf{x} \in \mathbb{P}^n \mid \langle \mathbf{h}, \mathbf{x} \rangle = h_0 x_0 + \cdots + h_n x_n = 0\}$.

Definition 2.16 The *conormal variety* $N_X \subset \mathbb{P}^n \times \mathbb{P}^n$ is the Zariski closure of the set of all pairs $(\mathbf{x}, \mathbf{h})$ of points in $\mathbb{P}^n \times \mathbb{P}^n$ such that $\mathbf{x}$ is a nonsingular point of X and $\mathbf{h}$ represents a hyperplane that is tangent to X at $\mathbf{x}$.

It is known that the conormal variety N_X has dimension $n - 1$, and if X is irreducible then so is N_X. The image of N_X under projection onto the second factor is the *dual variety* $X^\vee$; see Section 4.2. If X is a curve in the plane $\mathbb{P}^2$, then this yields the definition of dual curve from (1.7). The role of $\mathbf{x}$ and $\mathbf{h}$ in the definition of the conormal variety can be swapped. Theorem 4.11 from Chapter 4 implies that the following biduality relations hold:

$$N_X = N_{X^\vee} \quad \text{and} \quad (X^\vee)^\vee = X. \tag{2.15}$$

The conormal variety is an object of algebraic geometry that offers the theoretical foundations for various aspects of duality in optimization, including primal-dual algorithms.

Example 2.17 For a plane curve $X = V(f)$ in $\mathbb{P}^2$, the conormal variety N_X is a curve in $\mathbb{P}^2 \times \mathbb{P}^2$. Its ideal is derived from the ideal that is generated by f and the 2×2 minors of

$$\begin{bmatrix} h_0 & h_1 & h_2 \\ \partial f/\partial x_0 & \partial f/\partial x_1 & \partial f/\partial x_2 \end{bmatrix}.$$

By saturation, as in (2.13), we remove singularities and points on the isotropic quadric. The result is the bihomogeneous prime ideal in $\mathbb{R}[x_0, x_1, x_2, h_0, h_1, h_2]$ which defines N_X. The equation of the dual curve $X^\vee$ is obtained from this ideal by eliminating x_0, x_1, x_2.

For instance, if X is the homogeneous cardioid in (2.14) then its dual $X^\vee$ is the cubic curve defined by $16h_0^3 - 27h_0 h_1^2 - 24h_0^2 h_2 - 15h_0 h_2^2 - 2h_2^3$. The ideal of the conormal curve N_X has ten minimal generators. In addition to the above generators of bidegrees $(4, 0)$ and $(0, 3)$, we find the quadric $x_0 h_0 + x_1 h_1 + x_2 h_2$ of bidegree $(1, 1)$, three cubics of bidegree $(2, 1)$ like $x_1^2 h_1 - 3x_2^2 h_1 - x_0 x_1 h_2 + 4x_1 x_2 h_2$, and four cubics of bidegree $(1, 2)$. ◇

We now finally come to the polar degrees. The product of two projective spaces $\mathbb{P}^n \times \mathbb{P}^n$ serves as the ambient space for our primal-dual approach to the ED problem. We first consider its cohomology ring:

$$H^*(\mathbb{P}^n \times \mathbb{P}^n, \mathbb{Z}) \;=\; \mathbb{Z}[s,t]/\langle s^{n+1}, t^{n+1}\rangle.$$

The class of the conormal variety N_X in this cohomology ring is a binary form of degree $n + 1 = \mathrm{codim}(N_X)$ whose coefficients are nonnegative integers:

$$[N_X] \;=\; \delta_1(X)s^n t + \delta_2(X)s^{n-1}t^2 + \delta_3(X)s^{n-2}t^3 + \cdots + \delta_n(X)st^n. \qquad (2.16)$$

In `Macaulay2` one uses the command `multidegree` to compute this binary form.

Definition 2.18 The coefficients $\delta_i(X)$ in (2.16) are called the *polar degrees* of X.

Remark 2.19 The polar degrees satisfy $\delta_i(X) = \#(N_X \cap (L \times L'))$, where $L \subset \mathbb{P}^n$ and $L' \subset \mathbb{P}^n$ are general linear subspaces of dimensions $n + 1 - i$ and i respectively. We will see this in Chapter 4. This geometric interpretation implies that $\delta_i(X) = 0$ for $i < \mathrm{codim}(X^\vee)$ and for $i > \dim(X) + 1$. Moreover, the first and last polar degree are the classical degrees:

$$\delta_i(X) \;=\; \begin{cases} \mathrm{degree}(X) & \text{for } i = \dim(X) + 1, \\ \mathrm{degree}(X^\vee) & \text{for } i = \mathrm{codim}(X^\vee). \end{cases} \qquad (2.17)$$

Example 2.20 Let $X \subset \mathbb{P}^2$ be the cardioid in (2.14). The ideal of the curve $N_X \subset \mathbb{P}^2 \times \mathbb{P}^2$ was derived in Example 2.17. From this ideal we compute the cohomology class

$$[N_X] \;=\; \mathrm{degree}(X^\vee) \cdot s^2 t + \mathrm{degree}(X) \cdot st^2 \;=\; 3 \cdot s^2 t + 4 \cdot st^2.$$

Thus the polar degrees of the cardioid are 3 and 4. Their sum 7 is the ED degree.   ◇

Example 2.21 Let X be a general surface of degree d in $\mathbb{P}^3$. Its dual $X^\vee$ is a surface of degree $d(d-1)^2$ in $(\mathbb{P}^3)^*$. The conormal variety N_X is a surface in $\mathbb{P}^3 \times (\mathbb{P}^3)^*$, with class

$$[N_X] \;=\; d(d-1)^2 s^3 t + d(d-1) s^2 t^2 + d \, st^3.$$

The sum of the three polar degrees is $\mathrm{EDdegree}(X) = d^3 - d^2 + d$; see Proposition 2.4. ◇

Theorem 2.13 allows us to compute the ED degree for many interesting varieties, e.g. using Chern classes [60, Theorem 5.8]. This is relevant for many applications, including those in machine learning, our topic in Chapter 10. Frequently, these applications involve low-rank approximation of matrices and tensors with special structure [33, 142].

Example 2.22 (Determinantal varieties) Let $X = X_r \subset \mathbb{P}^{m^2-1}$ be the projective variety of $m \times m$ matrices $x = (x_{ij})$ of rank $\leq r$. We claim that its conormal variety N_X is cut out by matrix equations (here, the product symbol $\cdot$ denotes the multiplication of matrices):

$$N_X \;=\; \big\{ (\mathbf{x}, \mathbf{h}) \in \mathbb{P}^{m^2-1} \times \mathbb{P}^{m^2-1} \mid \mathrm{rank}(\mathbf{x}) \leq r, \ \mathrm{rank}(\mathbf{h}) \leq m - r, \ \mathbf{x} \cdot \mathbf{h} = 0 \text{ and } \mathbf{h} \cdot \mathbf{x} = 0 \big\}.$$

This follows from Lemma 9.12 in Chapter 9. In particular, among determinantal varieties the duality relation $(X_r)^\vee = X_{m-r}$ holds. We now type the above formula for N_X into Macaulay2 [73], for $r = 1$ and $m = 3$. With this, we compute the polar degrees as follows:

```
QQ[x11,x12,x13,x21,x22,x23,x31,x32,x33,
    h11,h12,h13,h21,h22,h23,h31,h32,h33,
    Degrees=> {{1,0},{1,0},{1,0},{1,0},{1,0},{1,0},{1,0},{1,0},{1,0},
               {0,1},{0,1},{0,1},{0,1},{0,1},{0,1},{0,1},{0,1},{0,1}}];
x = matrix {{x11,x12,x13},{x21,x22,x23},{x31,x32,x33}};
h = matrix {{h11,h12,h13},{h21,h22,h23},{h31,h32,h33}};
I = minors(2,x) + minors(3,h) + minors(1,x*h) + minors(1,h*x);
isPrime(I), codim(I), degree I
multidegree(I)
```

The code starts with the bigraded coordinate ring of $\mathbb{P}^8 \times \mathbb{P}^8$. It verifies that N_X has codimension 9 and that I is its prime ideal. The last command computes the polar degrees:

$$[N_X] = 3s^8 t + 6s^7 t^2 + 12s^6 t^3 + 12s^5 t^4 + 6s^4 t^5. \tag{2.18}$$

At this point, the reader may verify (2.17). Using Corollary 2.14, we conclude

$$\mathrm{EDdegree}_{\mathrm{gen}}(X_1) = 3 + 6 + 12 + 12 + 6 = 39.$$

Indeed, after changing coordinates, the EDdegree for 3×3-matrices of rank 1 equals 39. We saw this in Example 2.12, where 39 critical points were found by numerical computation. ◇

The primal-dual set-up of conormal varieties allows for a very elegant formulation of the critical equations for the ED problem (2.1). This will be presented in the next theorem. We now assume that X is an irreducible variety defined by homogeneous polynomials in n variables. We view X as an affine cone in $\mathbb{C}^n$. Its dual $Y = X^\vee$ is the affine cone over the dual of the projective variety given by X. Thus Y is also an affine cone in $\mathbb{C}^n$. In this setting, the conormal variety N_X is viewed as an affine variety of dimension n in $\mathbb{C}^{2n}$. The homogeneous ideals of these cones are precisely the ideals we discussed above.

Theorem 2.23 *The ED problems for X and Y coincide:* $\mathrm{EDdegree}(X) = \mathrm{EDdegree}(Y)$. *Given a general data point $\mathbf{u} \in \mathbb{R}^n$, the critical equations for this ED problem are:*

$$(\mathbf{x}, \mathbf{h}) \in N_X \quad \text{and} \quad \mathbf{x} + \mathbf{h} = \mathbf{u}. \tag{2.19}$$

Proof See [60, Theorem 5.2].   □

It is instructive to verify Theorem 2.23 for the low-rank variety in Example 2.22. For any data matrix $\mathbf{u}$ of size $m \times m$, the sum in (2.19) is a special decomposition of $\mathbf{u}$, namely as a matrix $\mathbf{x}$ of rank r plus a matrix $\mathbf{h}$ of rank $m - r$. By the Eckart–Young Theorem 2.9, this arises from zeroing out complementary singular values σ_i in the two matrices $\mathbf{x}$ and $\mathbf{h}$.

In general, there is no free lunch, even with a simple formulation like (2.19). The difficulty lies in computing the ideal of the conormal variety N_X. However, this should be thought of as a preprocessing step, to be carried out only once per model X. If an efficient presentation of N_X is available, then our task is to solve the system $\mathbf{x} + \mathbf{h} = \mathbf{u}$ of n linear equations in the $2n$ coordinates on N_X, now viewed as an n-dimensional affine variety.

The discussion in this section was restricted to the Euclidean norm. But, we can measure distances in $\mathbb{R}^n$ with any other norm $\|\cdot\|$. Our optimization problem (2.1) extends naturally:

$$\text{minimize } \|\mathbf{x} - \mathbf{u}\| \text{ subject to } \mathbf{x} \in X. \tag{2.20}$$

The unit ball $B = \{\mathbf{x} \in \mathbb{R}^n \mid \|\mathbf{x}\| \leq 1\}$ in the chosen norm is a centrally symmetric convex body. Conversely, every centrally symmetric convex body B in $\mathbb{R}^n$ defines a norm, and we can paraphrase the previous optimization problem as follows:

$$\text{minimize } \lambda \text{ subject to } \lambda \geq 0 \text{ and } (\mathbf{u} + \lambda B) \cap X \neq \emptyset.$$

We will meet minimization problems for unit balls that are polyhedra in Chapter 5.

If the boundary of the unit ball B is smooth and algebraic then we can express the critical equations for the corresponding norm as a polynomial system. This is derived as before, but we now replace the first row of the augmented Jacobian matrix $\mathcal{AJ}$ with the gradient vector of the map $\mathbb{R}^n \to \mathbb{R}$, $\mathbf{x} \mapsto \|\mathbf{x} - \mathbf{u}\|^2$. In conclusion, this chapter was dedicated to computing the minimal distance from a data point to a given variety for the Euclidean norm. The algebraic machinery we developed can be applied to much more general scenarios. For instance, Kubjas, Kuznetsova and Sodomaco [114] study this topic for p-norms.

Chapter 3
Computations

In this chapter, we study two computational approaches to solve a system of polynomial equations. A system of m polynomial equations in n variables can be written in the form

$$F(\mathbf{x}) := \begin{pmatrix} f_1(\mathbf{x}) \\ \vdots \\ f_m(\mathbf{x}) \end{pmatrix} = 0,$$

where

$$f_1, \ldots, f_m \in \mathbb{C}[\mathbf{x}] := \mathbb{C}[x_1, \ldots, x_n].$$

If $n = m$, we call $F(\mathbf{x})$ a *square system*. If $n > m$, then we call $F(\mathbf{x})$ *underdetermined*. If $n < m$, we call $F(\mathbf{x})$ *overdetermined*. Here, we will mostly focus on square systems.

Example 3.1 The previous chapter considered constrained optimization problems of the form $\min_{\mathbf{x} \in \mathbb{R}^n : g(\mathbf{x}) = 0} f(\mathbf{x})$, where f and g are polynomials in n variables $\mathbf{x} = (x_1, \ldots, x_n)$. To solve this problem, one can compute the solutions of the critical equations

$$g(\mathbf{x}) = \frac{\partial f}{\partial x_1}(\mathbf{x}) + \lambda \cdot \frac{\partial g}{\partial x_1}(\mathbf{x}) = \cdots = \frac{\partial f}{\partial x_n}(\mathbf{x}) + \lambda \cdot \frac{\partial g}{\partial x_n}(\mathbf{x}) = 0.$$

This is a square system in the $n + 1$ variables $(\mathbf{x}, \lambda)$, where λ is a Lagrange multiplier. ◇

Solving the system $F(\mathbf{x}) = 0$ means that we compute *all* points $\mathbf{z} = (z_1, \ldots, z_n) \in \mathbb{C}^n$ such that $F(\mathbf{z}) = 0$. The first step is to find an appropriate *data structure* to represent a solution. In fact, $F(\mathbf{z}) = 0$ is already *implicitly* represented by its equation. Some information can be read off from this representation. For instance, if F has rational coefficients and we know that $F(\mathbf{z}) = 0$ has only finitely many complex solutions, then each coordinate z_i of $\mathbf{z}$ is an algebraic number. On the other hand, further information, like whether or not there is a real solution $\mathbf{z} \in \mathbb{R}^n$, is not directly accessible from this implicit representation.

Our aim in this chapter is to introduce two data structures for representing solutions of systems of polynomial equations: the first is *Gröbner bases* and the second is *approximate zeros*. Readers who are familiar with these concepts can skip ahead to the next chapter.

© The Author(s) 2024
P. Breiding et al., *Metric Algebraic Geometry*, Oberwolfach Seminars 53,
https://doi.org/10.1007/978-3-031-51462-3_3

Remark 3.2 In Chapter 2 we formulated critical equations for optimization problems like the ED problem (2.1). In this chapter, we present methods for solving these equations. The field of *polynomial optimization* rests on an alternative approach, namely one employs relaxations based on moments and sums of squares, and these are solved using semidefinite programming. We refer to the book [143] for an introduction to polynomial optimization.

3.1 Gröbner Bases

We use the notation $\mathbf{x}^\alpha := x_1^{\alpha_1} \cdots x_n^{\alpha_n}$ for the monomial defined by the exponent vector $\alpha = (\alpha_1, \ldots, \alpha_n) \in \mathbb{N}^n$. We can identify monomials with their exponent vectors. A monomial order $>$ on $\mathbb{C}[\mathbf{x}]$ is then defined by a total order $>$ on $\mathbb{N}^n$ that satisfies (1) if $\alpha > \beta$, then $\alpha + \gamma > \beta + \gamma$ for every $\gamma \in \mathbb{N}^n$ and (2) every non-empty subset of $\mathbb{N}^n$ has a smallest element under $>$ (see, e.g., [52, Chapter 2 §2, Definition 1]).

Example 3.3 The Lex (Lexicographic) order on the set of monomials in n variables is defined by setting $\alpha > \beta$ if $\alpha_j - \beta_j > 0$ for $\alpha, \beta \in \mathbb{N}^n$, where $j := \min\{i \mid \alpha_i \neq \beta_i\}$ is the first index where α and β are not equal. For instance, $x_1^2 x_2 > x_1 x_2 x_3^3$. Intuitively speaking, the Lex order views a polynomial $f \in \mathbb{C}[\mathbf{x}]$ as a polynomial in x_1 with coefficients that are polynomials in x_2, which has coefficients that are polynomials in x_3, and so on.   ◇

A monomial order induces the notion of *leading term*. Let $f = c_\alpha \mathbf{x}^\alpha + c_\beta \mathbf{x}^\beta + \cdots + c_\gamma \mathbf{x}^\gamma$, where $\alpha > \beta > \cdots > \gamma$ and $c_\alpha \neq 0$. Then, the leading term of f is $\mathrm{LT}(f) := c_\alpha \mathbf{x}^\alpha$. The *leading term ideal* of an ideal I in $\mathbb{C}[\mathbf{x}]$ is defined as

$$\mathrm{LT}(I) := \langle \{\mathrm{LT}(f) \mid f \in I \setminus \{0\}\} \rangle .$$

Definition 3.4 (Gröbner basis) Fix an ideal $I \subset \mathbb{C}[\mathbf{x}]$ and a monomial order $>$. A finite subset $\{g_1, \ldots, g_s\} \subset I$ is a *Gröbner basis* for I with respect to $>$ if its leading terms generate the leading term ideal:

$$\langle \mathrm{LT}(g_1), \ldots, \mathrm{LT}(g_s) \rangle = \mathrm{LT}(I).$$

If G is a Gröbner basis for an ideal I, then G generates I; i.e., $I = \langle G \rangle$. This justifies the name "basis", here chosen to mean "generating set". (See [52, Chapter 2 §5, Corollary 6].)

Our next example is identical to that in [52, Chapter 2 §8, Example 2]. It illustrates how Gröbner bases can be used to solve systems of polynomial equations.

Example 3.5 Consider the following system of three equations in three variables:

$$F(x, y, z) = \begin{pmatrix} x^2 + y^2 + z^2 - 1 \\ x^2 + z^2 - y \\ x - z \end{pmatrix} = 0.$$

We compute a Gröbner basis of the ideal $I = \langle x^2 + y^2 + z^2 - 1, x^2 + z^2 - y, x - z \rangle$ with respect to the Lex order with $x > y > z$ using Macaulay2 [73]. The code is as follows:

```
R = QQ[x, y, z, MonomialOrder => Lex];
f = x^2 + y^2 + z^2 - 1;
g = x^2 + z^2 - y;
h = x - z;
I = ideal {f, g, h};
G = gb I;
gens G
```

This computes the Gröbner basis $G = \{x - z, \ y - 2z^2, \ 4z^4 + 2z^2 - 1\}$. Since $I = \langle G \rangle$, we can solve $F(x, y, z) = 0$ by solving the equations given by G. Notice that the third polynomial in G only depends on z, the second only on y and z, and the first only on x and z. Thus, solving $F(x, y, z) = 0$ reduces to solving three univariate equations. This gives the four solutions $(a, 2a^2, a)$, where a is one of the four roots of $4z^4 + 2z^2 - 1$.   ◇

The reason why using the Lex order in Example 3.5 works well is the *Elimination Theorem*. To state this, consider any ideal $I \subset \mathbb{C}[\mathbf{x}]$. For every $0 \le j < n$, the intersection $I_j := I \cap \mathbb{C}[x_{j+1}, \ldots, x_n]$ is an ideal in a polynomial subring. It consists of those polynomials in I that only contain the variables $x_{j+1}, \ldots, x_n$. We call I_j the j-th *elimination ideal* of I. For a proof of the next theorem see [52, Chapter 3, §1, Theorem 1].

Theorem 3.6 (The Elimination Theorem) *Let $I \subset \mathbb{C}[\mathbf{x}]$ be an ideal and let G be a Gröbner basis for I with respect to the Lex order with $x_1 > \cdots > x_n$. Then $G \cap \mathbb{C}[x_{j+1}, \ldots, x_n]$ is a Gröbner basis of the j-th elimination ideal of I.*

For us, the most important consequence of the Elimination Theorem is that, if a system of polynomial equations $F(\mathbf{x}) = (f_1(\mathbf{x}), \ldots, f_m(\mathbf{x})) = 0$ has finitely many solutions, then the j-th elimination ideal will not be empty for $0 \le j < n$. Consequently, we can solve $F(\mathbf{x}) = 0$ by computing a Gröbner basis for the Lex order and then sequentially solving univariate equations. We can compute zeros of univariate polynomials by computing eigenvalues λ of the associated *companion matrix*. If $f(x) = x^d + \sum_{i=0}^{d-1} c_i x^i$ is a univariate polynomial, we have $f(\lambda) = 0$ if and only if λ is an eigenvalue of the companion matrix

$$\begin{bmatrix} 0 & \cdots & 0 & -c_0 \\ 1 & \cdots & 0 & -c_1 \\ \vdots & \ddots & \vdots & \vdots \\ 0 & \cdots & 1 & -c_{d-1} \end{bmatrix} \in \mathbb{C}^{d \times d}.$$

Sometimes we are not interested in the solutions of $F(\mathbf{x}) = 0$ per se, but only in the number of solutions. Gröbner bases naturally carry this information: Suppose $I \subset \mathbb{C}[\mathbf{x}]$ is an ideal and $>$ is a monomial order. Recall that a monomial $\mathbf{x}^\alpha \notin \mathrm{LT}(I)$ is called a *standard monomial* of I relative to $>$. The next result shows how to get the number of solutions of $F(\mathbf{x}) = 0$ from a Gröbner basis; see [164, Proposition 2.1].

Proposition 3.7 *Let $I \subset \mathbb{C}[\mathbf{x}]$ be an ideal, $>$ a monomial order, and $\mathcal{B}$ the set of standard monomials of I relative to $>$. Then, $\mathcal{B}$ is finite if and only if $V(I)$ is finite, and #$\mathcal{B}$ equals the number of points in $V(I)$ counting multiplicities.*

Example 3.8 In Example 3.5, there are four standard monomials, namely, $1, z, z^2$ and z^3. That is why the system $F(x, y, z) = 0$ has four solutions. We can view Proposition 3.7 as an n-dimensional version of the Fundamental Theorem of Algebra, which says that a univariate polynomial of degree d has d zeros.   ◇

We have now understood how to solve systems of polynomials with finitely many zeros using Gröbner bases. What about systems whose variety has positive dimensional components? In that case, the n-th elimination ideal is necessarily the zero ideal. To cope with that case, one can remove positive dimensional components using ideal saturation. Let $I, J \subset \mathbb{C}[\mathbf{x}]$ be two ideals. The *saturation* of I by J is the ideal

$$I : J^{\infty} := \{f \in \mathbb{C}[\mathbf{x}] \mid \text{there is } \ell > 0 \text{ with } f \cdot g^{\ell} \in I \text{ for all } g \in J\}. \tag{3.1}$$

Saturation is the ideal analogue of removing components on the level of varieties. We have

$$V(I : J^{\infty}) = \overline{V(I) \setminus V(J)}; \tag{3.2}$$

see [52, Chapter 4 §4, Corollary 11].

Recall that a solution to a square system $F(\mathbf{x}) = (f_1(\mathbf{x}), \ldots, f_n(\mathbf{x})) = 0$ is called *regular* if the Jacobian determinant $\det \left(\frac{\partial f_i}{\partial x_j} \right)_{1 \le i, j \le n}$ does not vanish at that solution. There can only be finitely many regular zeros of a square system of polynomial equations, and a finite set of points is Zariski closed. Consequently, if we saturate $I = \langle f_1, \ldots, f_n \rangle$ by the ideal $J = \langle \det \left(\frac{\partial f_i}{\partial x_j} \right) \rangle$, then we can use the strategy from above to solve $F(\mathbf{x}) = 0$.

Example 3.9 Consider the following system of two polynomials in two variables

$$F(x, y) = \begin{pmatrix} (x - 1) \cdot (x - 2) \cdot (x^2 + y^2 - 1) \\ (y - 1) \cdot (y - 3) \cdot (x^2 + y^2 - 1) \end{pmatrix} = 0.$$

This system has four regular solutions, namely $(1, 1), (1, 3), (2, 1), (2, 3)$. In addition, the system has the circle $x^2 + y^2 - 1$ as a positive dimensional component. We use Macaulay2 [73] to saturate the ideal generated by F. This is done as follows:

```
R = QQ[x, y, MonomialOrder => Lex];
f = (x-1) * (x-2) * (x^2+y^2-1);
g = (y-1) * (y-3) * (x^2+y^2-1);
I = ideal {f, g};
Jac = matrix{{diff(x, f), diff(x, g)}, {diff(y, f), diff(y, g)}};
J = ideal det(Jac);
K = saturate(I,J)
```

This code returns the ideal $K = \langle y^2 - 4y + 3, x^2 - 3x + 2 \rangle$. The two generators form a Gröbner basis for K, with leading terms y^2 and x^2.   ◇

Next, we present two propositions related to elimination and saturation of ideals with parameters, and one lemma on Gröbner bases of parametric ideals. For this, let $\mathbf{x} = (x_1, \ldots, x_n)$ and $\mathbf{p} = (p_1, \ldots, p_k)$ be two sets of variables, and

$$\mathbb{C}[\mathbf{x}, \mathbf{p}] := \mathbb{C}[x_1, \ldots, x_n, p_1, \ldots, p_k].$$

We regard the coordinates p_i of $\mathbf{p}$ as *parameters*. For a fixed parameter vector $\mathbf{q} \in \mathbb{C}^k$, we consider the surjective ring homomorphism

$$\phi_{\mathbf{q}} : \mathbb{C}[\mathbf{x}, \mathbf{p}] \to \mathbb{C}[\mathbf{x}], \quad f(\mathbf{x}; \mathbf{p}) \mapsto f(\mathbf{x}; \mathbf{q}). \tag{3.3}$$

Proposition 3.10 *Consider an ideal $I \subset \mathbb{C}[\mathbf{x}, \mathbf{p}]$ and let $G = \{g_1, \ldots, g_s\}$ be a Gröbner basis for I relative to the Lex order $x_1 > \cdots > x_n > p_1 > \cdots > p_k$. For $1 \leq i \leq s$ with $g_i \notin \mathbb{C}[\mathbf{p}]$, write g_i in the form $g_i = c_i(\mathbf{p})\mathbf{x}^{\alpha_i} + h_i$, where all terms of h_i are strictly smaller than $\mathbf{x}^{\alpha_i}$ in Lex order. Let $\mathbf{q} \in V(I \cap \mathbb{C}[\mathbf{p}]) \subseteq \mathbb{C}^k$ such that $c_i(\mathbf{q}) \neq 0$ for all $g_i \notin \mathbb{C}[\mathbf{p}]$. Then, the following set is a Gröbner basis for the ideal $\phi_{\mathbf{q}}(I) \subset \mathbb{C}[\mathbf{x}]$:*

$$\phi_{\mathbf{q}}(G) = \{\phi_{\mathbf{q}}(g_i) \mid g_i \notin \mathbb{C}[\mathbf{p}]\}.$$

Proof See, e.g., [52, Chapter 4 §7, Theorem 2].   □

Example 3.11 (Cardioid revisited) As an illustration, fix $n = k = 2$ and let I be the ideal generated by the cubic (2.4) and the determinant of (2.2), with u_1, u_2 replaced by p_1, p_2. The lexicographic Gröbner basis G for this ideal has 15 elements. The 15 leading coefficients $c_i(\mathbf{p})$ are quite complicated. Three of them are

$$(4p_1^2 + 4p_2^2 + 4p_2 + 1)^2,$$
$$32p_2^2 r(3p_2 + 1)(3p_2 + 2)^3(2p_2 + 1)^3(4p_2^2 + 5p_2 + 2)^3(1 + p_2)^5,$$
$$p_1 p_2^2 r(8p_1^4 + 8p_1^2 p_2^2 - 10p_1^2 p_2 - 18p_2^3 - 7p_1^2 - 21p_2^2 - 8p_2 - 1)(p_1 + p_2 + 1)^3(p_1 - p_2 - 1)^3,$$

where $r = 4p_2 + 3$. Suppose that we replace the unknowns p_1, p_2 by any complex numbers q_1, q_2 such that $c_i(q_1, q_2) \neq 0$ for all i. Then the specialization $\phi_{\mathbf{q}}(G)$ remains a Gröbner basis with the same leading terms. In particular, the number of zeros (x_1, x_2), which is six when counted with multiplicities, is independent of the choice of parameters q_1, q_2.   ◇

In many applications, polynomial systems come with a non-degeneracy constraint. These are usually expressed in the form of a polynomial inequation $h \neq 0$. We now show how to incorporate such a constraint when solving a polynomial system.

Proposition 3.12 *Let $I \subset \mathbb{C}[\mathbf{x}, \mathbf{p}]$ be an ideal and let $J = \langle h \rangle$ be a principal ideal in the same polynomial ring. Let u be an additional variable and $K := \langle 1 - u \cdot h \rangle$. Then,*

$$I : J^{\infty} = (I + K) \cap \mathbb{C}[\mathbf{x}, \mathbf{p}].$$

If G is a Gröbner basis of $I + K$ relative to the Lex order $u > x_1 > \cdots > x_n > p_1 > \cdots > p_k$, then $G \cap \mathbb{C}[\mathbf{x}, \mathbf{p}]$ is a Gröbner basis of the saturation ideal $I : J^{\infty}$.

Proof See e.g. [52, Chapter 4 §4, Theorem 14].   □

Example 3.13 Fix I from Example 3.11. We take $K = \langle 1 - u \cdot h \rangle$ where $h = x_1 x_2$, since the cardioid (2.4) is singular at the origin. The Gröbner basis G has 27 elements, but with friendlier leading coefficients: $p_1 p_2^5(p_1 + p_2 + 1)(p_1 - p_2 - 1)$, $p_2^5(p_2 + 1)^2$, $p_1 p_2^3$, $p_2^2(9p_1^4 + p_2^4 + 2p_2^3 - 9p_1^2 + p_2^2)$, $p_2^2(p_2 + 1)$, ... Nonvanishing of these coefficients ensures that the system has 3 complex solutions, with multiplicities.   ◇

We now apply the technique above to identifying a discriminant Δ for a polynomial system that has parameters. The following result is proved in [23].

Lemma 3.14 *Let $I \subset \mathbb{C}[\mathbf{x}, \mathbf{p}]$ be an ideal and $J = \langle h \rangle \subset \mathbb{C}[\mathbf{x}, \mathbf{p}]$ be a principal ideal such that $(I : J^\infty) \cap \mathbb{C}[\mathbf{p}] = \{0\}$. Let $\{g_1, \ldots, g_s\}$ be a Gröbner basis of $I : J^\infty$ with respect to the Lex order $x_1 > \cdots > x_n > p_1 > \cdots > p_k$. There is a proper subvariety $\Delta \subsetneq \mathbb{C}^k$ such that, for all $\mathbf{q} \notin \Delta$, the set $\{\phi_\mathbf{q}(g_1), \ldots, \phi_\mathbf{q}(g_s)\}$ is a Gröbner basis for $\phi_\mathbf{q}(I) : \phi_\mathbf{q}(J)^\infty$ and none of the leading terms of $g_1, \ldots, g_s$ vanish when evaluated at $\mathbf{q}$. In particular, we have*

$$\phi_\mathbf{q}(I : J^\infty) = \phi_\mathbf{q}(I) : \phi_\mathbf{q}(J)^\infty \qquad \text{for all } \mathbf{q} \notin \Delta.$$

Proof Let u be an additional variable. As in Proposition 3.12, we consider the ideal $K := \langle 1 - u \cdot h \rangle$. Then, we have $I : J^\infty = (I + K) \cap \mathbb{C}[\mathbf{x}, \mathbf{p}]$. Using our hypothesis that $(I : J^\infty) \cap \mathbb{C}[\mathbf{p}] = \{0\}$, we conclude

$$V((I + K) \cap \mathbb{C}[\mathbf{p}]) = \mathbb{C}^k. \tag{3.4}$$

We may therefore apply Proposition 3.10 to $I + K$ without any restrictions on $\mathbf{q}$. As in Proposition 3.12, we augment the Lex order by letting u be the largest variable. Let $\overline{G} := \{g_1, \ldots, g_r\}$ be a Gröbner basis of $I + K$ relative to this order. By (3.4), we have that $g_1, \ldots, g_r \notin \mathbb{C}[\mathbf{p}]$. We write each g_i in the form $g_i = c_i(\mathbf{p})u^\beta \mathbf{x}^{\alpha_i} + h_i$, where all terms of h_i are strictly smaller than $u^\beta \mathbf{x}^{\alpha_i}$, and define the hypersurface

$$\Delta := \{\mathbf{q} \in \mathbb{C}^k \mid c_1(\mathbf{q}) \cdots c_r(\mathbf{q}) = 0\}. \tag{3.5}$$

Now let $\mathbf{q} \in \mathbb{C}^k \setminus \Delta$. By Proposition 3.10, the set $\phi_\mathbf{q}(\overline{G}) = \{\phi_\mathbf{q}(g_1), \ldots, \phi_\mathbf{q}(g_r)\}$ is a Gröbner basis for the ideal

$$\phi_\mathbf{q}(I + K) = \phi_\mathbf{q}(I) + \phi_\mathbf{q}(K) = \phi_\mathbf{q}(I) + (1 - u \cdot \phi_\mathbf{q}(h)).$$

Without loss of generality, suppose the first $s \leq r$ elements in $\overline{G}$ do not depend on the variable u. We define $G := \{g_1, \ldots, g_s\} = \overline{G} \cap \mathbb{C}[\mathbf{x}, \mathbf{p}]$. It follows from Proposition 3.12 that G is a Gröbner basis of $I : J^\infty$. Because $\mathbf{q} \notin \Delta$, none of the leading terms in $\overline{G}$ when evaluated at $\mathbf{q}$ vanish. Consequently, we have $\phi_\mathbf{q}(G) \cap \mathbb{C}[\mathbf{x}] = \phi_\mathbf{q}(\overline{G}) \cap \mathbb{C}[\mathbf{x}]$. Therefore, $\phi_\mathbf{q}(G) = \{\phi_\mathbf{q}(g_1), \ldots, \phi_\mathbf{q}(g_s)\}$ is a Gröbner basis of $\phi_\mathbf{q}(I) : \phi_\mathbf{q}(J)^\infty$, by Proposition 3.12.□

We refer to Δ as the *discriminant* of the pair (I, h). Equation (3.5) leads to the following:

Corollary 3.15 *Fix ideals $I = \langle f_1, \ldots, f_n \rangle \subset \mathbb{C}[\mathbf{x}, \mathbf{p}]$ and $J = \langle h \rangle \subset \mathbb{C}[\mathbf{x}, \mathbf{p}]$. The discriminant Δ in Lemma 3.14 is found by computing a Lex Gröbner basis G for $I + \langle 1 - u \cdot h \rangle$, where $u > x_1 > \cdots > x_n > p_1 > \cdots > p_k$. Namely, Δ is the product of the leading coefficients $c_i(\mathbf{p})$ in G.*

Example 3.16 (Hyperdeterminant) Let $k = 8, n = 2$ and consider a general pair of bilinear equations. Their ideal is $I = \langle f_1, f_2 \rangle$, where

$$f_1 = p_1 x_1 x_2 + p_2 x_1 + p_3 x_2 + p_4 \quad \text{and} \quad f_2 = p_5 x_1 x_2 + p_6 x_1 + p_7 x_2 + p_8.$$

Our non-degeneracy constraint is $h = \partial f_1 / \partial x_1 \cdot \partial f_2 / \partial x_2 - \partial f_1 / \partial x_2 \cdot \partial f_2 / \partial x_1$. This is the Jacobian determinant of f_1 and f_2. The Gröbner basis G in Corollary 3.15 has 11 elements.

The most interesting Gröbner basis element has leading coefficient

$$c(\mathbf{p}) = p_1^2 p_8^2 - 2p_1 p_2 p_7 p_8 - 2p_1 p_3 p_6 p_8 - 2p_1 p_4 p_5 p_8 + 4p_1 p_4 p_6 p_7 + p_2^2 p_7^2$$
$$+ \; 4p_2 p_3 p_5 p_8 - 2p_2 p_3 p_6 p_7 - 2p_2 p_4 p_5 p_7 + p_3^2 p_6^2 - 2p_3 p_4 p_5 p_6 + p_4^2 p_5^2.$$

This coefficient is the main factor in our discriminant Δ. This quartic is the *hyperdeterminant* of the $2 \times 2 \times 2$ tensor with entries $p_1, p_2, \ldots, p_8$.   ◇

3.2 The Parameter Continuation Theorem

The theorem to be proved in this section states that a square system of polynomial equations with parameters has a well-defined degree. This degree is the number of complex solutions for generic parameter choices. This result is the *Parameter Continuation Theorem*, due to Morgan and Sommese [135]. We shall present a proof that rests on the results on Gröbner bases in the previous section. For this, we consider again the polynomial ring $\mathbb{C}[\mathbf{x}, \mathbf{p}] = \mathbb{C}[x_1, \ldots, x_n, p_1, \ldots, p_k]$. We interpret $\mathbf{x}$ as variables and $\mathbf{p}$ as parameters.

Definition 3.17 Let $f_1(\mathbf{x}; \mathbf{p}), \ldots, f_n(\mathbf{x}; \mathbf{p}) \in \mathbb{C}[\mathbf{x}, \mathbf{p}]$. We consider the image of the map

$$\mathbb{C}^k \mapsto \mathbb{C}[\mathbf{x}]^n, \quad \mathbf{p}_0 \mapsto F(\mathbf{x}; \mathbf{p}_0) = \begin{pmatrix} f_1(\mathbf{x}; \mathbf{p}_0) \\ \vdots \\ f_n(\mathbf{x}; \mathbf{p}_0) \end{pmatrix}.$$

This image is denoted $\mathcal{F} = \{F(\mathbf{x}; \mathbf{p}) \mid \mathbf{p} \in \mathbb{C}^k\}$. Note that this is a *family* of square polynomial systems of size n. In other words, the family $\mathcal{F}$ consists of n polynomials in n variables that depend polynomially on k parameters.

In Examples 3.11, 3.13 and 3.16, we studied families of polynomial systems in $n = 2$ variables, where the number k of parameters was 2, 2 and 8, respectively. The degrees of these families are 6, 3 and 2. These degrees count the numbers of complex zeros for generic parameters $\mathbf{p}$, which satisfy $c_i(\mathbf{p}) \neq 0$ for all i.

We now fix a family $\mathcal{F}$ depending on k parameters $\mathbf{p} = (p_1, \ldots, p_k)$. Let $\mathbf{z} \in \mathbb{C}^n$ be a zero of $F(\mathbf{x}; \mathbf{p}) = (f_1(\mathbf{x}; \mathbf{p}), \ldots, f_n(\mathbf{x}; \mathbf{p})) \in \mathcal{F}$, for some specific parameters $\mathbf{p} \in \mathbb{C}^k$. We say that $\mathbf{z}$ is a *regular zero* if the Jacobian determinant $\det \left(\frac{\partial f_i}{\partial x_j} \right)_{1 \leq i, j \leq n}$ does not vanish at the point $\mathbf{z}$. The next theorem is the main result of this section. It shows that, for almost all parameters $\mathbf{p}$, the number of regular solutions is the same.

Theorem 3.18 (The Parameter Continuation Theorem) *Let $\mathcal{F}$ be a family of polynomial systems that consists of systems of n polynomials in n variables depending on k parameters. For $\mathbf{p} \in \mathbb{C}^k$, denote by $N(\mathbf{p})$ the number of regular zeros of $F(\mathbf{x}; \mathbf{p})$, and $N := \sup_{\mathbf{p} \in \mathbb{C}^k} N(\mathbf{p})$. Then, $N < \infty$, and there exists a proper algebraic subvariety $\Delta \subsetneq \mathbb{C}^k$, called discriminant of the system $\mathcal{F}$, such that $N(\mathbf{p}) = N$ for all $\mathbf{p} \notin \Delta$.*

Proof We recall the proof from [23]. Another proof can also be found in the textbook [161].
Suppose $\mathcal{F} = \{F(\mathbf{x}; \mathbf{p}) \mid \mathbf{p} \in \mathbb{C}^k\}$, where $F(\mathbf{x}; \mathbf{p}) = (f_1(\mathbf{x}; \mathbf{p}), \ldots, f_n(\mathbf{x}; \mathbf{p})) \in \mathcal{F}$. Let

$$I = \langle f_1, \ldots, f_n \rangle \quad \text{and} \quad J := \langle \det \left(\frac{\partial f_i}{\partial x_j} \right) \rangle.$$

If $N = 0$, then no system in $\mathcal{F}$ has regular zeros. In this case, the statement is true. We
now assume $N > 0$. By (3.2), the variety $V(I : J^\infty)$ consists of all pairs $(\mathbf{x}, \mathbf{q}) \in \mathbb{C}^n \times \mathbb{C}^k$
such that $\mathbf{x}$ is a regular zero of $F(\mathbf{x}; \mathbf{q})$. Since $N > 0$, we therefore have $V(I : J^\infty) \neq \emptyset$.
Let $(\mathbf{x}, \mathbf{q}) \in V(I : J^\infty)$. The Implicit Function Theorem ensures that there is a Euclidean
open neighborhood U of $\mathbf{q}$ such that $F(\mathbf{x}; \mathbf{q})$ has regular zeros for all $\mathbf{q} \in U$. Consequently,

$$(I : J^\infty) \cap \mathbb{C}[\mathbf{p}] = \{0\},$$

so we can apply Lemma 3.14 in our situation.

Set $I_\mathbf{q} = \phi_\mathbf{q}(I)$ and $J_\mathbf{q} = \phi_\mathbf{q}(J)$. Let $G = \{g_1, \ldots, g_s\}$ be a Gröbner basis of $I : J^\infty$ for
the Lex order $x_1 > \cdots > x_n > p_1 > \cdots > p_k$. By Lemma 3.14, there is a proper algebraic
subvariety $\Delta \subsetneq \mathbb{C}^k$ such that $\phi_\mathbf{q}(G) = \{\phi_\mathbf{q}(g_1), \ldots, \phi_\mathbf{q}(g_s)\}$ is a Gröbner basis for $I_\mathbf{q} : J_\mathbf{q}^\infty$
and none of the leading terms of $g_1, \ldots, g_s$ vanish when evaluated at $\mathbf{q}$. This implies that
the leading monomials of $I_\mathbf{q} : J_\mathbf{q}^\infty$ are constant on $\mathbb{C}^k \setminus \Delta$.

We consider the set of *standard monomials*, i.e. monomials not in the lexicographic
initial ideal. We denote this set by

$$\mathcal{B}_\mathbf{q} := \{\text{standard monomials of } I_\mathbf{q} : J_\mathbf{q}^\infty\}.$$

We have shown that $\mathcal{B}_\mathbf{q}$ is constant on $\mathbb{C}^k \setminus \Delta$. On the other hand, by (3.2) and since finite
sets of points are Zariski closed, we have $V(I : J^\infty) = V(I) \setminus V(J)$. Proposition 3.7 and
the fact that regular zeros have multiplicity one imply that the following holds for all
parameters $\mathbf{q}$ that are not in the discriminant Δ: $N(\mathbf{q}) = \# \mathcal{B}_\mathbf{q}$. This shows that $N(\mathbf{q})$ is
constant on $\mathbb{C}^k \setminus \Delta$. The Implicit Function Theorem implies that, for all $\mathbf{q} \in \mathbb{C}^k$, there exists a
Euclidean neighborhood U of $\mathbf{q}$ such that $N(\mathbf{q}) \leq N(\mathbf{q}')$ for all $\mathbf{q}' \in U$. Since Δ is a proper
subvariety of $\mathbb{C}^k$ and thus lower-dimensional, we have $N = N(\mathbf{q}) < \infty$ for $\mathbf{q} \in \mathbb{C}^k \setminus \Delta$. $\quad\square$

We can use the algorithm in Corollary 3.15 to compute the discriminant. However, this
does not necessarily yield the smallest hypersurface with the properties in Theorem 3.18.

Nevertheless, the algorithm in Corollary 3.15 also returns the discriminant when
$F(\mathbf{x}; \mathbf{p}) = 0$ has nonregular solutions for *all* parameters $\mathbf{p}$. Resultant-based methods for
computing the discriminant would fail in such cases because the resultant will be constant
and equal to zero. Here is a simple example to illustrate this phenomenon.

Example 3.19 We slightly modify the system from Example 3.9 and consider

$$F(x, y; a) = \begin{pmatrix} (x - 1) \cdot (x - 2) \cdot (x^2 + y^2 - 1) \\ (y - 1) \cdot (y - a) \cdot (x^2 + y^2 - 1) \end{pmatrix} = 0,$$

where $a \in \mathbb{C}$ is a parameter. If $a \notin \{0, 1, \pm\sqrt{-3}\}$, then we have $N = 4$ regular solutions.

Let us compute the discriminant of the polynomial system $F(x, y; a)$ using the algorithm in Corollary 3.15. We use the `Macaulay2` code from [23]:

```
R = QQ[u, x, y, a, MonomialOrder => Lex];
f = (x-1) * (x-2) * (x^2+y^2-1);
g = (y-1) * (y-a) * (x^2+y^2-1);
I = ideal {f, g};
Jac = matrix{{diff(x, f), diff(x, g)}, {diff(y, f), diff(y, g)}};
K = ideal {1 - u * det(Jac)};
G = gens gb (I+K);
E = (entries(G))#0;
P = QQ[a][u, x, y, MonomialOrder => Lex];
result = apply(E, t -> leadCoefficient(sub(t, P)));
factor(product result)
```

The result is the polynomial $3538944 \cdot a^4 \cdot (a-1)^4 \cdot (a^2+3)^2$.   ◇

Example 3.20 The critical equations for the ED problem in (2.1) have parameters **u**. In many situations, the critical equations form a square system, and Theorem 3.18 applies. The degree N is the Euclidean distance degree. For instance, in Example 2.3, we have a square system in $n = 3$ variables with $k = 3$ parameters, and the ED degree equals $N = d_1 d_2 (d_1 + d_2 + 1)$. What is the discriminant Δ in this case?   ◇

Example 3.21 (Tact invariant) We consider two general quadratic equations in $n = 2$ variables. Each equation has six coefficients, so there are $k = 12$ parameters in total:

```
R = QQ[x,y,a20,a11,a02,a10,a01,a00,b20,b11,b02,b10,b01,b00];
f = a20*x^2 + a11*x*y + a02 * y^2 + a01*x + a10*y + a00;
g = b20*x^2 + b11*x*y + b02 * y^2 + b01*x + b10*y + b00;
```

For general parameters **q**, the set of lexicographic standard monomials is $B_\mathbf{q} = \{1, y, y^2, y^3\}$. Therefore, the number of solutions to our equations is $N = 4$. The main factor in the discriminant Δ is a polynomial in the 12 coefficients that is known as the *tact invariant*. This polynomial can be computed with the following `Macaulay2` code:

```
I = ideal(f, g, diff(x,f)*diff(y,g) - diff(y,f)*diff(x,g));
tact = eliminate({x,y}, I)
```

The tact invariant has degree 12 and is the sum of 3210 monomials. See also [156, §96]. ◇

3.3 Polynomial Homotopy Continuation

In Section 3.1, we have seen how to use Gröbner bases to reduce solving $F(\mathbf{x}) = 0$ to the problem of sequentially computing zeros of univariate polynomials. Another approach is *polynomial homotopy continuation* (PHC). This is a numerical method for computing the regular zeros of a square system of polynomial equations. The textbook of Sommese and Wampler [161] provides a detailed introduction to the theory of polynomial homotopy continuation. We also refer to the overview article [15]. This subject area is known as *numerical algebraic geometry*, and the present section offers a lightning introduction.

The goal in polynomial homotopy continuation is to compute *approximate zeros*. The following definition goes back to Smale (see [21, §8, Definition 1]). It rests on Newton's method from numerical analysis.

Definition 3.22 (Approximate Zeros) Let $F(\mathbf{x})$ be a square system of polynomial equations in n variables, and write $JF(\mathbf{x})$ for its (square) Jacobian matrix. A point $\mathbf{z} \in \mathbb{C}^n$ is called an *approximate zero* of F if the sequence of Newton iterates

$$\mathbf{z}_{k+1} := \mathbf{z}_k - JF(\mathbf{z}_k)^{-1} F(\mathbf{z}_k)$$

starting at $\mathbf{z}_0 := \mathbf{z}$ converges to a zero of F.

An approximate zero $\mathbf{z}$ of a polynomial system F is in a precise sense close to an actual zero $\mathbf{x}$. Applying the Newton operator to $\mathbf{z}$, we can get as close to $\mathbf{x}$ as we want. We can approximate $\mathbf{x}$ to any desired accuracy.

Let $F(\mathbf{x}) = 0$ be the system of polynomial equations that we want to solve. The idea in homotopy continuation is to find a family $\mathcal{F}$ and parameters $\mathbf{p}, \mathbf{q} \in \mathbb{C}^k$ with the properties that $F(\mathbf{x}) = F(\mathbf{x}; \mathbf{p})$ and $G(\mathbf{x}) := F(\mathbf{x}; \mathbf{q})$ is a system whose solutions are known or can be computed by other means. For a piecewise smooth path $\gamma(t)$ in $\mathbb{C}^k$ with $\gamma(1) = \mathbf{q}$ and $\gamma(0) = \mathbf{p}$, we define the *parameter homotopy*

$$H(\mathbf{x}, t) := F(\mathbf{x}; \gamma(t)),$$

and we *track* the solutions of $F(\mathbf{x}; \mathbf{q}) = 0$ to $F(\mathbf{x}; \mathbf{p})$ along the homotopy H. This tracking involves an ordinary differential equation (ODE). Namely, we use numerical algorithms to solve the *ODE initial value problem*

$$\left(\frac{\mathrm{d}}{\mathrm{d}\mathbf{x}} H(\mathbf{x}, t) \right) \frac{\mathrm{d}\mathbf{x}}{\mathrm{d}t} + \frac{\mathrm{d}}{\mathrm{d}t} H(\mathbf{x}, t) = 0, \quad \mathbf{x}(1) = \mathbf{z}. \tag{3.6}$$

Here the initial value $\mathbf{z}$ is a zero of $G(\mathbf{x})$. In this setting, $G(\mathbf{x}) = F(\mathbf{x}; \mathbf{q})$ is called *start system* and $F(\mathbf{x}) = F(\mathbf{x}; \mathbf{p})$ is called *target system*. The output of the numerical solver is then an approximate zero of $F(\mathbf{x})$. In implementations, one often uses piecewise linear paths, such as that described below.

Remark 3.23 The left factor $\left(\frac{\mathrm{d}}{\mathrm{d}\mathbf{x}} H(\mathbf{x}, t) \right)$ in (3.6) is the $n \times n$ Jacobian matrix. Throughout the tracking process, it is essential that this matrix is invertible. This means geometrically that our path must stay away from the discriminant Δ of the polynomial system. This is possible because Δ is a proper subvariety of the parameter space $\mathbb{C}^k$. The dimension of Δ over the real numbers $\mathbb{R}$ is an even number that is less than the real dimension $2k$ of the ambient space $\mathbb{R}^{2k} = \mathbb{C}^k$. This ensures that the space $\mathbb{C}^k \backslash \Delta$ is connected.

There are several software packages for solving polynomial systems that are based on homotopy continuation. In this book we use the software HomotopyContinuation.jl [31].

Example 3.24 We solve the system of polynomial equations from Example 3.5. Recall that those equations are $x^2 + y^2 + z^2 - 1 = x^2 + z^2 - y = x - z = 0$.

The following are commands in the programming language `Julia` [20] on which the package `HomotopyContinuation.jl` is based:

```
using HomotopyContinuation
@var x y z;
f = x^2 + y^2 + z^2 - 1;
g = x^2 + z^2 - y;
h = x - z;
F = System([f; g; h], variables = [x; y; z]);
solve(F)
```

This code returns the four solutions (here displayed with only the 4 most significant digits):

$$(0.556 + 0.0\sqrt{-1}, \ 0.618 - 0.0\sqrt{-1}, \ 0.556 + 0.0\sqrt{-1}),$$
$$(-0.0 - 0.899\sqrt{-1}, \ -1.618 + 0.0\sqrt{-1}, \ -0.0 - 0.899\sqrt{-1}),$$
$$(-0.556 - 0.0\sqrt{-1}, \ 0.618 + 0.0\sqrt{-1}, \ -0.556 + 0.0\sqrt{-1}),$$
$$(0.0 + 0.899\sqrt{-1}, \ -1.618 + 0.0\sqrt{-1}, \ -0.0 + 0.899\sqrt{-1}).$$

These are numerical approximations of the solutions found in Example 3.5. ◇

Remark 3.25 The capabilities of `HomotopyContinuation.jl` were on display in Example 2.7: we solved the critical equations for the ED problem on some complete intersections.

Remark 3.26 Numerical computations are not exact computations and therefore can produce errors. This is hence also true for polynomial homotopy continuation (PHC). It is possible, though, to *certify* the output of PHC. Certification means that we obtain a computer proof that we have indeed computed an approximate zero. There are various certification methods. Current implementations are [30, 83, 122].

The next proposition explains why polynomial homotopy continuation works and when the initial value problem from (3.6) is well-posed. Its proof relies crucially on the Parameter Continuation Theorem (Theorem 3.18), and on the connectedness result in Remark 3.23.

Proposition 3.27 *Let $\mathcal{F}$ be a family of polynomial systems with k parameters. Let $N \geq 2$ and Δ be as in Theorem 3.18. Given $\mathbf{q} \in \mathbb{C}^k \backslash \Delta$ and $\mathbf{p} \in \mathbb{C}^k$, for almost all choices of $\mathbf{p}_{\mathrm{mid}} \in \mathbb{C}^k$, the piecewise linear path*

$$\gamma(t) = \begin{cases} (2t - 1)\mathbf{q} + 2(1 - t)\mathbf{p}_{\mathrm{mid}} & \text{if } \frac{1}{2} \leq t \leq 1, \\ 2t\mathbf{p}_{\mathrm{mid}} + (1 - 2t)\mathbf{p} & \text{if } 0 < t \leq \frac{1}{2}, \end{cases}$$

satisfies:

(a) $\gamma((0, 1]) \cap \Delta = \emptyset$.
(b) The homotopy $H(\mathbf{x}, t) := F(\mathbf{x}; \gamma(t))$ defines N smooth curves $\mathbf{x}(t)$ with the property that $H(\mathbf{x}(t), t) = 0$ for $0 < t \leq 1$. These curves are called solution paths.
(c) As $t \to 0$, the limits of the solution paths include all regular solutions of $F(\mathbf{x}; \mathbf{p}) = 0$.
(d) If moreover $\gamma(0) \notin \Delta$, then every solution path $\mathbf{x}(t)$ converges for $t \to 0$ to a regular zero of $F(\mathbf{x}; \mathbf{p})$.

Proof Since $N \geq 2$, the discriminant $\Delta \subsetneq \mathbb{C}^k$ is a proper complex subvariety. Therefore, Δ is of complex codimension at least 1, hence of real codimension at least 2. This was a recap of Remark 3.23. It implies that, for general parameters $\mathbf{p}_{\text{mid}} \in \mathbb{C}^k$, the path $\gamma(t)$ does not intersect Δ for $t \in (0, 1]$. This proves the first item.

Since $\mathbf{q} = \gamma(1) \notin \Delta$, the Implicit Function Theorem ensures that there exists a Euclidean neighborhood $\mathcal{U}_0 \subset \mathbb{C}^k$ of $\mathbf{q}$ and a smooth *solution map* $s_0 : \mathcal{U}_0 \to \mathbb{C}^n$ such that $F(s_0(\mathbf{p}), \mathbf{p}) = 0$ for all $\mathbf{p} \in \mathcal{U}_0$. Let

$$t_0 := \min\{t \in [0, 1] \mid \gamma(t) \in \overline{\mathcal{U}_0}\},$$

where $\overline{\mathcal{U}_0}$ is the Euclidean closure of $\mathcal{U}_0$. If $t_0 > 0$, then $\gamma(t_0) \notin \Delta$ and we can repeat the construction for the new start system $F(\mathbf{x}; \gamma(t_0))$. Eventually, we obtain an open cover

$$(0, 1] = \bigcup_{i \in I} \mathcal{U}_i,$$

for some index set I, together with smooth solution maps $s_i : \mathcal{U}_i \to \mathbb{C}^n$. Taking a partition of unity $(\rho_i(t))_{i \in I}$ relative to this cover (see [121, Chapter 2]), we set

$$\mathbf{x}(t) := \sum_{i \in I} \rho_i(t) \cdot (s_i \circ \gamma)(t), \quad t \in (0, 1].$$

The path $\mathbf{x}(t)$ is smooth and it satisfies $H(\mathbf{x}(t), t) = 0$. Furthermore, as $t \to 0$, the solution path $\mathbf{x}(t)$ either converges to a point $\mathbf{z}$ or it diverges as $\|\mathbf{x}_i(t)\| \to \infty$. In the first situation, by continuity, $\mathbf{z}$ is a zero of $F(\mathbf{x}, \gamma(0)) = F(\mathbf{x}, \mathbf{p})$, which is not necessarily regular.

By Theorem 3.18, the system $F(\mathbf{x}; \mathbf{q})$ has N regular zeros. The construction above yields N solution paths $\mathbf{x}_1(t), \ldots, \mathbf{x}_N(t)$. Smoothness implies that $\mathbf{x}_i(t) \neq \mathbf{x}_j(t)$ for $i \neq j$ and all $t \in (0, 1]$. This proves the second item. Furthermore, for every regular zero of $F(\mathbf{x}; \mathbf{q})$, we also find a (local) solution map, which connects to exactly one of the smooth paths $\mathbf{x}_1(t), \ldots, \mathbf{x}_N(t)$. This implies the third and fourth items. $\square$

Corollary 3.28 *A general system $F(\mathbf{x}) = (f_1(\mathbf{x}), \ldots, f_n(\mathbf{x}))$ of n polynomials in n variables with $d_i = \deg f_i$ has*

$$N = d_1 \cdots d_n$$

isolated zeros in $\mathbb{C}^n$. (The number $d_1 \cdots d_n$ is called the Bézout number of F.)

Proof We consider the family $\mathcal{F}_{\text{Bézout}}$ of polynomial systems $F(\mathbf{x}) = (f_1(\mathbf{x}), \ldots, f_n(\mathbf{x}))$ with $d_i = \deg f_i$. The parameters are the coefficients of the polynomials $f_1, \ldots, f_n$. Here, we can use the start system

$$G(\mathbf{x}) = \begin{pmatrix} x_1^{d_1} - 1 \\ \vdots \\ x_n^{d_n} - 1 \end{pmatrix}.$$

This system has $d_1 \cdots d_n$ distinct complex zeros, namely $(\xi_1^{k_1}, \ldots, \xi_n^{k_n})$, where $\xi_i := \exp(2\pi\sqrt{-1}/d_i)$ is the d_i-th root of unity and k_i ranges from 1 to d_i. One calls $G(x)$ the *total degree start system*. All its zeros are regular, and it has no zeros at infinity.

Together with Proposition 3.27 this implies that the system $G(\mathbf{x})$ has the maximal number $N = d_1 \cdots d_n$ of regular zeros in $\mathcal{F}_{\text{Bézout}}$. □

Remark 3.29 Corollary 3.28 only states that the number of solutions equals $d_1 \cdots d_n$ for systems outside the discriminant. The full version of Bézout's theorem also applies to systems $F \in \Delta$, where it states that the number of zeros counted with multiplicities is $d_1 \cdots d_n$. Corollary 3.28 does not prove this full version.

Corollary 3.28 implies that one can use the total degree start system for homotopy continuation in the family of systems of polynomials with fixed degree pattern.

Example 3.30 The system $F(x, y, z) = (x^2+y^2+z^2-1, x^2+z^2-y, x-z)$ from Example 3.24 consists of three polynomials of degrees $d_1 = 2, d_2 = 2$ and $d_3 = 1$. The number of zeros of F is the Bézout number $N = d_1 \cdot d_2 \cdot d_3 = 4$. In HomotopyContinuation.jl [31], we can use the total degree start system by setting the following flag:

$$\text{solve(F; start\_system = :total\_degree)}$$

The default option in HomotopyContinuation.jl is the *polyhedral homotopy*. Here the start system is constructed from Newton polytopes, and it respects the mixed volume. ◇

We now introduce sparse systems, their mixed volume, and the polyhedral homotopy.

Example 3.31 Let $A \subset \mathbb{N}^n$ be a finite set and denote $\mathcal{F}_A := \{\sum_{\alpha \in A} c_\alpha \mathbf{x}^\alpha \mid c_\alpha \in \mathbb{C}\}$. An element in $\mathcal{F}_A$ is called a *sparse polynomial*, since only the monomials with exponent vector in A appear. For finite subsets $A_1, \ldots, A_n \subset \mathbb{N}^n$, we consider the family

$$\mathcal{F}_{\text{sparse}} := \mathcal{F}_{A_1} \times \cdots \times \mathcal{F}_{A_n}.$$

The parameters in this family are the coefficients of the n sparse polynomials of a system in $\mathcal{F}_{\text{sparse}}$. For $1 \leq i \leq n$, let P_i be the convex hull of A_i. The polytope P_i is the *Newton polytope* of the sparse polynomials in $\mathcal{F}_{A_i}$. We write $\text{MV}(P_1, \ldots, P_n)$ for the *mixed volume* of these n polytopes. The BKK Theorem [19] asserts that a general $F \in \mathcal{F}_{\text{sparse}}$ has $\text{MV}(P_1, \ldots, P_n)$ many zeros in the torus $(\mathbb{C}^*)^n$. Assuming that such an F only has zeros with non-zero entries, the maximal number of regular zeros in $\mathcal{F}_{\text{sparse}}$ therefore is

$$N = \text{MV}(P_1, \ldots, P_n).$$

If the supports A_i are all the same, then there is only one Newton polytope. Let us denote it by $P := P_1 = \cdots = P_n$. In this situation, which occurs frequently in practice, the mixed volume $\text{MV}(P, \ldots, P)$ equals $n!$ times the volume of P. This product is a nonnegative integer, and it is referred to as the normalized volume of P. Thus the BKK bound for unmixed square systems is the normalized volume of the Newton polytope. ◇

Remark 3.32 The article [98] introduced a combinatorial algorithm for computing an explicit start system for $\mathcal{F}_{\text{sparse}}$, called *polyhedral start system*. See also [15, Section 3]. The use of the polyhedral start system is the default option in HomotopyContinuation.jl.

We close this chapter with a brief discussion for two quadratic equations in two variables. The general system appeared in Example 3.21, where we computed the tact invariant, which serves as the discriminant. The nonvanishing of the tact invariant ensures that the two equations have four distinct complex solutions. Suppose we begin with the total degree start system $x_1^2 = x_2^2 = 1$, which has four solutions $(\pm 1, \pm 1)$. The homotopy in Proposition 3.27 is guaranteed to find the four solutions of the system we wish to solve.

By contrast, suppose now that our two equations are sparse, and the system has the form

$$F(x, y) = \begin{pmatrix} a + bx + cy + dxy \\ \alpha + \beta x + \gamma y + \delta xy \end{pmatrix} \in \mathcal{F}_{\text{sparse}}.$$

Here $\mathbf{p} = (a, b, c, d, \alpha, \beta, \gamma, \delta) \in \mathbb{C}^8$ is a vector of parameters. Both polynomials in $F(x, y)$ have the same Newton polytope P, namely the unit square. The normalized volume of the unit square equals $MV(P) = 2$. Therefore, the BKK Theorem tells us that $F(x, y) = 0$ has $N = 2$ solutions for general parameters $\mathbf{p} \in \mathbb{C}^8$.

The total degree start system $x_1^2 = x_2^2 = 1$ is not appropriate for the sparse family $F(x, y)$ because it has too many solutions. Two of the start solutions $(\pm 1, \pm 1)$ lead to paths that diverge when running the homotopy. Instead, a polyhedral start system [98] has precisely two solutions. A polyhedral start system is obtained by dividing the square P into two triangles, each of normalized area 1. The two zeros of $F(x, y)$ are distinct when the discriminant does not vanish at $\mathbf{p}$. The discriminant of the system $F(x, y)$ is the hyperdeterminant of format $2 \times 2 \times 2$, which we saw in Example 3.16. In other words, the role of the tact invariant for two dense quadrics is now played by our hyperdeterminant. Hyperdeterminants of larger tensors, and the spectral theory of tensors, will be featured in Chapter 12.

Chapter 4
Polar Degrees

The notion of polar degrees is fundamental for assessing the algebraic complexity of polynomial optimization problems of a metric origin. We already recognized this point for Euclidean distance optimization in Section 2.3, and we will see it again in Theorem 5.5 for polyhedral norms, with focus on the Wasserstein metric from optimal transport theory. The punchline is that polar degrees govern linear programming over real algebraic varieties.

This chapter offers a self-contained introduction to *polar degrees*, at a more leisurely pace, and with an emphasis on geometric intuition. We compare three definitions: in terms of non-transversal intersections, Schubert varieties and the Gauss map, and conormal varieties. The latter was used in Definition 2.16. We discuss the behavior of polar degrees under projective duality, and we explain how polar degrees are related to Chern classes. Throughout this chapter, we work over an algebraically closed field of characteristic zero.

4.1 Polar Varieties

Our starting point is a geometric definition of polar degrees, as the degrees of *polar varieties*. Recall from Definition 2.18 that we introduced the polar degrees $\delta_j(X)$ of a projective variety X as the multidegrees of the conormal variety of X. Polar varieties offer an alternative realization. We will prove in Theorem 4.16 below that both multidegrees of the conormal variety and degrees of polar varieties yield the same definition. Until then, we use the symbol $\mu_i(X)$ to denote the i-th polar degree that comes from polar varieties.

To illustrate the concept of polar varieties, we examine the case of surfaces in 3-space.

Example 4.1 Imagine that you look at an algebraic surface $X \subset \mathbb{P}^3$ from a point $\mathbf{v} \in \mathbb{P}^3$. If you want to sketch the surface from your point of view, then you would draw its *contour curve* $P(X, \mathbf{v})$. This is illustrated in Figure 4.1. The contour curve consists of all points $\mathbf{p}$ on the surface X such that the line spanned by $\mathbf{p}$ and $\mathbf{v}$ is tangent to X at $\mathbf{p}$. The *first polar degree* $\mu_1(X)$ is the degree of the contour curve on X for a generic point $\mathbf{v}$.

Suppose that X is defined by a general homogeneous polynomial $f(x_0, x_1, x_2, x_3)$ of degree d, and the viewpoint is $\mathbf{v} = (v_0 : v_1 : v_2 : v_3)$. To compute the contour curve, we

© The Author(s) 2024
P. Breiding et al., *Metric Algebraic Geometry*, Oberwolfach Seminars 53,
https://doi.org/10.1007/978-3-031-51462-3_4

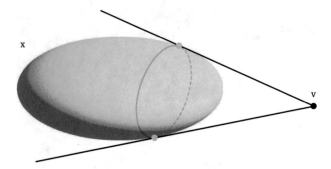

Fig. 4.1: The polar variety $P(X, \mathbf{v})$ is the contour curve when the ellipsoid X is viewed from the point $\mathbf{v}$.

consider the directional derivative $\partial_{\mathbf{v}} f = \sum_{i=0}^{3} v_i \, \partial f / \partial x_i$, which is a polynomial of degree $d - 1$. The contour curve $P(X, \mathbf{v})$ is the variety in $\mathbb{P}^3$ that is defined by f and $\partial_{\mathbf{v}} f$. So, by Bézout's Theorem, the degree of this curve is typically $\mu_1(X) = d(d - 1)$.

Now, we change the setting slightly and imagine that our viewing of the surface X is not centered at a point but at a line $V \subset \mathbb{P}^3$. This time our contour set $P(X, V)$ consists of all points $\mathbf{p}$ on the surface X such that the plane spanned by $\mathbf{p}$ and the line V is tangent at $\mathbf{p}$; see Figure 4.2. For a generic line V, the contour set $P(X, V)$ is finite. The cardinality of $P(X, V)$ is the *second polar degree* $\mu_2(X)$. The finite variety $P(X, V)$ is now defined by $f = \partial_{\mathbf{v}} f = \partial_{\mathbf{w}} f = 0$, where $\mathbf{v}$ and $\mathbf{w}$ are two distinct points on the line V. Using Bézout's Theorem again, we find that $\mu_2(X) = d(d - 1)^2$ for a general surface X of degree d.   ◇

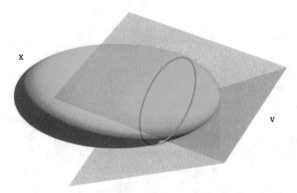

Fig. 4.2: The contour set consists of two points when the ellipsoid X is viewed from the line V.

The contour sets described in the example above are known as polar varieties. To define polar varieties in general, we need to fix some conventions and notations. For instance, the dimension of the empty set is considered to be -1. Given two projective subspaces V

and W in $\mathbb{P}^n$, their projective span (equivalently, their join) is denoted by $V + W \subseteq \mathbb{P}^n$. If the two subspaces V and W are disjoint in projective space then we have

$$\dim(V + W) \;=\; \dim(V) + \dim(W) + 1.$$

Consider any projective variety $X \subseteq \mathbb{P}^n$, with homogeneous ideal $I(X)$ in $\mathbb{C}[x_0, \ldots, x_n]$. We write $\mathrm{Reg}(X)$ for the regular locus of X. We recall that a projective subspace $W \subseteq \mathbb{P}^n$ is said to intersect X *non-transversely* at $\mathbf{p} \in \mathrm{Reg}(X)$ if $\mathbf{p} \in W$ and $\dim(W + T_{\mathbf{p}}X) < n$, where $T_{\mathbf{p}}X$ denotes the embedded tangent space of X at $\mathbf{p}$. For instance, if X is a smooth curve in $\mathbb{P}^3$, then every line that intersects it does so non-transversely, while the tangent planes of X are the only planes that meet X non-transversely.

Definition 4.2 The *polar variety* of a variety $X \subseteq \mathbb{P}^n$ with respect to a subspace $V \subseteq \mathbb{P}^n$ is

$$P(X, V) \;:=\; \overline{\{\,\mathbf{p} \in \mathrm{Reg}(X)\backslash V \;\mid\; V + \mathbf{p} \text{ intersects } X \text{ at } \mathbf{p} \text{ non-transversely}\}}.$$

Let $i \in \{0, 1, \ldots, \dim X\}$. If V is generic with $\dim V = \mathrm{codim}\, X - 2 + i$, then the degree of $P(X, V)$ is independent of V, and we can define

$$\mu_i(X) := \deg(P(X, V)).$$

The integer $\mu_i(X)$ is called the *i-th polar degree of X*.

Example 4.3 A surface $X \subset \mathbb{P}^3$ has three polar degrees. For $i = 0, 1, 2$, the generic subspace V in Definition 4.2 is empty, a point, or a line, respectively. We saw the latter two cases in Example 4.1. For the case $i = 0$, we observe that $P(X, \emptyset) = X$. So, the 0-th polar degree $\mu_0(X)$ is the degree of the surface X. $\diamond$

Example 4.4 The identity $\mu_0(X) = \deg(X)$ holds in general. If $i = 0$, then the dimension of the generic subspace V in Definition 4.2 is $\mathrm{codim}\, X - 2$. Hence, for every $\mathbf{p} \in \mathrm{Reg}(X)$,

$$\begin{aligned}
\dim\big((V + \mathbf{p}) + T_{\mathbf{p}}X\big) &= \dim\big(V + T_{\mathbf{p}}X\big) \\
&= \dim V + \dim X - \dim(V \cap T_{\mathbf{p}}X) \\
&\leq (\mathrm{codim}\, X - 2) + \dim X + 1 \\
&= n - 1.
\end{aligned}$$

This means that $V + \mathbf{p}$ intersects X at $\mathbf{p}$ non-transversely. We conclude that the polar variety for $i = 0$ is the variety itself. In symbols, we have $P(X, V) = X$, and so $\mu_0(X) = \deg(X)$. $\diamond$

Example 4.5 The polar varieties of a plane curve X are $P(X, \emptyset) = X$ and $P(X, \mathbf{v})$ for a point $\mathbf{v}$. The latter is the finite set of points on X whose tangent line passes through $\mathbf{v}$. Those points are contained in the vanishing locus of $\Delta_{\mathbf{v}}f$, where f is the defining polynomial of the curve X and $\Delta_{\mathbf{v}}$ is the differential operator in (1.6). Hence, the polar variety $P(X, \mathbf{v})$ is contained in the intersection of X with its polar curve of order $\deg(X) - 1$ as defined in Section 1.1. For a smooth curve X of degree d, that containment is an equality, and so $\mu_1(X) = |P(X, \mathbf{v})| = d(d - 1)$ by Bézout's Theorem. $\diamond$

We will now give a second definition of polar varieties in terms of the Gauss map and Schubert varieties. As before, we fix a projective subspace $V \subseteq \mathbb{P}^n$. We observe that $V + \mathbf{p}$

intersects X at $\mathbf{p} \in \mathrm{Reg}(X)$ non-transversely (i.e., $n > \dim((V + \mathbf{p}) + T_\mathbf{p}X)$) if and only if

$$\dim(V \cap T_\mathbf{p}X) > \dim V - \mathrm{codim}\, X. \tag{4.1}$$

Since $\dim V - \mathrm{codim}\, X$ is the expected dimension of the intersection of the two projective subspaces V and $T_\mathbf{p}X$, condition (4.1) means that the tangent space $T_\mathbf{p}X$ meets V in an unexpectedly large dimension.

The subspaces that meet V unexpectedly form special instances of *Schubert varieties*:

$$\Sigma_m(V) := \{T \in \mathrm{Gr}(m, \mathbb{P}^n) \mid \dim(V \cap T) > \dim V - n + m\}. \tag{4.2}$$

Here $\mathrm{Gr}(m, \mathbb{P}^n)$ is the *Grassmannian* of m-dimensional projective subspaces in $\mathbb{P}^n$. This is a smooth projective variety of dimension $(m + 1)(n - m)$, embedded in $\mathbb{P}^{\binom{n+1}{m+1}-1}$ via the *Plücker embedding*; see [133, Chapter 5]. For instance, if $m = 1$ and $n = 3$, we get the Grassmannian of lines in $\mathbb{P}^3$. As a projective variety, the Grassmannian $\mathrm{Gr}(1, \mathbb{P}^3)$ consists of all points $\mathbf{x} = (x_{01} : x_{02} : x_{03} : x_{12} : x_{13} : x_{23})$ in the 5-dimensional ambient space $\mathbb{P}^5$ that satisfy the *quadratic Plücker relation* $x_{01}x_{23} - x_{02}x_{13} + x_{03}x_{12} = 0$. Such a point $\mathbf{x}$ represents the line spanned by points $\mathbf{p}$ and $\mathbf{q}$ in $\mathbb{P}^3$ if $x_{ij} = p_i q_j - p_j q_i$ for $0 \leq i < j \leq 3$.

Every variety X of dimension m in $\mathbb{P}^n$ comes with a natural map to the Grassmannian:

$$\gamma_X : X \dashrightarrow \mathrm{Gr}(m, \mathbb{P}^n), \quad \mathbf{p} \mapsto T_\mathbf{p}X.$$

This is the *Gauss map*. It takes each regular point $\mathbf{p}$ on X to its tangent space $T_\mathbf{p}X$. If X is given by polynomial equations, then we compute γ_X by mapping $\mathbf{p}$ to the kernel of the Jacobian of the equations at $\mathbf{p}$. If X is given by a parametrization, then we can represent the Gauss map γ_X by taking derivatives in the parametrization.

Example 4.6 (Twisted cubic curve) Let X be the curve in $\mathbb{P}^3$ with the parametrization

$$\nu : \mathbb{P}^1 \to \mathbb{P}^3, \quad (s : t) \mapsto (s^3 : s^2t : st^2 : t^3).$$

The tangent space of X at the point $\mathbf{p} = \nu((s : t))$ is the line in $\mathbb{P}^3$ spanned by the rows of the Jacobian matrix

$$\begin{bmatrix} 3s^2 & 2st & t^2 & 0 \\ 0 & s^2 & 2st & 3t^2 \end{bmatrix}.$$

The Plücker coordinates of the tangent line $\gamma_X(\mathbf{p})$ are the 2×2 minors of this 2×4 matrix. After dividing the minors by 3, we obtain

$$x_{01} = s^4, \quad x_{02} = 2s^3t, \quad x_{03} = 3s^2t^2, \quad x_{12} = s^2t^2, \quad x_{13} = 2st^3, \quad x_{23} = t^4. \tag{4.3}$$

This shows that the image of the Gauss map of the twisted cubic curve X is a rational normal curve of degree four in the Grassmannian of lines $\mathrm{Gr}(1, \mathbb{P}^3) \subset \mathbb{P}^5$. ◇

We now turn to the Schubert varieties $\Sigma_m(V)$. These varieties are cut out by linear equations in Plücker coordinates.

Example 4.7 ($m = 1, n = 3$) We consider three types of Schubert varieties of lines. If V is a point in $\mathbb{P}^3$, then $\Sigma_1(V) = \mathrm{Gr}(1, \mathbb{P}^3)$. If V is a plane in $\mathbb{P}^3$, then $\Sigma_1(V)$ is the surface of all lines in that plane. Finally, suppose that V is a line in $\mathbb{P}^3$, with Plücker coordinates v_{ij}. Then $\Sigma_1(V)$ is defined by the linear equation

$$v_{01}x_{23} - v_{02}x_{13} + v_{03}x_{12} + v_{12}x_{03} - v_{13}x_{02} + v_{23}x_{01} = 0. \tag{4.4}$$

This Schubert variety is a threefold. It consists of all lines that intersect the given line V. ⋄

We now come to our punchline, which is the second definition of polar varieties.

Proposition 4.8 *Fix a variety X of dimension m in $\mathbb{P}^n$. For any subspace V of $\mathbb{P}^n$, the polar variety $P(X,V)$ is the pullback of the Schubert variety $\Sigma_m(V)$ under the Gauss map, i.e.*

$$P(X,V) = \overline{\gamma_X^{-1}\left(\Sigma_m(V)\right)}.$$

Proof The condition (4.1) is equivalent to $T_{\mathbf{p}}X \in \Sigma_m(V)$. □

Example 4.9 (Twisted cubic curve) In Example 4.7, if V is a point in $\mathbb{P}^3$, then we have $P(X,V) = \overline{\gamma_X^{-1}\left(\mathrm{Gr}(1,\mathbb{P}^3)\right)} = X$. For a line V, we compute $P(X,V) = \overline{\gamma_X^{-1}\left(\Sigma_1(V)\right)}$ by substituting (4.3) into (4.4). The result is a binary quartic in (s,t). This quartic has four zeros in $\mathbb{P}^1$. Their image under the map v is the polar variety $P(X,V)$. Thus $P(X,V)$ consists of four points. The polar degrees of our curve X are $\mu_0(X) = 3$ and $\mu_1(X) = 4$. ⋄

Corollary 4.10 *An inclusion of projective linear subspaces gives a reverse inclusion of polar varieties. To be precise, for any fixed variety X in $\mathbb{P}^n$ and subspaces $V \subset V'$, we have $P(X,V') \subset P(X,V)$.*

Proof Use Proposition 4.8 and the inclusion $\Sigma_m(V') \subset \Sigma_m(V)$ of Schubert varieties. □

4.2 Projective Duality

Duality is a fundamental concept in many fields of mathematics, including convexity, optimization, and algebraic geometry. Primal-dual approaches in polynomial optimization can be understood via the duality theory of projective algebraic geometry. We saw this in Chapter 2, in the context of polynomial optimization, where the conormal variety plays a key role for computing the ED degree in Theorem 2.13.

This section offers a second point of entry, by offering a self-contained introduction to duality theory of projective varieties, presented in a friendlier and more geometric way than in Section 2.3. There is a one-to-one correspondence between hyperplanes H in $\mathbb{P}^n$ and points $H^\vee$ in the dual projective space $(\mathbb{P}^n)^*$. We recall from our discussion in Section 2.3 that the dual variety $X^\vee$ of a projective variety $X \subseteq \mathbb{P}^n$ is the projection of the conormal variety N_X to the second factor. It parametrizes all tangent hyperplanes:

$$X^\vee := \overline{\left\{H^\vee \in (\mathbb{P}^n)^* \mid \exists \mathbf{p} \in \mathrm{Reg}(X) : T_{\mathbf{p}}X \subseteq H\right\}}.$$

This generalizes the dual of a plane curve in (1.7). In Chapter 2 we used the Euclidean inner product to identify projective space with its dual space. Here, we defined the dual variety without this identification. The following result is found in Section I.1.3 of the book by Gel'fand, Kapranov and Zelevinsky [71].

Theorem 4.11 (Biduality) *Fix a projective variety* $X \subseteq \mathbb{P}^n$, *and let* $\mathbf{p} \in \mathrm{Reg}(X)$ *and* $H^\vee \in \mathrm{Reg}(X^\vee)$. *The hyperplane* H *is tangent to the variety* X *at the point* $\mathbf{p}$ *if and only if the hyperplane* $\mathbf{p}^\vee$ *is tangent to the dual variety* $X^\vee$ *at the point* $H^\vee$. *In particular, we have*

$$(X^\vee)^\vee = X.$$

Example 4.12 If X is a projective subspace of $\mathbb{P}^n$, given as the row space of a matrix, then its dual $X^\vee$ is the subspace of $(\mathbb{P}^n)^*$ given by the kernel of that matrix, so $\dim X^\vee = n - 1 - \dim X$. Note that $(\mathbb{P}^n)^\vee = \emptyset$. ◇

Example 4.13 (Twisted cubic curve) Let $n = 3$ and fix coordinates $\mathbf{p} = (p_0 : p_1 : p_2 : p_3)$ on $\mathbb{P}^3$ and $H = (h_0 : h_1 : h_2 : h_3)$ on $(\mathbb{P}^3)^*$. The twisted cubic curve X in Example 4.6 is defined by the prime ideal

$$\langle\, p_0 p_2 - p_1^2, \ p_0 p_3 - p_1 p_2, \ p_1 p_3 - p_2^2 \,\rangle.$$

Its dual variety $X^\vee$ is a quartic surface in $(\mathbb{P}^3)^*$. The equation of $X^\vee$ is the *discriminant* of a binary cubic:

$$27 h_0^2 h_3^2 - 18 h_0 h_1 h_2 h_3 + 4 h_0 h_2^3 + 4 h_1^3 h_3 - h_1^2 h_2^2. \tag{4.5}$$

Metric geometry of discriminants will become important in Chapter 9. At this point, our readers are encouraged to check computationally that $(X^\vee)^\vee = X$. ◇

Example 4.14 Fix a surface $X \subset \mathbb{P}^3$. Figure 4.2 illustrates the polar variety $P(X, V)$ for a generic line V. The tangent planes through V correspond in $(\mathbb{P}^3)^*$ to points on the dual variety $X^\vee$ that lie on the line $V^\vee$; see Figure 4.3. Hence, if $X^\vee$ is a surface as well, then its degree is the second polar degree of X, i.e., $\mu_2(X) = \deg(X^\vee)$. Otherwise, if the dual variety $X^\vee$ is a curve, then the line $V^\vee$ misses it and $\mu_2(X) = 0$. This happens, for instance, when X is the quartic surface (4.5), now rewritten in primal coordinates $\mathbf{p}$.

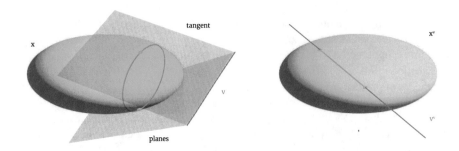

Fig. 4.3: A pair of dual surfaces X and $X^\vee$ in projective 3-space satisfies $\mu_2(X) = \deg(X^\vee)$.

Figure 4.1 shows the polar curve $P(X, \mathbf{v})$ of a surface $X \subset \mathbb{P}^3$ for a generic point $\mathbf{v}$. This consists of all points $\mathbf{p}$ on X whose tangent plane contains $\mathbf{v}$. We find $\mu_1(X) = \deg(P(X, \mathbf{v}))$

by intersecting the curve with a generic plane H. Hence, the first polar degree $\mu_1(X)$ counts all regular points $\mathbf{p} \in X$ such that

$$\mathbf{p} \in H \quad \text{and} \quad \mathbf{v} \in T_{\mathbf{p}}X. \tag{4.6}$$

These tangent planes correspond to points $\mathbf{q} := (T_{\mathbf{p}}X)^{\vee}$ on the dual variety $X^{\vee}$. Using the biduality in Theorem 4.11, we have $T_{\mathbf{q}}X^{\vee} = \mathbf{p}^{\vee}$ if the dual variety $X^{\vee}$ is a surface. In that case, the conjunction in (4.6) is equivalent to

$$H^{\vee} \in T_{\mathbf{q}}X^{\vee} \quad \text{and} \quad \mathbf{q} \in \mathbf{v}^{\vee}. \tag{4.7}$$

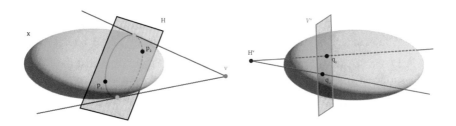

Fig. 4.4: The identity $\mu_1(X) = \mu_1(X^{\vee})$ holds for pairs of dual surfaces in projective 3-space.

Comparing (4.7) with (4.6), we see that the point-plane pair $(H^{\vee}, \mathbf{v}^{\vee})$ imposes the same conditions on points $\mathbf{q} \in X^{\vee}$ as the point-plane pair $(\mathbf{v}, H)$ imposes on points $\mathbf{p} \in X$; see Figure 4.4. By the genericity of $(\mathbf{v}, H)$, we conclude that $\mu_1(X) = \mu_1(X^{\vee})$ if $X^{\vee}$ is a surface. If $X^{\vee}$ is a curve, then $T_{\mathbf{q}}X^{\vee} \subsetneq \mathbf{p}^{\vee}$ and so the only condition imposed by (4.6) on the points $\mathbf{q} \in X^{\vee}$ is that they must lie in $\mathbf{v}^{\vee}$. Here, $\mu_1(X) = |X^{\vee} \cap \mathbf{v}^{\vee}| = \deg(X^{\vee}) = \mu_0(X^{\vee})$. Finally, if $X^{\vee}$ is a point (i.e., X is a plane), then $\mu_1(X) = 0$.   ◇

The observed relations between the polar degrees of a variety and its dual are true in great generality. The sequence of polar degrees of a projective variety X equals the sequence of polar degrees of its dual variety $X^{\vee}$ in reversed order. As seen in Example 4.4, the first non-zero entry is $\deg(X)$, and its last non-zero entry is the degree of the dual $X^{\vee}$.
We summarize these key properties of polar degrees:

Theorem 4.15 *Let X be an irreducible projective variety, and*

$$\alpha(X) := \dim X - \operatorname{codim} X^{\vee} + 1.$$

The following hold for the polar degrees of X:

(a) $\mu_i(X) > 0 \quad \Leftrightarrow \quad 0 \leq i \leq \alpha(X)$.
(b) $\mu_0(X) = \deg X$.
(c) $\mu_{\alpha(X)}(X) = \deg X^{\vee}$.
(d) $\mu_i(X) = \mu_{\alpha(X)-i}(X^{\vee})$.

The ideas discussed in Example 4.14 can be turned into formal proofs for almost all assertions in Theorem 4.15. The direction "$\Leftarrow$" in (a) is a bit tricky. For details, we refer to the article by Holme [91]. Another proof strategy is to first establish the relation of the polar degrees with the conormal variety N_X (see Definition 2.16). Recall from Definition 2.18 that we have defined the polar degrees of X to be the multidegrees of N_X. More precisely, for generic projective subspaces L_1 and L_2, we have

$$\delta_j(X) := |N_X \cap (L_1 \times L_2)|, \quad \text{where } \dim L_1 = n + 1 - j \text{ and } \dim L_2 = j.$$

We saw in Section 2.3 that the multidegree is the cohomology class of the conormal variety.

Theorem 4.16 *The multidegree agrees with the polar degrees. To be precise, we have* $\delta_j(X) = \mu_i(X)$, *where* $i := \dim X + 1 - j$.

This theorem and the biduality relation $N_X = N_{X^\vee}$ in (2.15) imply Theorem 4.15 (d). Using Example 4.4, they also imply (b) and (c) in Theorem 4.15. The direction "$\Rightarrow$" in Theorem 4.15 (a) can also be deduced directly from the definition of the $\delta_j(X)$.

Example 4.17 We see from Figure 4.4 and conditions (4.6) and (4.7) that the first polar degree of a surface X in $\mathbb{P}^3$ equals $\mu_1(X) = |N_X \cap (H \times V^\vee)| = \delta_2(X)$. Hence Theorem 4.16 holds for surfaces X in $\mathbb{P}^3$.   ◇

Proof (of Theorem 4.16)   We present the main ideas of the proof. For complete proofs, we refer to [106, Proposition 3 on page 187] or [70, Lemma 2.23 on page 169].

First, we remark that the formulation of the conormal variety in Chapter 2 used the Euclidean inner product to identify projective space with its dual. In this chapter, we do not use the Euclidean structure. Instead, we formulate the relevant notions using abstract duality. The conormal variety is

$$N_X = \overline{\left\{ (\mathbf{p}, H^\vee) \in \mathbb{P}^n \times (\mathbb{P}^n)^* \mid \mathbf{p} \in \mathrm{Reg}(X),\, T_\mathbf{p} X \subseteq H \right\}}.$$

The projection of the conormal variety N_X onto the first factor $\mathbb{P}^n$ is the variety X we started with. The projection onto the second factor $(\mathbb{P}^n)^*$ is the dual variety $X^\vee$.

We compute the multidegrees of N_X. Let $L_1 \subseteq \mathbb{P}^n$ and $L_2 \subseteq (\mathbb{P}^n)^*$ be generic subspaces of dimensions $n + 1 - j$ and j, respectively. Set $V := L_2^\vee$. Note that the subspace V has the correct dimension for the computation of the i-th polar degree, where $i = \dim X + 1 - j$. This follows from $\dim V = n - j - 1 = \mathrm{codim}\, X - 2 + i$. We now consider a generic point $(\mathbf{p}, H^\vee)$ in the intersection $N_X \cap (\mathbb{P}^n \times L_2)$. Then $\mathbf{p}$ is regular point of X. Both its tangent space $T_\mathbf{p} X$ and $V = L_2^\vee$ are contained in the hyperplane H. In particular, we have $\dim(V + T_\mathbf{p} X) < n$. Hence $\mathbf{p}$ is in the polar variety $P(X, V)$. In fact, the projection $N_X \cap (\mathbb{P}^n \times L_2) \to P(X, V)$ onto the first factor is birational. Hence,

$$\mu_i(X) = \deg(P(X, V)) = |P(X, V) \cap L_1| = |N_X \cap (L_1 \times L_2)| = \delta_j(X),$$

which yields the stated formula.   □

Many varieties in applications are defined by rank constraints on matrices or tensors. Determinantal varieties and Segre–Veronese varieties will make an appearance in several subsequent chapters. It is thus worthwhile to examine their polar varieties and polar degrees.

Example 4.18 (A Segre variety) We revisit Example 2.22, where an explicit multidegree was computed. Let X be the variety of 3×3 matrices of rank 1. This is a 4-dimensional smooth subvariety of $\mathbb{P}^8$. As an abstract variety, we have $X = \mathbb{P}^2 \times \mathbb{P}^2$. The prime ideal of X is generated by the nine 2×2 minors of the 3×3 matrix. The prime ideal of the conormal variety N_X is the ideal I defined in the Macaulay2 fragment in Example 2.22.

In light of Theorem 4.16, the polar degrees of the Segre variety X are

$$\mu_0(X) = 6, \ \mu_1(X) = 12, \ \mu_2(X) = 12, \ \mu_3(X) = 6, \ \mu_4(X) = 3.$$

Here $\alpha(X) = 4$. The dual variety $X^\vee$ is a cubic hypersurface in $\mathbb{P}^8$. Its defining polynomial is the 3×3-determinant. This was called minors(3,h) in Example 2.22. By Theorem 4.15 (d), we have

$$\mu_0(X^\vee) = 3, \ \mu_1(X^\vee) = 6, \ \mu_2(X^\vee) = 12, \ \mu_3(X^\vee) = 12, \ \mu_4(X^\vee) = 6.$$

Can you describe the polar varieties $P(X, V)$ and $P(X^\vee, V)$ in the language of linear algebra? Which matrices do they contain for a given linear subspace V of the matrix space $\mathbb{R}^{3 \times 3}$? How about symmetric 3×3 matrices?   ⋄

4.3 Chern Classes

In what follows we take a look at a concept that is ubiquitous in differential geometry, algebraic geometry, and algebraic topology. These three fields are now conspiring to furnish mathematical foundations for data science. It is thus natural for our book to devote one section to an important theoretical topic in their intersection.

Chern classes are topological invariants associated with vector bundles on smooth manifolds or varieties. If X is a smooth and irreducible projective variety, then the polar degrees $\mu_i(X)$ can be computed from the Chern classes of X. The formulas will be presented in Theorem 4.20. In light of Theorems 2.13 and 5.5, we can obtain formulas, in terms of Chern classes, for both ED degrees and Wasserstein degrees.

To a vector bundle $\mathcal{E}$ of rank r on a variety X, we associate the Chern classes $c_0(\mathcal{E}), \ldots, c_r(\mathcal{E})$. Formally, these are elements in the *Chow ring* of X. Chern classes are understood more easily when the vector bundle $\mathcal{E}$ is globally generated. Assuming that this holds for $\mathcal{E}$, we choose general global sections $\sigma_1, \ldots, \sigma_j : X \to \mathcal{E}$. These j global sections determine the following subvariety of the given variety X:

$$D(\sigma_1, \ldots, \sigma_j) := \{ x \in X \mid \sigma_1(x), \ldots, \sigma_j(x) \text{ are linearly dependent} \}. \qquad (4.8)$$

Experts refer to (4.8) as a *degeneracy locus* of the vector bundle $\mathcal{E}$. In concrete scenarios, $D(\sigma_1, \ldots, \sigma_j)$ is the determinantal variety given by the maximal minors of a certain $r \times j$ matrix with linear entries. We are interested in the class of this variety in the Chow ring.

Definition 4.19 The *Chern class* $c_{r+1-j}(\mathcal{E})$ is the class of (4.8) in the Chow ring of X.

For the purpose of this section, it is not necessary for the reader to master the definition of the Chow ring. It suffices for us to understand the *degree* of $c_{r+1-j}(\mathcal{E})$. This degree is a number, not a class. It is defined as the degree of the degeneracy locus $D(\sigma_1, \ldots, \sigma_j)$ as a projective variety, for general sections σ_i of $\mathcal{E}$. For instance, the degree of the *top Chern class* $c_r(\mathcal{E})$ is the degree of the vanishing locus of a single general global section.

There are some calculation rules that allow us to compute Chern classes of a vector bundle of interest in terms of simpler vector bundles. Most notable is the *Whitney sum formula* [69, Theorem 3.2]. This applies when we have a short exact sequence of vector bundles $0 \to \mathcal{E}' \to \mathcal{E} \to \mathcal{E}'' \to 0$. It states that

$$c_k(\mathcal{E}) = \sum_{i+j=k} c_i(\mathcal{E}') c_j(\mathcal{E}'').$$

Every smooth, irreducible variety X has a distinguished vector bundle, namely the *tangent bundle* $\mathcal{T}X$. The Chern class $c_k(X)$ of X is an abbreviation for the Chern class $c_k(\mathcal{T}X)$.

Theorem 4.20 ([91, Equation (3)]) *Let X be a smooth, irreducible projective variety, and let $m := \dim X$. Then,*

$$\mu_i(X) = \sum_{k=0}^{i} (-1)^k \binom{m-k+1}{m-i+1} \deg(c_k(X)). \tag{4.9}$$

This formula can be inverted to write degrees of Chern classes in terms of polar degrees:

$$\deg(c_k(X)) = \sum_{i=0}^{k} (-1)^i \binom{m-i+1}{m-k+1} \mu_i(X). \tag{4.10}$$

Remark 4.21 Both formulas also hold for singular varieties, after replacing the classical Chern classes with Chern–Mather classes. That result is due to Piene [145].

An important difference between polar degrees and Chern classes is the following: Polar degrees are projective invariants of the embedded variety $X \subseteq \mathbb{P}^n$. This holds also more generally for the *polar classes*, i.e., the rational equivalence classes (in the Chow ring of X) of the polar varieties. Chern classes are even *intrinsic invariants* of the variety X, i.e., they do not depend on the embedding of X in projective space.

Example 4.22 Here are some basic facts about the Chern classes of a smooth variety X.

(a) We see from (4.10) that $\deg(c_0(X)) = \mu_0(X) = \deg X$.
(b) The degree of the top Chern class of X equals its topological Euler characteristic:

$$\deg(c_m(X)) = \chi(X), \quad m = \dim X.$$

(c) If X is a curve of genus $g(X)$, then $\deg(c_1(X)) = \chi(X) = 2 - 2g(X)$ is independent of the embedding, while we see from (4.9) that

$$\mu_1(X) = 2 \deg X - \deg(c_1(X)) = 2(\deg X + g(X) - 1), \tag{4.11}$$

which does depend on the embedding of X.   ◇

Example 4.23 If $X \subset \mathbb{P}^n$ is a rational curve, then $\mu_1(X) = 2 \deg X - 2$ by (4.11). We can check this by examining the various cases:

- If X is a line, then its dual variety $X^\vee$ is never a hypersurface, and so $\mu_1(X) = 0$.
- If X is a conic (i.e., $\deg X = 2$) then $X^\vee$ is a (cone over a) conic, and so we have that $\mu_1(X) = \deg X^\vee = 2$.
- If X is a twisted cubic (i.e., $\deg X = 3$), then $X^\vee$ is (a cone over) the discriminant in (4.5). Therefore, $\mu_1(X) = \deg X^\vee$ is the degree of that discriminant, which is 4. See Example 4.13.
- If X is a rational normal curve of degree d, then its dual variety is (a cone over) the discriminant of a binary form of degree d. Its degree is $\mu_1(X) = \deg X^\vee = 2d - 2$. ⋄

We close with the expression for the ED degree in terms of Chern classes.

Corollary 4.24 *Let X be a smooth variety of dimension m in $\mathbb{P}^n$ which satisfies the hypotheses in Theorem 2.13. These always hold after a general linear change of coordinates. We have*

$$\mathrm{EDdegree}(X) = \sum_{i=0}^{m} (-1)^i \, (2^{m+1-i} - 1) \deg(c_i(X)).$$

Proof Equation (2.19) in Theorem 2.23 shows that the ED degree is the degree of the conormal variety N_X. By Theorem 4.16, this is the sum of the multidegrees, and hence the sum of the polar degrees. See Theorem 2.13. We now simply take the sum of the alternating sums in (4.9) for $i = 0, 1, \ldots, m$. □

As an application, we now compute the generic ED degree of the Veronese variety.

Example 4.25 Let $n = \binom{m+d}{d} - 1$ and let $X \subset \mathbb{P}^n$ be the d-th Veronese embedding of $\mathbb{P}^m$. The generic ED degree of X from Definition 2.8 satisfies

$$\mathrm{EDdegree}_{\mathrm{gen}}(X) = \frac{(2d-1)^{m+1} - (d-1)^{m+1}}{d}. \tag{4.12}$$

This is precisely the ED degree of the image of X under a generic linear change of coordinates in $\mathbb{P}^n$. Corollary 4.24 is a formula for the generic ED degree.

We shall now derive (4.12) from Corollary 4.24. Our argument will be taken from the proof of [60, Proposition 7.10]. The ith Chern class of the underlying projective space $\mathbb{P}^m$ is $c_i(\mathbb{P}^m) = \binom{m+1}{i} h^i$ where h is the hyperplane class in $\mathbb{P}^m$. Since X is the image under the dth Veronese embedding of $\mathbb{P}^m$, its hyperplane class is dh. The degree of its Chern class $c_i(X)$ is the integral of $(dh)^{m-i} c_i(\mathbb{P}^m)$ over $\mathbb{P}^m$. Since $h^{m-i} h^i = h^m = 1$, the numerical value of this formal integral (in the Chow ring of $\mathbb{P}^m$) equals $\deg(c_i(X)) = \binom{m+1}{i} d^{m-i}$. We plug this into Corollary 4.24. After some algebraic manipulations, we arrive at (4.12). For $m = 1$ the formula (4.12) evaluates to $d + (2d - 2) = \mu_0(X) + \mu_1(X)$, as desired. ⋄

In conclusion, Chern classes provide a conceptual framework for the degrees of optimization problems in metric algebraic geometry. This chapter offered a geometric introduction. Example 4.25 is a nice illustration of how Chern classes are used in practice.

Chapter 5
Wasserstein Distance

A fundamental problem in metric algebraic geometry is distance minimization. We seek a point in a variety X in $\mathbb{R}^n$ that is closest to a given data point $\mathbf{u} \in \mathbb{R}^n$. Thus, we wish to

$$\text{minimize } ||\mathbf{x} - \mathbf{u}|| \text{ subject to } \mathbf{x} \in X. \qquad (5.1)$$

In Chapter 2 we studied this problem for the Euclidean distance on $\mathbb{R}^n$. We here examine (5.1) for the case when the distance is given by a *polyhedral norm*. We note that the minimum in (5.1) is always attained because X is non-empty and closed. Hence, there exists at least one optimal solution. If that solution is unique then we denote it by $\mathbf{x}^*$. In Sections 5.2 and 5.3 we focus on a special class of polyhedral norms that arise from optimal transport theory. The corresponding distance is known as *Wasserstein distance*.

5.1 Polyhedral Norms

A norm $|| \cdot ||$ on the real vector space $\mathbb{R}^n$ is a *polyhedral norm* if its unit ball is polyhedral:

$$B = \{\mathbf{x} \in \mathbb{R}^n \mid ||\mathbf{x}|| \leq 1\}.$$

More precisely, B is a centrally symmetric convex polytope. Conversely, every centrally symmetric convex polytope B in $\mathbb{R}^n$ defines a polyhedral norm on $\mathbb{R}^n$. Using the unit ball, we can paraphrase the optimization problem (5.1) as follows:

$$\text{minimize } \lambda \text{ subject to } \lambda \geq 0 \text{ and } (\mathbf{u} + \lambda B) \cap X \neq \emptyset. \qquad (5.2)$$

Example 5.1 The unit ball $B \subset \mathbb{R}^n$ of the ∞-norm $||\mathbf{x}||_\infty = \max_{1 \leq i \leq n} |x_i|$ is the regular cube $[-1, +1]^n$. The unit ball of the dual norm $||\mathbf{x}||_1 := |x_1| + \cdots + |x_n|$ is the convex polytope $\text{conv}\{\pm\mathbf{e}_1, \pm\mathbf{e}_2, \ldots, \pm\mathbf{e}_n\} \subset \mathbb{R}^n$. The latter polytope is called the *crosspolytope*. It generalizes the octahedron from $n = 3$ to $n \geq 4$. ◇

© The Author(s) 2024
P. Breiding et al., *Metric Algebraic Geometry*, Oberwolfach Seminars 53,
https://doi.org/10.1007/978-3-031-51462-3_5

Polyhedral norms are very important in optimal transport theory, where one uses a Wasserstein norm on the space of probability distributions. Polytopes arise for distributions on finite state spaces. This will be our main application, to be developed later in this chapter.

We begin our discussion with a general polyhedral norm; that is, we allow the unit ball B to be an arbitrary n-dimensional centrally symmetric polytope in $\mathbb{R}^n$. As before, we use the Euclidean inner product $\langle \cdot, \cdot \rangle$. Recall that a subset F of the polytope B is called a *face* if there exists a vector $\mathbf{v} \in \mathbb{R}^n \setminus \{\mathbf{0}\}$ such that

$$F = \{\mathbf{x} \in B \mid \langle \mathbf{x}, \mathbf{v} \rangle \geq \langle \mathbf{y}, \mathbf{v} \rangle \text{ for all } \mathbf{y} \in B \}.$$

One says that the face F maximizes the linear functional $\ell(\mathbf{x}) := \langle \mathbf{x}, \mathbf{v} \rangle$. The vector $\mathbf{v}$ is a normal vector of F. The boundary of B consists of faces whose dimensions range from 0 to $n - 1$. Faces of maximal dimension are called *facets*. The set of all faces, ordered by inclusion, is a partially ordered set, called the *face poset* of B. For an introduction to polytopes see Ziegler's book [182]. An important combinatorial invariant is the f-*vector*.

Definition 5.2 Let $B \subset \mathbb{R}^n$ be a polytope. The f-vector of B is $f(B) = (f_0, f_1, \ldots, f_{n-1})$, where f_i denotes the number of i-dimensional faces of B, for $0 \leq i \leq n - 1$.

The dual of the unit ball B is also a centrally symmetric polytope, namely it is the set

$$B^* = \{\mathbf{v} \in \mathbb{R}^n \mid \langle \mathbf{x}, \mathbf{v} \rangle \leq 1 \text{ for all } \mathbf{x} \in B \}.$$

The norm $|| \cdot ||_*$ defined by the dual polytope B^* is dual to the norm $|| \cdot ||$ given by B. The f-vector of B^* is the reverse of the f-vector of B. More precisely, we have

$$f_i(B^*) = f_{n-1-i}(B) \quad \text{for} \quad i = 0, 1, \ldots, n - 1.$$

Example 5.3 Fix the unit cube $B = [-1, 1]^n$. Its dual is the crosspolytope

$$B^* = \text{conv}\{\pm \mathbf{e}_1, \pm \mathbf{e}_2, \ldots, \pm \mathbf{e}_n\} \subset \mathbb{R}^n.$$

Here $\mathbf{e}_j$ is the jth standard basis vector. The number of i-dimensional faces of the cube is $f_i(B) = \binom{n}{i} \cdot 2^{n-i}$. Consequently, its f-vector is $f(B) = (2^n, 2^{n-1} n, \ldots, 2n) \in \mathbb{R}^n$.

The 3-dimensional crosspolytope is the octahedron. The cube B has 8 vertices, 12 edges and 6 facets. By duality, the octahedron B^* has 6 vertices, 12 edges, and 8 facets.

Their f-vectors are $f(B) = (8, 12, 6)$ and $f(B^*) = (6, 12, 8)$. These numbers govern the combinatorial structure of the associated polyhedral norms $|| \cdot ||_\infty$ and $|| \cdot ||_1$ on $\mathbb{R}^3$.   ◇

We now turn to the optimization problem given in (5.1) or (5.2). To derive the critical equations, we shall use a combinatorial stratification of the problem by the face poset of the polytope B. The next lemma associates a unique face F of B to the optimal point $\mathbf{x}^*$.

We write L_F for the linear span of the face F in $\mathbb{R}^n$. If F has dimension $j < n$, then L_F has dimension $j + 1$, because the origin lies in the interior of B. For instance, if F is a vertex, then L_F is the line through F and the origin. If F is a facet, then $L_F = \mathbb{R}^n$.

Lemma 5.4 *Suppose that X is in general position. Given $\mathbf{u} \in \mathbb{R}^n$, let $\mathbf{x}^*$ be an optimal solution in (5.2) and let λ^* be the optimal value. Then $\mathbf{x}^*$ is unique, and the point $\frac{1}{\lambda^*}(\mathbf{x}^* - \mathbf{u})$ lies in the relative interior of a unique face F of the polytope B. Let ℓ_F be a linear functional whose maximum over B is attained on F. Then, the optimal point $\mathbf{x}^*$ in (5.1) can be recovered as the unique solution of the optimization problem*

$$\text{Minimize} \quad \ell_F(\mathbf{x}) \quad \text{subject to } \mathbf{x} \in (\mathbf{u} + L_F) \cap X. \tag{5.3}$$

Proof We defined λ^* to be the optimal value of (5.2). By construction, it is also the optimal value of (5.3). The general position hypothesis ensures that the affine space $\mathbf{u}+L_F$ intersects the real variety X transversally, and $\mathbf{x}^*$ is a smooth point of that intersection. The genericity assumption ensures that $(\mathbf{u} + \lambda^* B) \cap X = \{\mathbf{x}^*\}$ is a singleton. Hence, the optimal point $\mathbf{x}^*$ is unique, and it is also the minimizer of the linear functional ℓ_F on the variety $(\mathbf{u}+L_F)\cap X$. By our hypothesis, this linear function is generic relative to the variety, so the number of critical points is finite and the function values are distinct.   □

Lemma 5.4 motivates the following strategy for the distance minimization problem (5.1): For each of the finitely many faces F of B, solve the linear program (5.3) over X, and determine the distance from $\mathbf{u}$ to X from this finite amount of data. This splits into three subtasks: Combinatorial Preprocessing, Numerical Optimization, and (optionally) Algebraic Postprocessing. We give a high-level description of these steps in Algorithms 1–3.

Algorithm 1: Combinatorial Preprocessing

Output: A description of all faces F of the unit ball B.
1 Depending on how B is presented, apply appropriate tools from computational geometry to compute all its faces.
2 For each face F, fix a linear functional ℓ_F that is maximized on F, and let L_F be the linear span of the vertices of F.
3 **return** *the list of pairs (ℓ_F, L_F), one for each face of B.*

The complexity of computing the distance to a model X in the polyhedral norm has two components, seen clearly in (5.4). One is the combinatorial complexity of the unit ball B, which is encoded in the f-vector $f(B)$. This complexity affects Algorithm 1. Admittedly, we did not specify any details for the computation of the face lattice of B. Instead, we refer the reader to the vast literature on algorithms in polyhedral geometry.

The second component in the complexity of our problem (5.1) is the algebraic complexity of solving the linear optimization problem (5.3). It governs Algorithm 2. For an illustration see Figure 5.1 and the discussion in Examples 5.7 and 5.8.

We now determine the algebraic degree of the optimization problem (5.3) when F is a face of codimension i. To this end, we replace the affine variety $X \subset \mathbb{R}^n$ and the affine space $L = \mathbf{u} + L_F$ in $\mathbb{R}^n$ by their respective closures in complex projective space $\mathbb{P}^n$. We

---

Algorithm 2: Numerical Optimization

Input: Model X and a point $\mathbf{u} \in \mathbb{R}^n$; complete output from Algorithm 1.

Output: The optimal solution $\mathbf{x}^*$ of (5.1) along with its corresponding face G.

1 **for** *each face F of the unit ball B* **do**

2   Solve the optimization problem (5.3).

3   Store the optimal solution $\mathbf{x}$ and a basis for the linear subspace L_F of $\mathbb{R}^n$.

4 **end**

5 Among all candidate solutions, identify the solution $\mathbf{x}^*$ for which the distance to X to the given data point $\mathbf{u}$ is smallest. Record its face G.

6 **return** *the optimal solution $\mathbf{x}^*$, its associated linear space L_G, and the face normal ℓ_G.*

Algorithm 3: Algebraic Postprocessing

Input: The optimal solution $(\mathbf{x}^*, G)$ to (5.1) in the form found by Algorithm 2.

Output: The maximal ideal in the polynomial ring $\mathbb{Q}[x_1, \ldots, x_n]$ which has the zero $\mathbf{x}^*$.

1 Use Lagrange multipliers to give polynomial equations that characterize the critical points of the linear function ℓ_G on the subvariety $(\mathbf{u} + L_G) \cap X$ in the affine space $\mathbb{R}^n$.

2 Eliminate all variables representing Lagrange multipliers from the ideal in the previous step. This ideal lives in $\mathbb{Q}[x_1, \ldots, x_n]$.

3 **if** *the ideal in step 4 is maximal* **then**

4   Call the ideal M.

5 **else**

6   Remove extraneous primary components to get the maximal ideal M of $\mathbf{x}^*$.

7 **end**

8 Determine the degree of $\mathbf{x}^*$, which is the dimension of $\mathbb{Q}[x_1, \ldots, x_n]/M$ over $\mathbb{Q}$.

9 **return** *the generators for the maximal ideal M along with the degree.*

---

retain the same symbols X and L for the respective projective varieties. The following result formulates the algebraic degree of the linear program (5.3) in terms of the polar degrees that we have introduced in Definition 2.18 and again in Chapter 4.

Theorem 5.5 *Let $L \subset \mathbb{R}^n$ be a general affine-linear space of codimension $i - 1$ and ℓ be a general linear form. The number of critical points of ℓ on $L \cap X$ is the polar degree $\delta_i(X)$.*

Proof This result appears in [44, Theorem 5.1]. The number of critical points of a general linear form on the intersection $L \cap X$ in $\mathbb{R}^n$ is the degree of the dual variety $(L \cap X)^\vee$. That degree coincides with the polar degree $\delta_i(X)$. $\qquad\square$

Theorem 5.5 offers a direct interpretation of each polar degree $\delta_i(X)$ in terms of optimization on X. This interpretation can be used as a definition of polar degrees. Some readers might prefer this over the definitions given in Chapter 4.

Corollary 5.6 *If the variety X is in general position, then the total number of critical points of the optimization problem (5.1) arises from the f-vector of B and the polar degrees of X. That number equals*

$$\delta_n(X) \cdot f_0(B) + \delta_{n-1}(X) \cdot f_1(B) + \cdots + \delta_1(X) \cdot f_{n-1}(B). \tag{5.4}$$

Proof The total number of critical points is $\alpha_0 \cdot f_0(B) + \cdots + \alpha_{n-1} \cdot f_{n-1}(B)$, where α_j is the number of critical points of (5.3) for a face of dimension j. Fix a face F of dimension j.

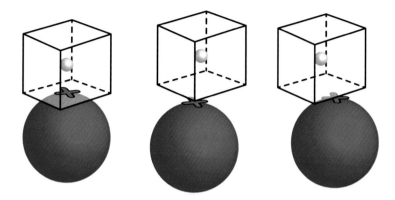

Fig. 5.1: The green sphere is the given variety X. The data point $\mathbf{u}$ is white. We solve the problem (5.1) for the norm $||\cdot||_\infty$. Balls in this norm are cubes. The contact point $\mathbf{x}^*$ is marked with a cross. The optimal face F is a facet, vertex, or edge.

The affine linear space $L := \mathbf{u} + L_F$ has codimension $n - j - 1$ in $\mathbb{R}^n$. Since X is assumed to be in general position, L is general relative to X and there exists a general linear functional ℓ that attains its maximum over B at F. Therefore, we can apply Theorem 5.5, and we find that the number of critical points of (5.3) is $\alpha_j = \delta_{n-j}(X)$. $\square$

We have learned in Chapter 2 that the polar degrees $\delta_i(X)$ of a variety X determine its ED degree and hence the algebraic complexity of Euclidean distance minimization for X. Corollary 5.6 highlights that this applies not just to the Euclidean distance, but also to the analogous problem for polyhedral norms. The two extreme cases in Theorem 5.5 are $i = 1$ and $i = n$. Touching at a vertex ($i = n$) can only happen when X is a hypersurface, and here $\delta_n(X) = \mathrm{degree}(X)$. Touching at a facet ($i = 1$) can happen for varieties of any dimension, as long as the dual variety $X^\vee$ is a hypersurface. Here, we have $\delta_1(X) = \mathrm{degree}(X^\vee)$.

Example 5.7 (Touching at a facet) Suppose that the face F is a facet of the unit ball B. Then $L_F = \mathbb{R}^n$, and ℓ_F is an outer normal vector to that facet, which is unique up to scaling. Here, the optimization problem (5.3) asks for the minimum of ℓ_F over X. This situation corresponds to the left diagram in Figure 5.1.   ◇

Example 5.8 (Touching at a vertex) Suppose F is a vertex of the unit ball B. This case arises when X is a hypersurface. It corresponds to the middle diagram in Figure 5.1. Here, the affine space $\mathbf{u} + L_F$ is the line that connects $\mathbf{u}$ and $\mathbf{x}^*$. That line intersects X in a finite set of cardinality $\mathrm{degree}(X)$. The optimum $\mathbf{x}^*$ is the real point in that finite set at which the value of the linear form ℓ is minimal.   ◇

Next, we work out Lemma 5.4 and Theorem 5.5 when X is a general surface in $\mathbb{R}^3$.

Example 5.9 Consider the problem in (5.1) and (5.2) for a general surface X of degree d in $\mathbb{R}^3$. The optimal face F of the unit ball B depends on the location of the data point $\mathbf{u}$. The algebraic degree of the solution $\mathbf{x}^*$ is $\delta_3(X) = d$ if $\dim(F) = 0$, it is $\delta_2(X) = d(d-1)$ if $\dim(F) = 1$, and it is $\delta_1(X) = d(d-1)^2$ if $\dim(F) = 2$. Here $\mathbf{x}^*$ is the unique point in $(\mathbf{u} + \lambda^* B) \cap X$, where λ^* is the optimal value in (5.2).

Figure 5.1 visualizes these three cases for $d = 2$ and $\|\cdot\|_\infty$. The variety X is the green sphere, which is a surface of degree $d = 2$. The unit ball for the norm $\|\cdot\|_\infty$ is the cube $B = [-1, 1]^3$. The picture shows the smallest λ^* such that $\mathbf{u} + \lambda^* B$ touches the sphere X. The cross marks the point of contact. This is the point $\mathbf{x}^*$ in X which is closest in ∞-norm to the white point $\mathbf{u}$ in the center of the cube. The point of contact is either on a facet, or on an edge, or it is a vertex. The algebraic degree of $\mathbf{x}^*$ is two in all three cases, i.e. we can write the solution $\mathbf{x}^*$ in terms of the data $\mathbf{u}$ by solving the quadratic formula. If the green surface X were a cubic surface ($d = 3$) then these polar degrees would be 3, 6 and 12. $\diamond$

Our geometric discussion can be translated into piecewise-algebraic formulas for the optimal point $\mathbf{x}^*$ and the optimal value λ^*. This rests on the algebraic postprocessing in Algorithm 3. It is carried out explicitly for a statistical scenario in Theorem 5.14. In that scenario, X is also a quadratic surface in 3-space, just like the green ball in Figure 5.1.

5.2 Optimal Transport and Independence Models

We now come to the title of this chapter, namely the Wasserstein distance to a variety X. For us, X will be an independence model in a probability simplex, given by matrices or tensors of low rank (see Chapter 12). We measure distances using Wasserstein metrics on that simplex. This is a class of polyhedral norms of importance in optimal transport theory.

We now present the relevant definitions. A probability distribution on the finite set $[n] = \{1, 2, \ldots, n\}$ is a point ν in the $(n-1)$-dimensional probability simplex

$$\Delta_{n-1} := \left\{ (\nu_1, \ldots, \nu_n) \in \mathbb{R}^n_{\geq 0} \mid \nu_1 + \cdots + \nu_n = 1 \right\}.$$

We shall turn this simplex into a metric space, by means of the *Wasserstein distance*. To define this notion, we first turn the finite state space $[n]$ into a finite metric space. The metric on $[n]$ is given by fixing a symmetric $n \times n$ matrix $d = (d_{ij})$ with nonnegative entries. These entries satisfy $d_{ii} = 0$ and $d_{ik} \leq d_{ij} + d_{jk}$ for all i, j, k.

Definition 5.10 Given two probability distributions $\mu, \nu \in \Delta_{n-1}$, we consider the following linear programming problem, where $\mathbf{z} = (z_1, \ldots, z_n)$ is the decision variable:

$$\text{Maximize } \sum_{i=1}^n (\mu_i - \nu_i) z_i \text{ subject to } |z_i - z_j| \leq d_{ij} \text{ for all } 1 \leq i < j \leq n. \quad (5.5)$$

The optimal value of (5.5), denoted $W_d(\mu, \nu)$, is the *Wasserstein distance* between μ and ν.

The optimal solution $\mathbf{z}^*$ to problem (5.5) is known as the *optimal discriminator* for the two probability distributions μ and ν. It satisfies $W_d(\mu, \nu) = \langle \mu - \nu, \mathbf{z}^* \rangle$, where $\langle \cdot, \cdot \rangle$ is the Euclidean inner product on $\mathbb{R}^n$. The coordinates z_i^* of the optimal discriminator $\mathbf{z}^*$ are the weights on the state space $[n]$ that best tell μ and ν apart. The linear program (5.5) is the *Kantorovich dual* of the *optimal transport problem*. The feasible region of the linear program (5.5) is unbounded because it is invariant under translation by

$$\mathbf{1} = (1, 1, \ldots, 1) \in \mathbb{R}^n.$$

It is compact after taking the quotient modulo the line $\mathbb{R}\mathbf{1}$. This motivates the following.

Definition 5.11 The *Lipschitz polytope* of the finite metric space $([n], d)$ equals

$$P_d = \{\mathbf{z} \in \mathbb{R}^n/\mathbb{R}\mathbf{1} \mid |z_i - z_j| \leq d_{ij} \text{ for all } 1 \leq i < j \leq n\}. \tag{5.6}$$

Note that $\dim(P_d) \leq n - 1$. Lipschitz polytopes arise prominently in tropical geometry (see [133, Chapter 7]), where they are called *polytropes*. A polytrope is a subset of $\mathbb{R}^n/\mathbb{R}\mathbf{1}$ that is convex both classically and tropically. The dual polytope P_d^* lies in the hyperplane perpendicular to the line $\mathbb{R}\mathbf{1}$. We call P_d^* a *root polytope* because its vertices are, up to scaling, the elements $\mathbf{e}_i - \mathbf{e}_j$ in the root system of Lie type A_{n-1}. The root polytope equals

$$P_d^* = \{\mathbf{x} \in \mathbb{R}^n \mid \max_{\mathbf{z} \in P_d} \langle \mathbf{x}, \mathbf{z} \rangle \leq 1\}$$
$$= \operatorname{conv}\left\{\tfrac{1}{d_{ij}}(\mathbf{e}_i - \mathbf{e}_j) \mid 1 \leq i, j \leq n\right\}. \tag{5.7}$$

This is a centrally symmetric polytope since the metric d satisfies $d_{ij} = d_{ji}$.

Proposition 5.12 *The Wasserstein metric W_d on the probability simplex Δ_{n-1} is equal to the polyhedral norm whose unit ball is the root polytope P_d^*. Hence, all results in Section 5.1 apply to Wasserstein metrics.*

Proof Fix the polyhedral norm with unit ball P_d^*. The distance between μ and ν in this norm is the smallest real number λ such that $\mu \in \nu + \lambda P_d^*$, or, equivalently, $\frac{1}{\lambda}(\mu - \nu) \in P_d^*$. By definition of the dual polytope, this minimal λ is the maximum inner product $\langle \mu - \nu, \mathbf{z} \rangle$ over all points $\mathbf{z}$ in the dual $(P_d^*)^*$ of the unit ball. But this dual is precisely the Lipschitz polytope because the biduality $(P_d^*)^* = P_d$ holds for polytopes. Hence the distance between μ and ν is equal to $W_d(\mu, \nu)$, which is the optimal value in (5.5). □

Example 5.13 Let $n = 4$ and fix the finite metric from the graph distance on the 4-cycle

$$d = \begin{bmatrix} 0 & 1 & 1 & 2 \\ 1 & 0 & 2 & 1 \\ 1 & 2 & 0 & 1 \\ 2 & 1 & 1 & 0 \end{bmatrix}. \tag{5.8}$$

The induced metric on the tetrahedron Δ_3 is given by the Lipschitz polytope

$$P_d = \{(x_1, x_2, x_3, x_4) \in \mathbb{R}^4/\mathbb{R}\mathbf{1} \mid |x_1 - x_2| \leq 1, |x_1 - x_3| \leq 1, |x_2 - x_4| \leq 1, |x_3 - x_4| \leq 1\}$$

$$= \operatorname{conv}\left\{\pm (1, 0, 0, -1), \pm(\tfrac{1}{2}, -\tfrac{1}{2}, -\tfrac{1}{2}, \tfrac{1}{2}), \pm(0, 1, -1, 0)\right\}.$$

Note that this 3-dimensional polytope is an octahedron. Therefore, its dual is a cube:

$$P_d^* = \{ (y_1, y_2, y_3, y_4) \in (\mathbb{R}\mathbf{1})^\perp \mid |y_1 - y_4| \leq 1, |y_2 - y_3| \leq 1, |y_2 + y_3| \leq 1 \}$$

$$= \mathrm{conv}\{\pm(1, -1, 0, 0), \pm(1, 0, -1, 0), \pm(0, 1, 0, -1), \pm(0, 0, 1, -1)\}.$$

This is the unit ball for the Wasserstein metric on the tetrahedron Δ_3 that is induced by d. Figure 5.1 illustrates the distance from a point to a surface with respect to this metric.   ◇

We wish to compute the Wasserstein distance from a given probability distribution μ to a fixed *discrete statistical model*. We now denote this model by

$$\mathcal{M} \subset \Delta_{n-1}.$$

This is the scenario studied in [43, 44]. The remainder of this chapter is based on the presentation in these two articles. As is customary in algebraic statistics [167], we assume that $\mathcal{M}$ is defined by polynomials in the unknowns $v = (v_1, \ldots, v_n)$. Thus

$$\mathcal{M} = X \cap \Delta_{n-1}$$

for some algebraic variety X in $\mathbb{R}^n$. Our task is to solve the following optimization problem:

$$W_d(\mu, \mathcal{M}) \quad := \quad \min_{v \in \mathcal{M}} W_d(\mu, v) \quad = \quad \min_{v \in \mathcal{M}} \max_{x \in P_d} \langle \mu - v, x \rangle. \tag{5.9}$$

Computing this quantity means solving a non-convex optimization problem. Our aim is to study this problem and propose solution strategies, using methods from geometry, algebra and combinatorics. We summarized these in Algorithms 1–3. Similar strategies for the Euclidean metric and the Kullback–Leibler divergence are found in Chapters 2 and 11.

We conclude this section with a detailed case study for the tetrahedron Δ_3 whose points are joint probability distributions of two binary random variables. The *2-bit independence model* $\mathcal{M} \subset \Delta_3$ consists of all nonnegative 2×2 matrices of rank one whose entries sum to one. This model has the parametric representation

$$\begin{bmatrix} v_1 & v_2 \\ v_3 & v_4 \end{bmatrix} = \begin{bmatrix} pq & p(1-q) \\ (1-p)q & (1-p)(1-q) \end{bmatrix}, \qquad (p, q) \in [0, 1]^2. \tag{5.10}$$

Thus, $\mathcal{M}$ is the surface in the tetrahedron Δ_3 defined by the equation $v_1 v_4 = v_2 v_3$. The next theorem gives the optimal value function and the solution function for this independence model. We use the Wasserstein metric W_d from Example 5.13. For the proof of Theorem 5.14 and a simpler example we refer to the paper [43] by Çelik, Jamneshan, Montúfar, Sturmfels and Venturello.

Theorem 5.14 involves a distinction into eight cases. This division of Δ_3 is shown in Figure 5.2. Each of the last four cases breaks into two subcases, since the numerator in the formulas is the absolute value of $\mu_1 \mu_4 - \mu_2 \mu_3$. The sign of this 2×2 determinant matters for the pieces of our piecewise algebraic function. Thus, the tetrahedron Δ_3 is divided into 12 regions. On each region the optimal value map $\mu \mapsto W_d(\mu, \mathcal{M})$ is one algebraic function.

Theorem 5.14 *The Wasserstein distance from a distribution* $\mu \in \Delta_3$ *to the model* $\mathcal{M}$ *equals*

$$
W_d(\mu, \mathcal{M}) = \begin{cases}
2\sqrt{\mu_1}(1 - \sqrt{\mu_1}) - \mu_2 - \mu_3, & \text{if } \mu_1 \geq \mu_4, \ \sqrt{\mu_1} \geq \mu_1 + \mu_2, \ \sqrt{\mu_1} \geq \mu_1 + \mu_3, \\
2\sqrt{\mu_2}(1 - \sqrt{\mu_2}) - \mu_1 - \mu_4, & \text{if } \mu_2 \geq \mu_3, \ \sqrt{\mu_2} \geq \mu_1 + \mu_2, \ \sqrt{\mu_2} \geq \mu_2 + \mu_4, \\
2\sqrt{\mu_3}(1 - \sqrt{\mu_3}) - \mu_1 - \mu_4, & \text{if } \mu_3 \geq \mu_2, \ \sqrt{\mu_3} \geq \mu_1 + \mu_3, \ \sqrt{\mu_3} \geq \mu_3 + \mu_4, \\
2\sqrt{\mu_4}(1 - \sqrt{\mu_4}) - \mu_2 - \mu_3, & \text{if } \mu_4 \geq \mu_1, \ \sqrt{\mu_4} \geq \mu_2 + \mu_4, \ \sqrt{\mu_4} \geq \mu_3 + \mu_4, \\
|\mu_1\mu_4 - \mu_2\mu_3|/(\mu_1 + \mu_2), & \text{if } \mu_1 \geq \mu_4, \ \mu_2 \geq \mu_3, \ \mu_1 + \mu_2 \geq \sqrt{\mu_1}, \ \mu_1 + \mu_2 \geq \sqrt{\mu_2}, \\
|\mu_1\mu_4 - \mu_2\mu_3|/(\mu_1 + \mu_3), & \text{if } \mu_1 \geq \mu_4, \ \mu_3 \geq \mu_2, \ \mu_1 + \mu_3 \geq \sqrt{\mu_1}, \ \mu_1 + \mu_3 \geq \sqrt{\mu_3}, \\
|\mu_1\mu_4 - \mu_2\mu_3|/(\mu_2 + \mu_4), & \text{if } \mu_4 \geq \mu_1, \ \mu_2 \geq \mu_3, \ \mu_2 + \mu_4 \geq \sqrt{\mu_4}, \ \mu_2 + \mu_4 \geq \sqrt{\mu_2}, \\
|\mu_1\mu_4 - \mu_2\mu_3|/(\mu_3 + \mu_4), & \text{if } \mu_4 \geq \mu_1, \ \mu_3 \geq \mu_2, \ \mu_3 + \mu_4 \geq \sqrt{\mu_4}, \ \mu_3 + \mu_4 \geq \sqrt{\mu_3}.
\end{cases}
$$

The solution function $\Delta_3 \to \mathcal{M}, \ \mu \mapsto \nu^*(\mu)$ *is given (with the same case distinction) by*

$$
\nu^*(\mu) = \begin{cases}
\left(\mu_1, \ \sqrt{\mu_1} - \mu_1, \ \sqrt{\mu_1} - \mu_1, \ -2\sqrt{\mu_1} + \mu_1 + 1\right), \\
\left(\sqrt{\mu_2} - \mu_2, \ \mu_2, \ -2\sqrt{\mu_2} + \mu_2 + 1, \ \sqrt{\mu_2} - \mu_2\right), \\
\left(\sqrt{\mu_3} - \mu_3, \ -2\sqrt{\mu_3} + \mu_3 + 1, \ \mu_3, \ \sqrt{\mu_3} - \mu_3\right), \\
\left(-2\sqrt{\mu_4} + \mu_4 + 1, \ \sqrt{\mu_4} - \mu_4, \ \sqrt{\mu_4} - \mu_4, \ \mu_4\right), \\
\left(\mu_1, \ \mu_2, \ \mu_1(\mu_3 + \mu_4)/(\mu_1 + \mu_2), \ \mu_2(\mu_3 + \mu_4)/(\mu_1 + \mu_2)\right), \\
\left(\mu_1, \ \mu_1(\mu_2 + \mu_4)/(\mu_1 + \mu_3), \ \mu_3, \ \mu_3(\mu_2 + \mu_4)/(\mu_1 + \mu_3)\right), \\
\left(\mu_2(\mu_1 + \mu_3)/(\mu_2 + \mu_4), \ \mu_2, \ \mu_4(\mu_1 + \mu_3)/(\mu_2 + \mu_4), \ \mu_4\right), \\
\left(\mu_3(\mu_1 + \mu_2)/(\mu_3 + \mu_4), \ \mu_4(\mu_1 + \mu_2)/(\mu_3 + \mu_4), \ \mu_3, \ \mu_4\right).
\end{cases}
$$

The boundaries separating the various cases are given by the surfaces

$$
\{\mu \in \Delta_3 \mid \mu_1 - \mu_4 = 0, \mu_1 + \mu_2 \geq \sqrt{\mu_1}, \mu_1 + \mu_3 \geq \sqrt{\mu_1}\}
$$
$$
and \ \{\mu \in \Delta_3 \mid \mu_2 - \mu_3 = 0, \mu_1 + \mu_2 \geq \sqrt{\mu_2}, \mu_2 + \mu_4 \geq \sqrt{\mu_2}\}.
$$

5.3 Wasserstein meets Segre–Veronese

At the end of the previous section, we studied the Wasserstein distance from a probability distribution μ in the three-dimensional probability simplex to the variety of 2×2 matrices of rank one. We now turn to the general case. Let $\mathcal{M}$ be an arbitrary smooth variety in $\Delta_{n-1} \subset \mathbb{R}^n$. For the moment, we do not specify a statistical model $\mathcal{M}$; later in this section, it will be an independence model. Furthermore, let $d = (d_{ij}) \in \mathbb{R}^{n \times n}$ induce a metric on the finite state space $[n]$. Here are three examples of metrics $([n], d)$:

- The discrete metric on any finite set $[n]$ where $d_{ij} = 1$ for distinct i, j.
- The L_0-metric on the Cartesian product $[m_1] \times \cdots \times [m_k]$ where $d_{ij} = \#\{l \mid i_l \neq j_l\}$. Here $i = (i_1, \ldots, i_k)$ and $j = (j_1, \ldots, j_k)$ are elements in that Cartesian product.
- The L_1-metric on the Cartesian product $[m_1] \times \cdots \times [m_k]$ where $d_{ij} = \sum_{l=1}^{k} |i_l - j_l|$.

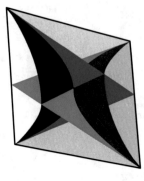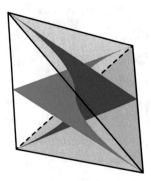

Fig. 5.2: The optimal value function of Theorem 5.14 subdivides the tetrahedron (left). The red surface consists of four pieces that, together with the blue surface, separate the eight cases in Theorem 5.14. Four convex regions are enclosed between the red surfaces and the edges they meet. These regions represent the first four cases in Theorem 5.14. The remaining four regions are each bounded by two red and two blue pieces, and correspond to the last four cases. Each of these four regions is further split in two by the model. We do not depict this in our visualization. The two sides are determined by the sign of the determinant $\mu_1\mu_4 - \mu_2\mu_3$. The two blue surfaces (right) specify the points $\mu \in \Delta_3$ with more than one optimal solution.

For the last two metrics, the number of states of the relevant independence models is $n = m_1 \cdots m_k$. All three metrics above are *graph metrics*. This means that there exists an undirected simple graph G with vertex set $[n]$ such that d_{ij} is the length of the shortest path from i to j in G. The corresponding Wasserstein balls are called *symmetric edge polytopes*. The combinatorics of these polytopes is investigated in [44, Section 4].

For any $\mu \in \Delta_{n-1}$, we now seek the Wasserstein distance $W_d(\mu, M)$ to the model M. Recall from Proposition 5.12 that the unit ball of the Wasserstein metric is the root polytope $P_d^* = \text{conv}\left\{\frac{1}{d_{ij}}(\mathbf{e}_i - \mathbf{e}_j) \mid 1 \le i, j \le n\right\}$. As before, for computing $W_d(\mu, M)$ we iterate through the faces of the unit ball $B = P_d^*$, and we solve the optimization problem in Lemma 5.4. That is, for a fixed face F of the polytope B we solve:

$$\text{Minimize } \ell_F = \ell_F(\nu) \text{ subject to } \nu \in (\mu + L_F) \cap M, \qquad (5.11)$$

where ℓ_F is any linear functional on $\mathbb{R}^n$ that attains its maximum over B at F.

Let $\mathcal{F}$ be the set of pairs (i, j) such that the point $\frac{1}{d_{ij}}(\mathbf{e}_i - \mathbf{e}_j)$ is a vertex of B and it lies in F. The linear space spanned by the face F is

$$L_F = \left\{\textstyle\sum_{(i,j)\in\mathcal{F}} \lambda_{ij}(\mathbf{e}_i - \mathbf{e}_j) \mid \lambda_{ij} \in \mathbb{R}\right\}.$$

With this notation, the decision variables for (5.11) are the multipliers λ_{ij} for $(i, j) \in \mathcal{F}$. The algebraic complexity of this problem is given by the polar degree (Theorem 5.5). The combinatorial complexity is governed by the facial structure of the Wasserstein ball $B = P_d^*$. They are combined in Corollary 5.6.

We now work this out for the case when $M \subset \Delta_{n-1}$ is an independence model for discrete random variables, given by tensors of rank one. We denote by $(m)_r$ a multinomial distribution with m possible outcomes and r trials, which can be interpreted as an unordered set of r identically distributed random variables on $[m] = \{1, 2, ..., m\}$. The subscript r is omitted if $r = 1$. For integers $m_1, \ldots, m_k$ and $r_1, \ldots, r_k$ we consider the model M whose elements are given by k independent multinomial distributions $(m_1)_{r_1}, \ldots, (m_k)_{r_k}$. We denote this independence model by

$$M = ((m_1)_{r_1}, \ldots, (m_k)_{r_k}).$$

The number of states of the model M equals

$$n = \prod_{i=1}^{k} \binom{m_i + r_i - 1}{r_i}.$$

Example 5.15 The variety $M = (2_2, 2)$ is the independence model for three binary random variables where the first two are identically distributed. This model has $n = 6$ states. Note that M is the image of the map from the square $[0, 1]^2$ into the simplex Δ_5 given by

$$(p, q) \mapsto \left(p^2 q, \ 2p(1-p)q, \ (1-p)^2 q, \ p^2(1-q), \ 2p(1-p)(1-q), \ (1-p)^2(1-q) \right).$$

Our parameterization lists the $n = 6$ states in the order $00, 10, 20, 01, 11, 21$. These are the vertices of the associated graph G, which is the product of a 3-chain and a 2-chain. ◇

Example 5.16 The following four models are used for the case studies in [44, Section 6]: the 3-bit model $(2, 2, 2)$ with the L_0-metric on $[2]^3$, the model $(3, 3)$ for two ternary variables with the L_1-metric on $[3]^2$, the model (2_6) for six identically distributed binary variables with the discrete metric on $[7]$, and the model $(2_2, 2)$ in Example 5.15 with the L_1-metric on $[3] \times [2]$. In Table 5.1, we report the f-vectors of the Wasserstein balls for each of these models, thus hinting at combinatorial complexity. ◇

M	n	$\dim(M)$	Metric d	f-vector of the $(n-1)$-polytope P_d^*
$(2, 2, 2)$	8	3	$L_0 = L_1$	$(24, 192, 652, 1062, 848, 306, 38)$
$(3, 3)$	9	4	L_1	$(24, 216, 960, 2298, 3048, 2172, 736, 82)$
(2_6)	7	1	discrete	$(42, 210, 490, 630, 434, 126)$
$(2_2, 2)$	6	2	L_1	$(14, 60, 102, 72, 18)$

Table 5.1: f-vectors of the Wasserstein balls for the four models in Example 5.16.

As is customary in algebraic statistics, we replace the independence model introduced above, which is a semialgebraic set inside a simplex, by its complex Zariski closure in a projective space. This allows us to compute the algebraic degrees of our optimization problem. Independence models correspond in algebraic geometry to *Segre–Veronese varieties*. The Segre–Veronese variety $M = ((m_1)_{r_1}, \ldots, (m_k)_{r_k})$ is the embedding of $\mathbb{P}^{m_1-1} \times \cdots \times \mathbb{P}^{m_k-1}$ in the projective space of partially symmetric tensors, $\mathbb{P}(S^{r_1}(\mathbb{R}^{m_1}) \otimes \cdots \otimes S^{r_k}(\mathbb{R}^{m_k}))$.

That projective space equals $\mathbb{P}^{n-1}$ where $n = \prod_{i=1}^{k} \binom{m_i + r_i - 1}{r_i}$. By definition, the Segre–Veronese variety $\mathcal{M}$ is the set of all tensors of rank one inside this projective space. The dimension of $\mathcal{M}$ is $\mathbf{m} := m_1 + \cdots + m_k - k$. Tensors of rank one are discussed in more detail in Chapter 12. See Section 12.3 for their volumes.

Example 5.17 Let $k = 2$. The Segre–Veronese variety $\mathcal{M} = (2_2, 2)$ is an embedding of $\mathbb{P}^1 \times \mathbb{P}^1$ into $\mathbb{P}^5$. This is a toric surface of degree four. Its points are rank one tensors of format $2 \times 2 \times 2$ which are symmetric in the first two indices. This model appears in the last row of Table 5.3.   ◇

The computation of the Chern classes of the Segre–Veronese variety $\mathcal{M}$ first appeared in the doctoral dissertation of Luca Sodomaco [160, Chapter 5]. We obtain the polar degrees using the formula that connects Chern classes with polar degrees in Theorem 4.20. We here state the result of this computation, which was published in [113, Proposition 6.11].

Theorem 5.18 *With the notation above, the polar degrees of the Segre–Veronese variety are given by the following formula:*

$$\delta_{j+1}(\mathcal{M}) = \sum_{i=0}^{\mathbf{m}-j} (-1)^i \binom{\mathbf{m}+1-i}{j+1} (\mathbf{m}-i)! \sum_{\alpha \in \mathcal{A}} \prod_{\ell=1}^{k} \frac{\binom{m_\ell}{\alpha_\ell} r_\ell^{m_\ell - 1 - \alpha_\ell}}{(m_\ell - 1 - \alpha_\ell)!}, \quad (5.12)$$

where $\mathcal{A} = \{\alpha \in \mathbb{N}^k \mid \alpha_1 + \cdots + \alpha_k = i \text{ and } \alpha_i \leq m_i - 1 \text{ for } i = 1, \ldots, k\}$.

We next examine this formula for various special cases, starting with the binary case.

Corollary 5.19 *Let M be the k-bit independence model (this is the case $r_1 = \cdots = r_k = 1$ and $m_1 = \cdots = m_k = 2$). The formula (5.12) specializes to*

$$\delta_{j+1}(\mathcal{M}) = \sum_{i=0}^{k-j} (-1)^i \binom{k+1-i}{j+1} (k-i)! \binom{k}{i} 2^i. \quad (5.13)$$

In algebraic geometry language, our model $\mathcal{M}$ is the Segre embedding of $(\mathbb{P}^1)^k$ into $\mathbb{P}^{2^k - 1}$. This is the toric variety associated with the k-dimensional unit cube. Its degree is the normalized volume of that cube, which is $k!$. The dual variety $\mathcal{M}^\vee$ is a hypersurface of degree δ_1, known as the *hyperdeterminant* of format 2^k. For instance, for $k = 3$, this hypersurface in $\mathbb{P}^7$ is the $2 \times 2 \times 2$-hyperdeterminant. This hyperdeterminant is a homogeneous polynomial of degree four in eight unknowns. We computed it in Example 3.16. The polar degrees for the k-bit independence model in (5.13) are shown for $k \leq 7$ in Table 5.2.

We now briefly discuss the independence models (m_1, m_2) for two random variables. These are the classical determinantal varieties of $m_1 \times m_2$ matrices of rank one. Here, $n = m_1 m_2$ and $\mathbf{m} = m_1 + m_2 - 2$.

Corollary 5.20 *The Segre variety $\mathcal{M} = \mathbb{P}^{m_1 - 1} \times \mathbb{P}^{m_2 - 1}$ in $\mathbb{P}^{n-1}$ has the polar degrees*

$$\delta_{j+1}(\mathcal{M}) = \sum_{i=0}^{m_1 + m_2 - 2 - j} (-1)^i \binom{m_1 + m_2 - 1 - i}{j+1} (m_1 + m_2 - 2 - i)! \, \sigma(i),$$

where $\sigma(i) = \sum_s \frac{\binom{m_1}{s}}{(m_1 - 1 - s)!} \cdot \frac{\binom{m_2}{i-s}}{(m_2 - 1 - i + s)!}$ and the sum is over the set of integers s such that both $m_1 - 1 - s$ and $m_2 - 1 - i + s$ are nonnegative.

j	$k = 2$	$k = 3$	$k = 4$	$k = 5$	$k = 6$	$k = 7$
7						5040
6					720	30240
5				120	3600	80640
4			24	480	7920	124320
3		6	72	840	9840	120960
2	2	12	96	800	7440	75936
1	2	12	64	440	3408	30016
0	2	4	24	128	880	6816

Table 5.2: The table shows the polar degrees $\delta_{j+1}(M)$ of the k-bit independence model for $k \leq 7$. The indices j with $\delta_{j+1}(M) \neq 0$ range from 0 to k. The bottom row, labeled 0, contains the degree of the hyperdeterminant $M^\vee$. On the antidiagonal ($j = k$) we find degree$(M) = k!$. The entries in the first column ($k = 2$) correspond to the three scenarios in Figure 5.1, where each algebraic degree equals 2.

The polar degrees above serve as upper bounds for any particular Wasserstein distance problem. For a fixed model M, the equality in Theorem 5.5 holds only when the data (ℓ, L) is generic. However, for the optimization problem in (5.11), the linear space $L = L_F$ and the linear functional $\ell = \ell_F$ are very specific. They depend on the Lipschitz polytope P_d and the position of the face F relative to the model M.

Proposition 5.21 *Consider the optimization problem (5.11) for the independence model* $M = ((m_1)_{r_1}, \ldots, (m_k)_{r_k})$ *with a given face F of the Wasserstein ball $B = P_d^*$. Suppose that F has codimension i. The number of critical points of (5.11) is bounded above by the polar degree $\delta_i(M)$.*

Proof The critical points of (5.11) are given as the solutions of a system of polynomial equations that depends on parameters (μ, ℓ, L). It follows from Theorem 5.5 that the number of critical points for general parameters is the polar degree $\delta_i(M)$. The Parameter Continuation Theorem (Theorem 3.18) implies that the number of solutions can only go down when we pass from general parameters to special parameters.   □

Example 5.22 We investigate the drop in algebraic degree experimentally for the four independence models in Example 5.16. In the language of algebraic geometry, these models are the Segre threefold $\mathbb{P}^1 \times \mathbb{P}^1 \times \mathbb{P}^1$ in $\mathbb{P}^7$, the variety $\mathbb{P}^2 \times \mathbb{P}^2$ of rank one 3×3 matrices in $\mathbb{P}^8$, the rational normal curve $\mathbb{P}^1$ in $\mathbb{P}^6 = \mathbb{P}(S^6(\mathbb{R}^2))$, and the Segre–Veronese surface $\mathbb{P}^1 \times \mathbb{P}^1$ in $\mathbb{P}^5 = \mathbb{P}(S^2(\mathbb{R}^2) \otimes \mathbb{R}^2)$. The finite metrics d are specified in the fourth column of Table 5.1. The fifth column in Table 5.1 records the combinatorial complexity of our optimization problem. The algebraic complexity is recorded in Table 5.3.

The second column in Table 5.3 gives the vector $(\delta_0, \delta_1, \ldots, \delta_{n-2})$ of polar degrees. The third and fourth columns are the results of a computational experiment. For each model, we take 1000 uniform samples μ with rational coordinates from Δ_{n-1}, and we solve the optimization problem (5.9). The output is an exact representation of the optimal solution v^*. This includes the optimal face F that specifies v^*, along with its maximal ideal over $\mathbb{Q}$. The algebraic degree of the optimal solution v^* is computed as the number of

M	Polar degrees	Maximal degree	Average degree
$(2,2,2)$	$(0,0,0,6,12,12,4)$	$(0,0,0,4,12,6,0)$	$(0,0,0,2.138,6.382,3.8,0)$
$(3,3)$	$(0,0,0,6,12,12,6,3)$	$(0,0,0,2,8,6,6,0)$	$(0,0,0,1.093,3.100,4.471,6.0,0)$
(2_6)	$(0,0,0,0,6,10)$	$(0,0,0,0,6,5)$	$(0,0,0,0,6,5)$
$(2_2,2)$	$(0,0,4,6,4)$	$(0,0,3,5,2)$	$(0,0,2.293,3.822,2.0)$

Table 5.3: The algebraic degrees of the problem (5.9) for the four models in Example 5.16.

complex zeros of that maximal ideal. This number is bounded above by the polar degree (cf. Proposition 5.21). The fourth column in Table 5.3 shows the average of the algebraic degrees we found. For example, for the 3-bit model $(2,2,2)$ we have $\delta_3 = 6$, corresponding to P_d^* touching M at a 3-face F. However, the maximum degree we saw in our computations was 4, with an average degree of 2.138. For 4-faces F, we have $\delta_4 = 12$. This degree was attained in some runs. The average of the degrees for 4-faces was found to be 6.382.   ◇

In this chapter, we measured the distance to a real algebraic variety with a polyhedral norm. We focused on the important case when the norm is a Wasserstein norm and the variety is an independence model. We emphasized the distinction between combinatorial complexity, given by the f-vector of the unit ball, and algebraic complexity, given by the polar degrees of the model. This distinction was made completely explicit in Theorem 5.14.

Chapter 6
Curvature

The notion of curvature is central to differential geometry and its numerous applications. The aim of this chapter is to offer a first introduction to curvature. Our main point is to explore how curvature connects to algebraic geometry. We start out with plane curves, and we then turn to more general real algebraic varieties. The third section addresses the fundamental question of how to compute the volume of a tubular neighborhood of a variety.

6.1 Plane Curves

We consider a smooth algebraic curve $C \subset \mathbb{R}^2$. This is given to us in its implicit representation, as the zero set of an irreducible polynomial of degree $d \geq 1$ in two real variables:

$$f(x_1, x_2) \in \mathbb{R}[x_1, x_2].$$

We are interested in the *curvature* of the curve C at one of its points $\mathbf{x} = (x_1, x_2)$. Geometrically, the curvature is defined as the rate of change at $\mathbf{x}$ of a unit normal vector traveling along the curve C. To be precise, we first define

$$N(\mathbf{x}) := \frac{1}{\|\nabla f(\mathbf{x})\|} \nabla f(\mathbf{x}), \tag{6.1}$$

where the column vector $\nabla f(\mathbf{x}) = (\partial f/\partial x_1, \partial f/\partial x_2)^\top$ is the gradient of the polynomial f. For all $\mathbf{x} \in C$, the formula for $N(\mathbf{x})$ returns a normal vector of C at $\mathbf{x}$. One calls $N(\mathbf{x})$ a *unit normal field*. Similarly, a *unit tangent field* is given by the row vector

$$T(\mathbf{x}) := \big(N(\mathbf{x})_2, -N(\mathbf{x})_1\big). \tag{6.2}$$

The curvature of C at $\mathbf{x}$ is defined as the (signed) magnitude of the derivative of the unit normal field $N(\mathbf{x})$ in tangent direction. Thus, the curvature is the following scalar quantity:

$$c(\mathbf{x}) := \left\langle T(\mathbf{x}), \, T(\mathbf{x})_1 \cdot \frac{\partial N}{\partial x_1}(\mathbf{x}) + T(\mathbf{x})_2 \cdot \frac{\partial N}{\partial x_2}(\mathbf{x}) \right\rangle. \tag{6.3}$$

© The Author(s) 2024
P. Breiding et al., *Metric Algebraic Geometry*, Oberwolfach Seminars 53,
https://doi.org/10.1007/978-3-031-51462-3_6

This measures the rate of change of $N(\mathbf{x})$ as it travels along C. Since the derivative of a unit normal field at a curve always points in tangent direction, the definition of the curvature $c(x)$ in (6.3) is equivalent to the following identity:

$$T(\mathbf{x})_1 \cdot \frac{\partial N}{\partial x_1}(\mathbf{x}) + T(\mathbf{x})_2 \cdot \frac{\partial N}{\partial x_2}(\mathbf{x}) \; = \; c(\mathbf{x}) \cdot T(\mathbf{x}).$$

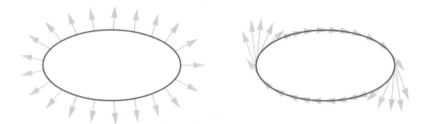

Fig. 6.1: The left picture shows the unit normal field of an ellipse, and the right picture shows how the unit normal field changes when traveling along the ellipse. The lengths of the tangent vectors in the right picture indicate the curvature of the ellipse at its various points.

Example 6.1 Consider the ellipse defined by $f(\mathbf{x}) = x_1^2 + 4x_2^2 - 1$. The left picture in Figure 6.1 shows this ellipse in green and its unit normal field $N(\mathbf{x})$ in yellow. The right picture displays the curvature via $c(\mathbf{x}) \cdot T(\mathbf{x})$. The magnitude of a yellow vector attached to a point $\mathbf{x}$ in the right picture gives the curvature at $\mathbf{x}$. On the top and bottom, where the ellipse is rather flat, the normal vectors do not change much, hence the curvature is small. On the sides, the normal vectors change more rapidly, so there the curvature is larger.   ◇

The inverse of the signed curvature is denoted $r(\mathbf{x}) := c(\mathbf{x})^{-1}$. This quantity is called the (signed) *radius of curvature*. Indeed, the curve C contains an infinitesimally small arc of a circle with radius $|r(\mathbf{x})|$ and center $\mathbf{x} - r(\mathbf{x}) \cdot N(\mathbf{x})$. This is the circle that best approximates C at $\mathbf{x}$. The center of this circle is called a *focal point* or *center of curvature* of the curve C at the point $\mathbf{x}$. The reason for the negative sign in this formula is that a normal vector pointing towards the focal point changes towards the direction that is opposite to $T(\mathbf{x})$; see Figure 6.2 for an illustration. We now connect to our historical discussion.

Proposition 6.2 *The Zariski closure of the set of all centers of curvature of a plane curve C is the evolute of C, as defined in Section 1.3.*

Proof We consider a local parametrization $\gamma(t)$ of C with $\gamma(0) = \mathbf{x}$ and $\dot{\gamma}(0) = T(\mathbf{x})$. Let E denote the curve that is traced out by the centers of curvature of C. Then, the curve that is defined by $\varepsilon(t) := \gamma(t) - r(\gamma(t)) \cdot N(\gamma(t))$ gives a local parametrization of E. The derivative of this parametrization at $t = 0$ equals

$$\dot{\varepsilon}(0) \; = \; T(\mathbf{x}) - \langle \nabla r(\mathbf{x}), T(\mathbf{x}) \rangle \cdot N(\mathbf{x}) - r(\mathbf{x}) \cdot \left(T(\mathbf{x})_1 \cdot \tfrac{\partial N}{\partial x_1}(\mathbf{x}) + T(\mathbf{x})_2 \cdot \tfrac{\partial N}{\partial x_2}(\mathbf{x}) \right).$$

Since $T(\mathbf{x})_1 \cdot \frac{\partial N}{\partial x_1}(\mathbf{x}) + T(\mathbf{x})_2 \cdot \frac{\partial N}{\partial x_2}(\mathbf{x}) = c(\mathbf{x}) \cdot T(\mathbf{x}) = r(\mathbf{x})^{-1} \cdot T(\mathbf{x})$, we get

$$\dot{\varepsilon}(0) = -\langle \nabla r(\mathbf{x}), T(\mathbf{x}) \rangle \cdot N(\mathbf{x}). \tag{6.4}$$

Hence the tangent line of E at $\varepsilon(0) = \mathbf{x} - r(\mathbf{x}) \cdot N(\mathbf{x})$ is the normal line of C at $\mathbf{x}$. The curve E is the envelope of the normal lines. In other words, the curve E is the evolute of C. □

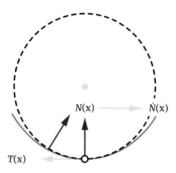

Fig. 6.2: The green curve contains an infinitesimally small arc of the dashed circle. The grey center of the circle is a focal point of the green curve. The red normal vector $N(\mathbf{x})$, pointing towards the focal point, changes into a normal vector that is slightly tilted in the direction opposite to $T(\mathbf{x})$. Hence $\dot{N}(\mathbf{x}) := T(\mathbf{x})_1 \frac{\partial N(\mathbf{x})}{\partial x_1} + T(\mathbf{x})_2 \frac{\partial N(\mathbf{x})}{\partial x_2}$ is a negative multiple of $T(\mathbf{x})$.

The previous proof shows that the absolute value of the curvature $|c(\mathbf{x})|$ is the inverse distance from $\mathbf{x}$ to its corresponding point on the evolute. Indeed, the latter point is $\mathbf{x} - r(\mathbf{x}) \cdot N(\mathbf{x})$, and so its distance from $\mathbf{x}$ is $|r(\mathbf{x})| = |c(\mathbf{x})|^{-1}$. In the remainder of this section, we will study two types of points: *inflection points* and *points of critical curvature*. These points exhibit special curvature of C. Inflection points are points where C is locally flat, and critical curvature points are points where the curvature has a local extremum.

Definition 6.3 Let $\mathbf{x} \in C$. We call $\mathbf{x}$ an *inflection point* if the curvature is zero, i.e. $c(\mathbf{x}) = 0$. We call $\mathbf{x}$ a *critical curvature point* if $\mathbf{x}$ is a critical point of the function $C \to \mathbb{R}$, $\mathbf{x} \mapsto c(\mathbf{x})$.

Example 6.4 We consider the Trott curve $f(\mathbf{x}) = 144(x^4 + y^4) - 225(x^2 + y^2) + 350x^2y^2 + 81$, as we did in Figure 2.1. This has degree $d = 4$. Figure 6.3 shows the curve with its points from Definition 6.3. We first compute inflection points using the numerical software HomotopyContinuation.jl [31]. This is based on the formulation in Lemma 6.7.

```
using HomotopyContinuation, LinearAlgebra
@var x y z; v = [x; y; z];
F = 144*(x^4 + y^4) - 225*(x^2 + y^2) + 350*x^2*y^2 + 81*z^4;
dF = differentiate(F, v);
H0 = differentiate(dF, v);
f = subs(F, z=>1);
h = subs(-det(H0), z=>1);
inflection_points = solve([f; h])
```

By Theorem 6.8, a general curve of degree d has $3d(d-2) = 24$ complex inflection points. Felix Klein [107] proved that at most $d(d-2) = 8$ can be real. Indeed, for the Trott curve, we find 8 real inflection points. They are the yellow points in Figure 6.3.

Next, we compute critical curvature points using the equations in Lemma 6.10.

```
f1, f2 = dF[1:2]
f11, f12, f12, f22 = H0[1:2,1:2];
hx, hy = differentiate(h,[x; y]);
g = f1 * f2 * (f11-f22) + f12 * (f2^2- f1^2);
c = subs(f2 * hy - f1 * hx - 3 * h * g, z=>1);
crit_curv = solve([f; c])
```

By Theorem 6.11, there are $2d(3d - 5) = 56$ complex critical curvature points. We find that 24 of them are real; out of these, 8 are close (but not equal) to the 8 inflection points, which is why they are not visible in the picture. The other 16 critical curvature points are shown in red in Figure 6.3. By Proposition 6.5, the critical curvature points correspond to cusps on the evolute, which is illustrated in Figure 6.4.   ◇

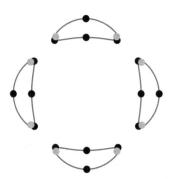

Fig. 6.3: The Trott curve (2.7) with its inflection points (yellow) and critical curvature points (red).

Proposition 6.5 *Let $C \subset \mathbb{R}^2$ be a smooth algebraic curve and $E \subset \mathbb{R}^2$ its evolute. For any point $\mathbf{x} \in C$, let $r(\mathbf{x})$ be the radius of curvature and let $\Gamma(\mathbf{x}) := \mathbf{x} - r(\mathbf{x}) \cdot N(\mathbf{x}) \in E$ be the corresponding point on the evolute. Then*

(a) $\mathbf{x}$ is an inflection point if and only if $\Gamma(\mathbf{x})$ is a point at infinity, and
(b) $\mathbf{x}$ is a point of critical curvature if and only if E has a cusp at $\Gamma(\mathbf{x})$.

Proof Recall that $c(\mathbf{x}) = r(\mathbf{x})^{-1}$. The point $\Gamma(\mathbf{x})$ is at infinity if and only if $c(\mathbf{x}) = 0$, which means that $\mathbf{x}$ is an inflection point. This proves the first item. For the second item, we consider a local parametrization $\gamma(t)$ of C with $\gamma(0) = \mathbf{x}$ and $\dot{\gamma}(0) = T(\mathbf{x})$. As in the proof of Proposition 6.2, the function $\varepsilon(t) := \Gamma(\gamma(t))$ gives a local parametrization of the evolute E. The evolute has a cusp at $\Gamma(\mathbf{x}) = \varepsilon(0)$ if and only if $\dot{\varepsilon}(0) = 0$. By (6.4), the latter is equivalent to $\langle \nabla r(\mathbf{x}), T(\mathbf{x}) \rangle = 0$, where $T(\mathbf{x})$ is the unit tangent field (6.2). This identity holds exactly when the curvature $c(\mathbf{x})$ is critical at $\mathbf{x}$ since $\nabla c(\mathbf{x}) = -\nabla r(\mathbf{x})/r(\mathbf{x})^2$.   □

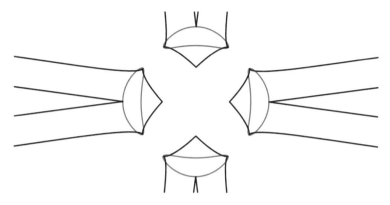

Fig. 6.4: The picture shows the Trott curve (2.7) in blue and its evolute in red. The cusps on the evolute correspond to the red critical curvature points in Figure 6.3. The Trott curve has 24 real critical curvature points. Out of those, 8 have a radius of curvature that exceeds the boundary of this picture. This is why we only see 16 cusps. We thank Emil Horobet and Pierpaola Santarsiero for helping us by computing the equation for the evolute.

Our next goal is to count the number of complex inflection and critical curvature points for a curve given by a general polynomial f of degree d. For this, let us first understand the curvature $c(\mathbf{x})$ better. In the following, we denote partial derivatives by $f_i := \frac{\partial f}{\partial x_i}$ and $f_{i,j} := \frac{\partial^2 f}{\partial x_i x_j}$. With this, the *Hessian* equals

$$H := \begin{bmatrix} f_{1,1} & f_{1,2} \\ f_{1,2} & f_{2,2} \end{bmatrix}.$$

Lemma 6.6 *The curvature of C at* $\mathbf{x}$ *equals*

$$c(\mathbf{x}) = \frac{1}{\|\nabla f(\mathbf{x})\|} \cdot T(\mathbf{x}) H(\mathbf{x}) T(\mathbf{x})^\top.$$

Proof By applying the product rule to (6.1), we obtain

$$T(\mathbf{x})_1 \cdot \frac{\partial N}{\partial x_1}(\mathbf{x}) + T(\mathbf{x})_2 \cdot \frac{\partial N}{\partial x_2}(\mathbf{x}) = \|\nabla f(\mathbf{x})\|^{-1} \cdot H(\mathbf{x}) T(\mathbf{x})^\top + a(\mathbf{x}) \cdot \nabla f(\mathbf{x})$$

for some scalar function $a(\mathbf{x})$. Since $\langle T(\mathbf{x}), \nabla f(\mathbf{x}) \rangle = 0$, Equation (6.3) gives the assertion.□

We can write the formula from Lemma 6.6 more explicitly as

$$c(\mathbf{x}) = \frac{f_{1,1} \cdot f_2^2 - 2 f_{1,2} \cdot f_1 \cdot f_2 + f_{2,2} \cdot f_1^2}{(f_1^2 + f_2^2)^{\frac{3}{2}}}(\mathbf{x}). \tag{6.5}$$

This formula is appealing to algebraic geometers because it writes curvature in terms of the polynomial $f(\mathbf{x})$. To make it even more appealing, we now embed C into the complex projective plane. We write

$$F(x_0, x_1, x_2) := x_0^d f\left(\frac{x_1}{x_0}, \frac{x_2}{x_0}\right)$$

for the homogenization of f. The Hessian of the ternary form F is the 3×3 matrix

$$H_0 = \begin{bmatrix} F_{0,0} & F_{0,1} & F_{0,2} \\ F_{0,1} & F_{1,1} & F_{1,2} \\ F_{0,2} & F_{1,2} & F_{2,2} \end{bmatrix}. \tag{6.6}$$

But, we now view the entries $F_{i,j}$ as inhomogeneous polynomials in the two variables $\mathbf{x} = (x_1, x_2)$ by setting $x_0 = 1$. We can rewrite the curvature of C in terms of F. The next lemma goes back to Salmon [156].

Lemma 6.7 *The curvature of the degree d curve C at the point $\mathbf{x}$ is equal to*

$$c(\mathbf{x}) = \frac{-\det H_0}{(d-1)^2 \cdot (f_1^2 + f_2^2)^{\frac{3}{2}}}(\mathbf{x}).$$

Proof By homogenizing the polynomials in (6.5), we get $c = \frac{P}{Q}$, where

$$P = F_{1,1} \cdot F_2^2 - 2F_{1,2} \cdot F_1 \cdot F_2 + F_{2,2} \cdot F_1^2 \quad \text{and} \quad Q = (F_1^2 + F_2^2)^{\frac{3}{2}}.$$

By Euler's formula for homogeneous polynomials, we have

$$(d-1) \cdot F_j = x_0 F_{0,j} + x_1 F_{1,j} + x_2 F_{2,j}, \quad 0 \le j \le 2. \tag{6.7}$$

Substituting this into P gives the following equation:

$$(d-1)^2 \cdot P = (F_{1,1}F_{2,2} - F_{1,2}^2) \cdot (x_1^2 F_{1,1} + x_2^2 F_{2,2} + 2(x_0 x_1 F_{0,1} + x_0 x_2 F_{0,2} + x_1 x_2 F_{1,2}))$$
$$+ x_0^2 (F_{0,1}^2 F_{2,2} - 2F_{1,2}F_{0,1}F_{0,2} + F_{1,1}F_{0,2}^2).$$

Note that the equation $0 = dF = x_0 F_0 + x_1 F_1 + x_2 F_2$ holds on the curve C. Substituting (6.7) into this equation, we obtain $x_1^2 F_{1,1} + x_2^2 F_{2,2}^2 + 2(x_0 x_1 F_{0,1} + x_0 x_2 F_{0,2} + x_1 x_2 F_{1,2}) = -x_0^2 F_{0,0}$. This identity holds on the curve C. From this we conclude

$$P = \frac{x_0^2}{(d-1)^2} \cdot (-(F_{1,1}F_{2,2} - F_{1,2}^2)F_{0,0} + F_{0,1}^2 F_{2,2} - 2F_{1,2}F_{0,1}F_{0,2} + F_{1,1}F_{0,2}^2).$$

The latter is $-x_0^2 \cdot \det H_0/(d-1)^2$. Setting $x_0 = 1$ finishes the proof.   □

We can now count the inflection points of a plane curve. This is due to Felix Klein [107].

Theorem 6.8 *The number of complex inflection points of a general curve C of degree d is $3d(d-2)$.*

Klein also proved that the number of real inflection points is at most $d(d-2)$. So, the real bound is one third of the complex count. Here, we give a short proof for Theorem 6.8.

Proof By Lemma 6.7, the inflection points on C are defined by the equations $f = \det H_0 = 0$. This is a system of two polynomial equations in two variables $\mathbf{x} = (x_1, x_2)$. The degree of f is d and the degree of $\det H_0$ is $3(d - 2)$. Bézout's theorem implies that the number of inflection points is at most $3d(d - 2)$. To show that the number is also at least $3d(d - 2)$, we use the Parameter Continuation Theorem (Theorem 3.18), and we show that there exist degree d curves with this number of inflection points.

Consider a univariate polynomial $g(x_1) \in \mathbb{R}[x_1]$ of degree d and let $G(x_0, x_1)$ be its homogenization. We write $M := \begin{bmatrix} G_{0,0} & G_{0,1} \\ G_{0,1} & G_{1,1} \end{bmatrix}$ for the Hessian of the binary form $G(x_0, x_1)$. We assume (1) that g has d regular zeros, (2) that $\det M = 0$ has only regular zeros, and (3) that $G = \det M = 0$ has no solutions in $\mathbb{P}^1$. All three are Zariski open conditions, so almost all polynomials g satisfy this assumption. Define

$$f(x_1, x_2) := x_2^d - g(x_1).$$

The Hessian (6.6) of the plane curve f satisfies $\det H_0 = d(d-1) \cdot x_2^{d-2} \cdot \det M$. Consequently, $\det H_0 = 0$ if and only if either $x_2 = 0$, or x_1 is among the $2(d - 2)$ zeros of $\det M$. This means that $\mathbf{x}$ is an inflection point if either $\mathbf{x} = (x_1, 0)$ and x_1 is a zero of g, or $\mathbf{x} = (x_1, x_2)$ where x_1 is a zero of $\det M$ and $x_2^d = g(x_1)$.

In the first case, we find d inflection points, and each has multiplicity $d-2$. In the second case, since $g(x_1) \neq 0$, we find $2d(d - 2)$ many regular inflection points with multiplicity one. Now, if we perturb f slightly, then the d points with multiplicity $d - 2$ will split into $d(d - 2)$ inflection points, while the other $2d(d - 2)$ inflection points will remain distinct. In total, this gives $3d(d - 2)$ inflection points. □

Corollary 6.9 *For a general plane curve C of degree d, the evolute has degree $3d(d - 1)$.*

Proof We compute the degree by intersecting the evolute with the line at infinity. For that, we consider the Zariski closure $\bar{C}$ of the curve C in the complex projective plane $\mathbb{P}^2_{\mathbb{C}}$. By Proposition 6.2, the evolute is the image of $\Gamma : \bar{C} \to \mathbb{P}^2_{\mathbb{C}}, \mathbf{x} \mapsto \mathbf{x} - r(\mathbf{x}) \cdot N(\mathbf{x})$. A point $\Gamma(\mathbf{x})$ on the evolute can be at infinity for two reasons: Either $\mathbf{x} \in \bar{C}$ is at infinity, or $\mathbf{x}$ is a finite point and a complex inflection point of the curve C by Proposition 6.5. Since C is a general curve of degree d, there are d points $\mathbf{x}$ of the first kind and $3d(d - 2)$ points $\mathbf{x}$ of the second kind, by Theorem 6.8. For each of the d points $\mathbf{x} \in \bar{C}$ at infinity, Salmon [156, §119] shows that $\Gamma(\mathbf{x})$ is a cusp whose tangent line is the line at infinity. Hence, when intersecting the evolute with the line at infinity, there are d cusps (that count with multiplicity three each) plus $3d(d - 2)$ points that correspond to the complex inflection points of C. All in all, the degree of the evolute is $3d + 3d(d - 2) = 3d(d - 1)$. □

Let us now find polynomial equations for critical curvature.

Lemma 6.10 *The points of critical curvature on the curve $C = \{f(\mathbf{x}) = 0\}$ are defined by*

$$(f_1^2 + f_2^2) \cdot \left(f_2 \cdot \frac{\partial \det H_0}{\partial x_1} - f_1 \cdot \frac{\partial \det H_0}{\partial x_2} \right) - 3 \det H_0 \cdot g = 0,$$

where $g := f_1 f_2 \cdot (f_{1,1} - f_{2,2}) + f_{1,2}(f_2^2 - f_1^2)$.

Proof Critical curvature points on C are defined by the equations

$$f(\mathbf{x}) = f_2(\mathbf{x}) \cdot \frac{\partial c(\mathbf{x})}{\partial x_1} - f_1(\mathbf{x}) \cdot \frac{\partial c(\mathbf{x})}{\partial x_2} = 0.$$

By Lemma 6.7 and the product rule, we have

$$-(d-1)^2 (f_1^2 + f_2^2)^{\frac{5}{2}} \cdot \frac{\partial c(\mathbf{x})}{\partial x_i} = \frac{\partial \det H_0}{\partial x_i} \cdot (f_1^2 + f_2^2) - 3 \det H_0 \cdot (f_1 \cdot f_{1,i} + f_2 \cdot f_{2,i}).$$

This implies the polynomial equation stated above.   □

Theorem 6.11 *A general plane curve C of degree d has $2d(3d-5)$ critical curvature points over the complex numbers $\mathbb{C}$.*

Proof Recall that critical curvature points correspond to finite cusps of the evolute by Proposition 6.5. Piene, Riener, and Shapiro prove in [146, Proposition 3.3] that, counting in the complex projective plane, the number of cusps on the evolute for a general plane curve C of degree d is $6d^2 - 9d$. As explained in the proof of Corollary 6.9 above, Salmon [156, §119] shows that d of these cusps lie at infinity. Therefore, the curve C has $6d^2 - 9d - d = 2d(3d-5)$ complex critical curvature points.   □

6.2 Algebraic Varieties

We now turn to smooth algebraic varieties in $\mathbb{R}^n$ of any dimension. Our aim is to study their curvature. This will lead us to the notions of the *second fundamental form* and the *Weingarten map*. These are fundamental concepts in Riemannian geometry. In standard textbooks, these concepts are presented in much more general contexts; see, for instance, [58, 120, 140]. Our main goal in this section is to formulate the second fundamental form and Weingarten map in terms of the polynomial equations that define the variety.

Let $X \subset \mathbb{R}^n$ be a smooth algebraic variety of dimension m. As in the case of plane curves, we consider a unit normal field $N(\mathbf{x})$ for X and we differentiate it along a tangent field $T(\mathbf{x})$. The main difference to the case of plane curves is that there are usually many tangent directions and many normal directions. In fact, every choice of normal and tangent direction defines a curvature. Similar to (6.3), we define the curvature of X at a point $\mathbf{x}$ in tangent direction $T(\mathbf{x})$ and in normal direction $N(\mathbf{x})$ to be $\langle T(\mathbf{x}), T(\mathbf{x})_1 \cdot \frac{\partial N}{\partial x_1}(\mathbf{x}) + \cdots + T(\mathbf{x})_n \cdot \frac{\partial N}{\partial x_n}(\mathbf{x}) \rangle$. We will see that, as for plane curves, this only depends on the values of $T(\mathbf{x})$ and $N(\mathbf{x})$ at a fixed point $\mathbf{x}$, but not on how those fields behave locally around $\mathbf{x}$.

Let us work this out. We assume that the Zariski closure of X is irreducible. Its ideal

$$I(X) = \langle f_1, \ldots, f_k \rangle$$

is prime. We denote the gradients of the polynomials f_i by ∇f_i and their Hessians by H_i for $i = 1, \ldots, k$. The (transpose of the) Jacobian of $f_1, \ldots, f_k$ at $\mathbf{x}$ is denoted by

$$J(\mathbf{x}) := \begin{bmatrix} \nabla f_1 & \ldots & \nabla f_k \end{bmatrix} \in \mathbb{R}^{n \times k}.$$

A smooth unit normal field on X is given by

$$N(\mathbf{x}) = \frac{J(\mathbf{x})\,w(\mathbf{x})}{\|J(\mathbf{x})\,w(\mathbf{x})\|} = \frac{1}{\|J(\mathbf{x})\,w(\mathbf{x})\|} \sum_{i=1}^{k} w_i(\mathbf{x}) \cdot \nabla f_i(\mathbf{x}), \tag{6.8}$$

where $w : X \to \mathbb{R}^k$ is a smooth function with $w(\mathbf{x}) \notin \ker J(\mathbf{x})$ for all $\mathbf{x}$. Differentiating (6.8) leads to the following equation, for some matrix-valued function $R(\mathbf{x})$:

$$\frac{\mathrm{d}}{\mathrm{d}\mathbf{x}} N(\mathbf{x}) = \frac{1}{\|J(\mathbf{x})\,w(\mathbf{x})\|} \sum_{i=1}^{k} w_i(\mathbf{x}) \cdot H_i(\mathbf{x}) + J(\mathbf{x})\,R(\mathbf{x}).$$

Let us now fix a point $\mathbf{x} \in X$. We denote the tangent vector by $\mathbf{t} := T(\mathbf{x}) \in T_{\mathbf{x}}X$ and the normal vector by $\mathbf{v} := J(\mathbf{x})w(\mathbf{x}) \in N_{\mathbf{x}}X$. Since $\mathbf{t}^\top J(\mathbf{x}) = 0$, this implies

$$\left\langle \mathbf{t}, \mathbf{t}_1 \cdot \frac{\partial N}{\partial x_1} + \cdots + \mathbf{t}_n \cdot \frac{\partial N}{\partial x_n} \right\rangle = \frac{1}{\|\mathbf{v}\|} \cdot \mathbf{t}^\top \left(\sum_{i=1}^{k} w_i H_i \right) \mathbf{t}.$$

The right-hand side only depends on the values of T and N at $\mathbf{x}$.

Definition 6.12 The *curvature* of a real algebraic variety X at a smooth point $\mathbf{x}$ in tangent direction $\mathbf{t} \in T_{\mathbf{x}}X$ and in normal direction $\mathbf{v} \in N_{\mathbf{x}}X$ is the scalar

$$c(\mathbf{x}, \mathbf{t}, \mathbf{v}) := \frac{1}{\|\mathbf{v}\|} \mathbf{t}^\top \left(\sum_{i=1}^{k} w_i \cdot H_i \right) \mathbf{t},$$

where $\mathbf{v} = \sum_{i=1}^{k} w_i \nabla f_i \in N_{\mathbf{x}}X$. In this formula, we use the notation introduced above.

Remark 6.13 If the codimension of X is greater than k, then the w_i are not unique. Still, the formula for $c(\mathbf{x}, \mathbf{t}, \mathbf{v})$ is well-defined. Indeed, if $\sum_{i=1}^{k} w_i \cdot \nabla f_i = 0$, then $\mathbf{t}^\top (\sum_{i=1}^{k} w_i \cdot H_i) \mathbf{t} = 0$.

In fact, for $\mathbf{x} \in X$ and a fixed normal vector $\mathbf{v} \in N_{\mathbf{x}}X$, the curvature $c(\mathbf{x}, \mathbf{t}, \mathbf{v})$ is a quadratic form on $T_{\mathbf{x}}X$. This quadratic form is called the *second fundamental form* of X at $\mathbf{x}$ and $\mathbf{v}$. It is often denoted by

$$\mathrm{II}_{\mathbf{v}}(\mathbf{t}) := c(\mathbf{x}, \mathbf{t}, \mathbf{v}). \tag{6.9}$$

The linear map associated with this quadratic form is the *Weingarten map*. We denote it by

$$L_{\mathbf{v}} : T_{\mathbf{x}}X \to T_{\mathbf{x}}X, \quad L_{\mathbf{v}}(\mathbf{t}) = P_{\mathbf{x}} \left(\sum_{i=1}^{k} w_i \cdot H_i \cdot \mathbf{t} \right), \quad \text{where } \mathbf{v} = \sum_{i=1}^{k} w_i \nabla f_i \tag{6.10}$$

and $P_{\mathbf{x}} : \mathbb{R}^n \to T_{\mathbf{x}}X$ is the orthogonal projection onto the tangent space of X at $\mathbf{x}$. Since $L_{\mathbf{v}}$ is a self-adjoint operator given by a real symmetric matrix, its eigenvalues are all real. If $\|\mathbf{v}\| = 1$, the eigenvalues of $L_{\mathbf{v}}$ are called the *principal curvatures* of X at $\mathbf{x}$ and in normal direction $\mathbf{v}$. The product of the principal curvatures is called the *Gauss curvature*; the arithmetic mean of the principal curvatures is called *mean curvature*.

Since the principal curvatures are the critical points of the quadratic form $\mathrm{II}_\mathbf{v}(\mathbf{t})$, the *maximal curvature*

$$C(X) := \max_{\mathbf{x} \in X,\, \mathbf{t} \in T_\mathbf{x} X,\, \mathbf{v} \in N_\mathbf{x} X} c(\mathbf{x}, \mathbf{t}, \mathbf{v}) = \max_{\mathbf{x} \in X,\, \mathbf{v} \in N_\mathbf{x} X} \max_{\mathbf{t} \in T_\mathbf{x} X} c(\mathbf{x}, \mathbf{t}, \mathbf{v}) \qquad (6.11)$$

is the maximum over all principal curvatures for varying $(\mathbf{x}, \mathbf{v})$.

Example 6.14 (Hypersurfaces) If X is defined by one polynomial $f(\mathbf{x})$, then we have only one normal direction (up to sign). Here, the formula in Definition 6.12 can be written as

$$c(\mathbf{x}, \mathbf{t}) := \frac{1}{\|\nabla f(\mathbf{x})\|}\, \mathbf{t}^\top H \mathbf{t},$$

where H is the Hessian of f. This generalizes the formula in Lemma 6.6.   ◇

Let us now focus on surfaces in $\mathbb{R}^3$. Let $S \subset \mathbb{R}^3$ be a smooth algebraic surface and $\mathbf{x} \in S$. When the two principal curvatures of S at $\mathbf{x}$ are equal, the point $\mathbf{x}$ is called an *umbilic* or *umbilical point* of the surface S. Equivalently, the best second-order approximation of S at $\mathbf{x}$ is a 2-sphere. Umbilical points can be formulated as the zeros of a system of polynomial equations, whose complex zeros are called *complex umbilics* of the surface S. Salmon [157] computed the number of complex umbilics of a general surface.

Theorem 6.15 *A general surface of degree d in* $\mathbb{R}^3$ *has* $10d^3 - 28d^2 + 22d$ *complex umbilics.*

In the case of surfaces of degree $d = 2$, we have results on the number of real umbilics and critical curvature points. Observe that rotations and translations do not affect the curvature, and that after a rotation and translation every quadric surface in $\mathbb{R}^3$ has the form

$$S = \{a_1 x_1^2 + a_2 x_2^2 + a_3 x_3^2 = 1\}.$$

By Theorem 6.15, the surface S has 12 complex umbilics. The next theorem is proved in [29]. We shall assume that $a_1 a_2 a_3 (a_1 - a_2)(a_1 - a_3)(a_2 - a_3) \neq 0$.

Theorem 6.16 *The number of real umbilics of the quadratic surface S equals*

- *4 if S is an ellipsoid (a_1, a_2, a_3 are positive) or a two-sheeted hyperboloid (one of the a_i is positive and two are negative);*
- *0 if S is a one-sheeted hyperboloid (two of the a_i are positive and one is negative).*

Similar to the case of plane curves, we call a point $\mathbf{x}$ on a surface S a *critical curvature point* if one of the two principal curvatures of S attains a critical value at $\mathbf{x}$. The first observation is that umbilics are always critical curvature points. This was shown in [29]. The following result from [29] covers the case of quadrics.

Theorem 6.17 *A general quadric surface* $S \subset \mathbb{R}^3$ *has* 18 *complex critical curvature points. The number of real critical curvature points equals*

- *10 if S is an ellipsoid (a_1, a_2, a_3 are positive);*
- *4 if S is a one-sheeted hyperboloid (two of the a_i are positive and one is negative);*
- *6 if S is a two-sheeted hyperboloid (one of the a_i is positive and two are negative).*

Fig. 6.5: The pictures illustrate Theorems 6.16 and 6.17. The figure on the left shows an ellipsoid with 4 red umbilics and 6 green critical curvature points. The umbilics are also critical curvature points, so there are 10 in total. Similarly, the figure in the middle shows a one-sheeted hyperboloid with 4 green critical curvature points, and the figure on the right shows a two-sheeted hyperboloid with 4 red umbilics and 2 green critical curvature points (so 6 critical curvature points in total).

6.3 Volumes of Tubular Neighborhoods

In this section, we study the volume of a *tubular neighborhood* of a real algebraic variety. This is closely connected to curvature, as we will see. Methods for computing volumes of semialgebraic sets numerically will be presented in Chapter 14.

The tubular neighborhood of radius ε of a variety $X \subset \mathbb{R}^n$ is the set

$$\text{Tube}(X, \varepsilon) := \{ \mathbf{u} \in \mathbb{R}^n \mid d(\mathbf{u}, X) < \varepsilon \},$$

where $d(\mathbf{u}, X) = \min_{\mathbf{x} \in X} \|\mathbf{u} - \mathbf{x}\|$ is the Euclidean distance from $\mathbf{u}$ to X. There are several general formulas for upper bounds on the volume of $\text{Tube}(X, \varepsilon)$ in the literature. For instance, Lotz [125] studied the case of a general complete intersection. Bürgisser, Cucker, and Lotz studied the case of a (possibly) singular hypersurface [36] in the sphere.

We now state the most general formula, due to Basu and Lerario [14]. Their theorem also holds for singular varieties. The proof of the theorem is based on approximating X in the Hausdorff topology by a sequence of smooth varieties $(X_k)_{k \in \mathbb{N}}$ and showing that the volume of the tubular neighborhood of X_k can be controlled as $k \to \infty$.

Theorem 6.18 *Let $X \subset \mathbb{R}^n$ be a real variety of dimension m, defined by polynomials of degree $\leq d$. Fix $\mathbf{u} \in \mathbb{R}^n$ and let $B_r(\mathbf{u})$ denote the ball of radius $r > 0$ around $\mathbf{u}$. For every $0 < \varepsilon \leq r/(4dm + m)$, we have*

$$\frac{\text{vol}(\text{Tube}(X, \varepsilon) \cap B_r(\mathbf{u}))}{\text{vol}(B_r(\mathbf{u}))} \leq 4e \left(\frac{4nd\varepsilon}{r} \right)^{n-m}.$$

The volume of tubular neighborhoods of smooth varieties (in fact, of smooth submanifolds of $\mathbb{R}^n$) is given by *Weyl's tube formula*. We shall now derive this formula. For a more detailed derivation and discussion, we refer to Weyl's original paper [175].

Let $X \subset \mathbb{R}^n$ be smooth and NX be the normal bundle of X. The ε-normal bundle is

$$\mathcal{N}_\varepsilon X := \{(\mathbf{x}, \mathbf{v}) \in NX \mid \|\mathbf{v}\| < \varepsilon\}.$$

The (normal) *exponential map* is the following parametrization of the tubular neighborhood:

$$\varphi_\varepsilon : \mathcal{N}_\varepsilon X \to \text{Tube}(X, \varepsilon), \quad (\mathbf{x}, \mathbf{v}) \mapsto \mathbf{x} + \mathbf{v}. \tag{6.12}$$

Definition 6.19 The *reach* of X is defined as

$$\tau(X) := \sup\{\varepsilon > 0 \mid \varphi_\varepsilon \text{ is a diffeomorphism}\}. \tag{6.13}$$

If X is smooth and compact, then the set in (6.13) is non-empty, and hence the reach $\tau(X)$ is a positive real number; see, e.g., [121, Theorem 6.24]. See Chapters 7 and 15 for further properties of the reach. In what follows, we assume that X is compact. If not, then we replace X by the semialgebraic set $X \cap B$, where B is a ball.

The desired volume is the integral of the constant function 1 over the tube. If $\varepsilon < \tau(X)$, then the exponential map φ_ε is a diffeomorphism and we can pull that integral back to the normal bundle $\mathcal{N}_\varepsilon X$. To be precise, let A be the matrix that represents the derivative of φ_ε with respect to orthonormal bases. We have

$$\text{vol}(\text{Tube}(X, \varepsilon)) = \int_{\text{Tube}(X, \varepsilon)} d\mathbf{u} = \int_{\mathbf{x} \in X} \int_{\mathbf{v} \in N_{\mathbf{x}} X: \|\mathbf{v}\| < \varepsilon} |\det(A(\mathbf{x}, \mathbf{v}))| \, d\mathbf{v} \, d\mathbf{x}. \tag{6.14}$$

We compute the matrix $A := A(\mathbf{x}, \mathbf{v})$. It represents a linear map

$$T_{(\mathbf{x},\mathbf{v})} \mathcal{N}_\varepsilon X \cong T_{\mathbf{x}} X \oplus N_{\mathbf{x}} X \to \mathbb{R}^n \cong T_{\mathbf{x}} X \oplus N_{\mathbf{x}} X.$$

Let B_1 be an orthonormal basis for the tangent space $T_{\mathbf{x}} X$ and B_2 one for the normal space $N_{\mathbf{x}} X$. An orthonormal basis for $T_{\mathbf{x}} X \oplus N_{\mathbf{x}} X$ is $\{(\mathbf{t}, 0) \mid \mathbf{t} \in B_1\} \cup \{(0, \mathbf{z}) \mid \mathbf{z} \in B_2\}$. We compute the image of these basis vectors under A. First, let $\mathbf{t} \in B_1$. If $\mathbf{v} = \lambda \cdot \mathbf{w}, \lambda = \|\mathbf{v}\|$, and $L_{\mathbf{w}}$ denotes the Weingarten map from (6.10), then we have

$$A \begin{bmatrix} \mathbf{t} \\ 0 \end{bmatrix} = \mathbf{t} + (\lambda \cdot L_{\mathbf{w}}) \mathbf{t}.$$

This is because $(\mathbf{t}, 0)$ is the tangent vector of a curve $(\mathbf{x}(t), \mathbf{v}(t)) \in \mathcal{N}_\varepsilon X$ passing through $(\mathbf{x}(0), \mathbf{v}(0)) = (\mathbf{x}, \mathbf{v})$, such that $\|\mathbf{v}(t)\|$ is constant and equal to λ. If the derivative of $\mathbf{x}(t)$ at $t = 0$ is $\mathbf{t}$, then the derivative of $\mathbf{v}(t)$ at $t = 0$ is $(\lambda \cdot L_{\mathbf{w}}) \mathbf{t}$, since the image of $\mathbf{t}$ under the Weingarten map $L_{\mathbf{w}}$ is the derivative of the unit normal vector $\lambda^{-1} \cdot \mathbf{v}(t)$. Linearity of differentiation implies that $A(\mathbf{t}, 0)^\top$ is the sum of these two terms.

Next, we consider a basis vector $(0, \mathbf{z})$ for $\mathbf{z} \in B_2$. It is the tangent vector of a curve $(\mathbf{x}(t), \mathbf{v}(t))$ in $\mathcal{N}_\varepsilon X$, where $\mathbf{x}(t)$ is constant and $\mathbf{v}(t)$ is a curve in the normal space $N_{\mathbf{x}} X$, whose derivative at $t = 0$ is $\mathbf{z}$. This shows the equation

$$A \begin{bmatrix} 0 \\ \mathbf{z} \end{bmatrix} = \mathbf{z}.$$

Therefore, a matrix representation with respect to orthonormal bases is

$$A(\mathbf{x}, \mathbf{v}) = \begin{bmatrix} I_m + \lambda \cdot L_{\mathbf{w}} & 0 \\ 0 & I_{n-m} \end{bmatrix}, \qquad m = \dim X. \tag{6.15}$$

Since $\lambda = \|\mathbf{v}\| < \tau(X)$, the eigenvalues of $\lambda \cdot L_{\mathbf{w}}$ have absolute value at most 1. This implies that the determinant of $A(\mathbf{x}, \mathbf{v})$ is positive, and therefore

$$|\det(A(\mathbf{x}, \mathbf{v}))| = \det(I_m + \lambda \cdot L_{\mathbf{w}}).$$

The transformation $\mathbf{v} \to (\mathbf{w}, \lambda)$ has Jacobian determinant λ^{n-m-1}. Plugging all this into the inner integral in (6.14), we arrive at the following integral for the volume of the tube:

$$\mathrm{vol}(\mathrm{Tube}(X, \varepsilon)) = \int_{\mathbf{x} \in X} \int_{\lambda=0}^{\varepsilon} \int_{\mathbf{w} \in N_{\mathbf{x}}X: \|\mathbf{w}\|=1} \lambda^{n-m-1} \cdot \det(I_m + \lambda \cdot L_{\mathbf{w}}) \, d\mathbf{w} \, d\lambda \, d\mathbf{x}.$$

Expanding the characteristic polynomial inside this integral, we see that $\mathrm{vol}(\mathrm{Tube}(X, \varepsilon))$ is a polynomial in ε of degree n whose coefficients are integrals of the principal minors of the Weingarten map $L_{\mathbf{v}}$ over X. Since $L_{-\mathbf{v}} = -L_{\mathbf{v}}$, the integrals over the odd-dimensional minors of $L_{\mathbf{v}}$ vanish. All this leads to:

Theorem 6.20 (Weyl's tube formula [175]) *In the notation above, we have*

$$\mathrm{vol}(\mathrm{Tube}(X, \varepsilon)) = \sum_{0 \le 2i \le m} \kappa_{2i}(X) \cdot \varepsilon^{n-m+2i}.$$

The coefficients $\kappa_{2i}(X)$ of this polynomial are called curvature coefficients *of X. Explicitly,*

$$\kappa_{2i}(X) = \frac{1}{n - m + 2i} \int_{\mathbf{x} \in X} \int_{\mathbf{w} \in N_{\mathbf{x}}X: \|\mathbf{w}\|=1} m_{2i}(L_{\mathbf{w}}) \, d\mathbf{w} \, d\mathbf{x},$$

where $m_{2i}(\cdot)$ denotes the sum of the principal minors of format $2i \times 2i$.

In fact, the curvature coefficient $\kappa_0(X)$ is always equal to the m-dimensional volume of X times the volume of the unit ball $B^{n-m} = \{\mathbf{x} \in \mathbb{R}^{n-m} \mid \|\mathbf{x}\| \le 1\}$. This yields:

Corollary 6.21 *The volume of a compact variety X of dimension m in $\mathbb{R}^n$ is equal to*

$$\mathrm{vol}(X) = \lim_{\varepsilon \to 0} \frac{\mathrm{vol}(\mathrm{Tube}(X, \varepsilon))}{\varepsilon^{n-m} \cdot \mathrm{vol}(B^{n-m})}.$$

We close this chapter by discussing Weyl's tube formula in two low-dimensional cases.

Example 6.22 (Curves) If $X \subset \mathbb{R}^n$ is a smooth algebraic curve, then

$$\mathrm{vol}(\mathrm{Tube}(X, \varepsilon)) = \varepsilon^{n-1} \cdot \mathrm{vol}(B^{n-1}) \cdot \mathrm{length}(X).$$

For instance, the volume of the ε-tube around a plane curve C is $2\varepsilon \cdot \mathrm{length}(C)$. $\diamond$

Example 6.23 (Surfaces in 3-space) Let $S = \{f(\mathbf{x}) = 0\} \subset \mathbb{R}^3$ be a compact smooth algebraic surface with Euler characteristic $\chi(S)$. For $\varepsilon < \tau(S)$, the volume of its ε-tube is

$$2\varepsilon \,\text{area}(S) + \varepsilon^3 \,\text{vol}(B^3)\,\chi(S).$$

Let us prove this. Weyl's tube formula implies

$$\text{vol}(\text{Tube}(S, \varepsilon)) = 2\varepsilon \,\text{area}(S) + \varepsilon^3 \kappa_2(S),$$

where the coefficient $\kappa_2(S)$ is the integral of the Gauss curvature:

$$\kappa_2(S) = \frac{2}{3}\int_{\mathbf{x}\in S} \det(L_{N(\mathbf{x})})\,d\mathbf{x},$$

where $N(\mathbf{x}) = \|\nabla f(\mathbf{x})\|^{-1} \cdot \nabla f(\mathbf{x})$ denotes the normal field of S at $\mathbf{x}$ given by the gradient of f. The surface S is orientable since it lives in $\mathbb{R}^3$. Indeed, the orientation is given by the normal field $N(\mathbf{x})$. The Gauss–Bonnet theorem (see, e.g., [120, Theorem 9.3]) implies that the integral over the Gauss curvature is $2\pi \cdot \chi(S)$. Moreover, the volume of the three-dimensional unit ball is $\text{vol}(B^3) = 4\pi/3$. ◇

Chapter 7
Reach and Offset

In this chapter, we study the medial axis, bottlenecks, and offset hypersurfaces. These notions are intuitive and important for many applications. They will also lead to a better understanding of the geometry of the reach, which was introduced in Definition 6.19. The last section is devoted to the offset discriminant of a variety. This offers a direct link between the ED problem in Chapter 2 and differential geometric concepts we saw in the previous chapter, namely the second fundamental form (6.9) and the Weingarten map (6.10).

7.1 Medial Axis and Bottlenecks

The *medial axis* $\mathrm{Med}(X)$ of a set $X \subset \mathbb{R}^n$ is the set of points $\mathbf{u} \in \mathbb{R}^n$ such that there exist at least two distinct points on X at which the distance from X to $\mathbf{u}$ is attained. In other words, the medial axis is the locus of points in $\mathbb{R}^n$ where the Euclidean projection to X is not well-defined. We shall later focus on the case when X is a smooth variety in $\mathbb{R}^n$.

Proposition 7.1 *If X is a semialgebraic set in $\mathbb{R}^n$ then its medial axis $\mathrm{Med}(X)$ is also a semialgebraic set.*

Proof The sentence in the introductory paragraph that defines the medial axis can be expressed using polynomial inequalities, together with the existential quantifier in *"there exist at least two points"*. By Tarski's Theorem on Quantifier Elimination, there exists a quantifier-free formula for $\mathrm{Med}(X)$. The medial axis is, therefore, semialgebraic. □

The reach $\tau(X)$ was defined as the supremum over all ε such that the exponential map of X restricted to normal vectors of length $< \varepsilon$ is a diffeomorphism. This implies that all points $\mathbf{u} \in \mathbb{R}^n$ whose distance to X is less than $\tau(X)$ have a unique closest point on X. Consequently, the distance between X and its medial axis must be at least $\tau(X)$. This shows that the reach is intimately linked with the medial axis. We make this connection more precise in the following result. Proposition 7.2 can be used as a definition for the reach of singular varieties. Yet another characterization of the reach will be given in Theorem 7.8.

© The Author(s) 2024
P. Breiding et al., *Metric Algebraic Geometry*, Oberwolfach Seminars 53,
https://doi.org/10.1007/978-3-031-51462-3_7

Proposition 7.2 *The distance from a smooth real variety X to its medial axis* Med(X) *is the reach* $\tau(X)$.

Proof The exponential map $\varphi_\varepsilon : N_\varepsilon X \to$ Tube(X, ε) was defined in (6.12) by the formula $(\mathbf{x}, \mathbf{v}) \mapsto \mathbf{x} + \mathbf{v}$. Suppose $\varepsilon < \tau(X)$. Then φ_ε is a diffeomorphism, and the tubular neighborhood Tube(X, ε) is disjoint from the medial axis Med(X) since otherwise φ_ε cannot be injective. Hence, the distance from X to Med(X) is at least $\tau(X)$. To show the reverse inequality, let $\varepsilon > 0$ be less than the distance from X to Med(X). Fix a point $\mathbf{u}$ in Tube(X, ε). The Euclidean closure of Med(X) does not contain $\mathbf{u}$. Thus, in an open neighborhood $U \subset$ Tube(X, ε) of $\mathbf{u}$, the points have a unique closest point on X. This defines a smooth map $U \to X, \mathbf{u} \mapsto \mathbf{x}(\mathbf{u})$. The map $\psi_\mathbf{u} : U \to N_\varepsilon X, \mathbf{u} \mapsto (\mathbf{x}(\mathbf{u}), \mathbf{u} - \mathbf{x}(\mathbf{u}))$ is then smooth with smooth inverse $\psi_\mathbf{u}^{-1} = \varphi_\varepsilon|_{\psi_\mathbf{u}(U)}$. Via a partition of unity of Tube(X, ε) (see e.g. [121, Chapter 2]), we can obtain a global smooth inverse of φ_ε from the local inverses $\psi_\mathbf{u}$. Hence, φ_ε is a diffeomorphism and $\varepsilon < \tau(X)$. $\qquad\square$

Example 7.3 The reach of the parabola X in Example 7.4 is $\tau(X) = \frac{1}{2}$. It is realized as the distance between $(0, 0) \in X$ and $(0, \frac{1}{2})$, which is a point in the Euclidean closure of Med$(X) = \{(0, u_2) \in \mathbb{R}^2 \mid u_2 > \frac{1}{2}\}$. $\qquad\diamond$

We define the *algebraic medial axis* as the Zariski closure of the medial axis. In symbols,

$$M_X := \overline{\text{Med}(X)}.$$

The variety M_X is our algebraic proxy for the medial axis Med(X). By Proposition 7.1, the variety M_X and the semialgebraic set Med(X) have the same dimension.

Example 7.4 Consider the parabola $X = V(x_2 - x_1^2)$. We compute the algebraic medial axis of X. The result is shown in Figure 7.4 below. If $\mathbf{x} = (x_1, x_2) \in X$ minimizes the distance to a point $\mathbf{u} = (u_1, u_2) \in \mathbb{R}^2$, then we have $\langle \mathbf{x} - \mathbf{u}, \mathbf{t} \rangle = 0$, where $\mathbf{t}$ spans the tangent space $T_\mathbf{x} X$. We use Macaulay2 [73] to compute M_X algebraically:

```
R = QQ[x1, x2, y1, y2, u1, u2];
fx = x2 - x1^2; fy = y2 - y1^2;
Jx = matrix {{x1-u1, x2-u2}, {diff(x1, fx), diff(x2, fx)}};
Jy = matrix {{y1-u1, y2-u2}, {diff(y1, fy), diff(y2, fy)}};
distxu = (x1 - u1)^2 + (x2 - u2)^2;
distyu = (y1 - u1)^2 + (y2 - u2)^2;
I = ideal {fx, fy, det(Jx), det(Jy), distxu - distyu};
K = saturate(I, ideal {x1-y1, x2-y2});
eliminate({x1, x2, y1, y2}, K)
```

This returns the ideal $\langle u_1 \rangle$, so the algebraic medial axis is $M_X = \{u_1 = 0\}$. $\qquad\diamond$

Remark 7.5 Propositions 7.1 and 7.2 imply that the reach of a real variety is an algebraic notion. More concretely, if X is smooth and defined by polynomials with rational coefficients, then its reach $\tau(X)$ is an algebraic number over $\mathbb{Q}$. This was shown by Horobeţ and Weinstein in [92, Proposition 3.14].

Our next goal is to characterize the reach of a smooth variety $X \subset \mathbb{R}^n$ in terms of maximal curvature and bottlenecks. Curvature is discussed in detail in the previous chapter. We now introduce bottlenecks. Let $\mathbf{x}, \mathbf{y} \in X$ be two distinct points. If $\mathbf{x} - \mathbf{y}$ is normal to $T_{\mathbf{x}}X$ (i.e., $\langle \mathbf{t}, \mathbf{x} - \mathbf{y} \rangle = 0$ for all $\mathbf{t} \in T_{\mathbf{x}}X$) and also normal to $T_{\mathbf{y}}X$, then we call $\{\mathbf{x}, \mathbf{y}\}$ a *bottleneck*. Complex solutions to the corresponding polynomial equations are *complex bottlenecks*. Di Rocco, Eklund and Weinstein [57] expressed the number of complex bottlenecks of a variety in terms of polar classes. The following theorem is their result for planar curves.

Theorem 7.6 *A general plane curve of degree d has $\frac{1}{2}(d^4 - 5d^2 + 4d)$ complex bottlenecks.*

Example 7.7 We compute the bottlenecks of the Trott curve from (2.7) using the numerical software `HomotopyContinuation.jl` [31]:

```
using HomotopyContinuation, LinearAlgebra
@var x y u v
f = 144*(x^4 + y^4) - 225*(x^2 + y^2) + 350*x^2*y^2 + 81;
g = subs(f, x=>u, y=>v);
df = differentiate(f, [x; y]); dg = differentiate(g, [u; v]);
N = [x-u; y-v];
bottlenecks = solve([f; g; det([N df]); det([N dg])])
```

By Theorem 7.6 there are $\frac{1}{2}(d^4 - 5d^2 + 4d) = 96$ complex bottlenecks. Our computation reveals that 36 of the bottlenecks are real. They are marked in Figure 7.1. We note here that one point can appear in more than one bottleneck.   ◇

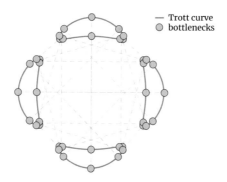

Fig. 7.1: Bottlenecks of the Trott curve (2.7) are displayed as grey normal lines with yellow endpoints.

The *width* of a bottleneck is $b(\mathbf{x}, \mathbf{y}) := \frac{1}{2}\|\mathbf{x} - \mathbf{y}\|$. We denote the width of the smallest bottleneck of the variety X by

$$B(X) \;:=\; \min_{\{\mathbf{x},\mathbf{y}\}\text{ bottleneck of } X} b(\mathbf{x}, \mathbf{y}).$$

The next theorem links the reach of X to its bottlenecks and its maximal curvature $C(X)$. Recall that $C(X)$ was defined in Equation (6.11).

Theorem 7.8 *Let X be a smooth variety in $\mathbb{R}^n$. Then the reach of X equals*

$$\tau(X) \;=\; \min\left\{ B(X), \; \frac{1}{C(X)} \right\}.$$

Proof Recall from (6.13) that the reach $\tau(X)$ is the supremum over all $\varepsilon > 0$ such that the exponential map $\varphi_\varepsilon : N_\varepsilon X \to \mathrm{Tube}(X, \varepsilon)$, $(\mathbf{x}, \mathbf{v}) \mapsto \mathbf{x} + \mathbf{v}$ is a diffeomorphism. Let

$$\varepsilon := \tau(X). \tag{7.1}$$

For $\varepsilon' > \varepsilon$, the exponential map $\varphi_{\varepsilon'}$ is not a diffeomorphism. The map is either not an immersion or it is not injective. Thus, there is a point $\mathbf{u} = \mathbf{x} + \mathbf{v} \in \mathbb{R}^n$, where $(\mathbf{x}, \mathbf{v}) \in N X$, at distance $\varepsilon = \|\mathbf{v}\|$ from X such that, for $\varepsilon' > \varepsilon$, either the derivative of $\varphi_{\varepsilon'}$ at $(\mathbf{x}, \mathbf{v})$ is not injective or $\varphi_{\varepsilon'}^{-1}(\mathbf{u})$ has at least two elements.

Suppose first that the derivative of each $\varphi_{\varepsilon'}$ at $(\mathbf{x}, \mathbf{v})$ is not injective. Then it follows from (6.15) that ε^{-1} is a principal curvature at $\mathbf{x}$ in normal direction $-\varepsilon^{-1}\mathbf{v}$. According to (7.1), ε is the smallest positive number with that property. Therefore, its inverse ε^{-1} must be the maximal curvature $C(X)$.

The remaining case to analyze is when each fiber $\varphi_{\varepsilon'}^{-1}(\mathbf{u})$ contains at least two points and the derivative of $\varphi_{\varepsilon'}$ is injective at these points. We have two distinct points $\mathbf{x}, \mathbf{y} \in X$ such that $\mathbf{u} = \mathbf{x} + \mathbf{v} = \mathbf{y} + \mathbf{w}$, where $(\mathbf{y}, \mathbf{w}) \in N X$ and $\delta := \|\mathbf{w}\| \le \varepsilon$. We distinguish two subcases. First, we assume that $\mathbf{u}$ lies on the line spanned by $\mathbf{x}$ and $\mathbf{y}$. Then, $\{\mathbf{x}, \mathbf{y}\}$ is a bottleneck. Moreover, $\mathbf{u}$ must be the midpoint between $\mathbf{x}$ and $\mathbf{y}$; otherwise there is a σ with $b(\mathbf{x}, \mathbf{y}) < \sigma < \varepsilon$ and the fiber of the midpoint $\frac{1}{2}(\mathbf{x} + \mathbf{y})$ under φ_σ would contain at least two points, but the latter implies $\tau(X) \le \sigma < \varepsilon$; a contradiction to (7.1). Hence, $\varepsilon = b(\mathbf{x}, \mathbf{y})$. By (7.1), there cannot be any smaller bottleneck, so that $\varepsilon = B(X)$ and we are done.

Second, we assume that $\mathbf{x}, \mathbf{y}, \mathbf{u}$ form a triangle. Since the derivative of $\varphi_{\varepsilon'}$ is injective at both $(\mathbf{x}, \mathbf{v})$ and $(\mathbf{y}, \mathbf{w})$, the Inverse Function Theorem implies the existence of two locally defined and smooth maps $\mathbf{u} \mapsto \mathbf{x}(\mathbf{u})$ and $\mathbf{u} \mapsto \mathbf{y}(\mathbf{u})$ that project locally around $\mathbf{u}$ to X. These define two local smooth functions $d_{\mathbf{x}}(\mathbf{u}) := \|\mathbf{u} - \mathbf{x}(\mathbf{u})\|$ and $d_{\mathbf{y}}(\mathbf{u}) := \|\mathbf{u} - \mathbf{y}(\mathbf{u})\|$ that locally measure the distance to X. Their gradients are

$$\nabla d_{\mathbf{x}}(\mathbf{u}) = \varepsilon^{-1}(\mathbf{u} - \mathbf{x}) \quad \text{and} \quad \nabla d_{\mathbf{y}}(\mathbf{u}) = \delta^{-1}(\mathbf{u} - \mathbf{y});$$

see e.g. [72, Lemma 2.11] and also Remark 7.9.

Let $\mathbf{a}$ be a unit norm vector that starts at $\mathbf{u}$ and points inside the triangle with vertices $\mathbf{x}, \mathbf{y}, \mathbf{u}$ such that $\langle \mathbf{u} - \mathbf{x}, \mathbf{a} \rangle < 0$ and $\langle \mathbf{u} - \mathbf{y}, \mathbf{a} \rangle < 0$. Here is an illustration:

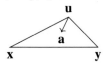

The partial derivatives of $d_{\mathbf{x}}(\mathbf{u})$ and $d_{\mathbf{y}}(\mathbf{u})$ in direction $\mathbf{a}$ satisfy $\frac{\partial d_{\mathbf{x}}(\mathbf{u})}{\partial \mathbf{a}} < 0$ and $\frac{\partial d_{\mathbf{y}}(\mathbf{u})}{\partial \mathbf{a}} < 0$. When we move from $\mathbf{u}$ in direction $\mathbf{a}$, the local distances from $\mathbf{u}$ to X both decrease. Hence, there is a $\sigma < \varepsilon$ such that φ_σ is not injective. Thus, $\tau(X) \le \sigma < \varepsilon$. This contradicts (7.1) and so $\mathbf{x}, \mathbf{y}, \mathbf{u}$ cannot form a triangle. $\qquad\square$

Remark 7.9 As above, let d be a (locally) defined projection map to a variety X, such that $\mathbf{x} = d(\mathbf{u})$. Write $\mathbf{v} := (\mathbf{u} - \mathbf{x})/\|\mathbf{u} - \mathbf{x}\|$ for the unit normal direction. We claim that $\nabla d(\mathbf{u}) = \mathbf{v}$. An informal proof is as follows: If we move from $\mathbf{u}$ infinitesimally in a direction that is perpendicular to $\mathbf{v}$, then the distance $d(\mathbf{u})$ does not change. This means that the derivative of $d(\mathbf{u})$ in a direction perpendicular to $\mathbf{v}$ is zero, hence the gradient of $d(\mathbf{u})$ must be a multiple of $\mathbf{v}$. On the other hand, $d(\mathbf{u} + t \cdot \mathbf{v}) = d(\mathbf{u}) + t$, which shows that the derivative of $d(\mathbf{u})$ in direction $\mathbf{v}$ is 1. Consequently, $\nabla d(\mathbf{u}) = \mathbf{v}$. In particular, the Weingarten map $L_{\mathbf{v}}$ can be obtained via the second derivative of $d(\mathbf{u})$. This is worked out in (7.5) below.

7.2 Offset Hypersurfaces

This section is based on the article [92] by Horobeţ and Weinstein. We fix an irreducible variety X in $\mathbb{R}^n$, and we identify X with its Zariski closure in $\mathbb{C}^n$. The *ED correspondence* $\mathcal{E}_X$ of X is the Zariski closure of the set of pairs $(\mathbf{x}, \mathbf{u}) \in \mathrm{Reg}(X) \times \mathbb{R}^n$ such that $\mathbf{x}$ is an ED critical point for $\mathbf{u}$. We recall from Theorem 2.23 that

$$\mathcal{E}_X = \overline{\{(\mathbf{x}, \mathbf{x} + \mathbf{h}) \mid \mathbf{x} \in \mathrm{Reg}(X), (\mathbf{x}, \mathbf{h}) \in N_X\}} \subseteq X \times \mathbb{C}^n.$$

The branch locus of the projection $\mathcal{E}_X \to \mathbb{C}^n$ is called the *ED discriminant* or evolute. For plane curves, this coincides with the definition of the evolute in Section 1.3. We denote the ED discriminant of the given variety X by $\Sigma_X \subset \mathbb{C}^n$.

For $\varepsilon \in \mathbb{C}$ and $\mathbf{u} \in \mathbb{C}^n$, the ε-*sphere around* $\mathbf{u}$ is the variety $S(\mathbf{u}, \varepsilon) := V(\|\mathbf{x} - \mathbf{u}\|^2 - \varepsilon^2)$.

Definition 7.10 The *offset correspondence* of X is the variety

$$\mathrm{OC}_X = (\mathcal{E}_X \times \mathbb{C}) \cap \{(\mathbf{x}, \mathbf{u}, \varepsilon) \in \mathbb{C}^n \times \mathbb{C}^n \times \mathbb{C} \mid \mathbf{x} \in S(\mathbf{u}, \varepsilon)\}.$$

Hence, OC_X is the complex Zariski closure of the set of triples $(\mathbf{x}, \mathbf{u}, \varepsilon)$ such that $\mathbf{x}$ is an ED critical point for $\mathbf{u}$, and ε^2 is the squared Euclidean distance (over $\mathbb{R}$) between $\mathbf{x}$ and $\mathbf{u}$.

We consider the two projections $\pi_1 : \mathrm{OC}_X \to X$ and $\pi_2 : \mathrm{OC}_X \to \mathbb{C}^n \times \mathbb{C}$. The map π_1 is dominant, i.e. $\overline{\pi_1(\mathrm{OC}_X)} = X$, because $(\mathbf{x}, \mathbf{x}, 0) \in \mathrm{OC}_X$ for all $\mathbf{x} \in X$. However, the other projection is not dominant. The next definition specifies the image of that projection.

Definition 7.11 The *offset hypersurface* of X is the variety $\mathrm{Off}_X := \overline{\pi_2(\mathrm{OC}_X)} \subset \mathbb{C}^n \times \mathbb{C}$.

The next lemma justifies the name.

Lemma 7.12 $\mathrm{codim} \, \mathrm{Off}_X = 1$.

Proof The ED correspondence $\mathcal{E}_X$ is the Zariski closure of a vector bundle of rank $\mathrm{codim}(X)$ over $\mathrm{Reg}(X)$, which shows that $\dim \mathcal{E}_X = n$. Since X is irreducible, also $\mathcal{E}_X$ is irreducible. The offset correspondence OC_X is the intersection of $\mathcal{E}_X \times \mathbb{C}$, which is irreducible as well, with a hypersurface. This implies $\dim \mathrm{OC}_X = n$. Since the ED degree of X is finite, the projection π_2 has finite fibers generically, which implies $\dim \mathrm{Off}_X = n. \square$

Remark 7.13 For a fixed radius $r > 0$, the level set $\mathrm{Off}_{X,r} \subset \mathbb{C}^n$ is the intersection of the offset hypersurface Off_X with the hyperplane $\varepsilon = r$ in $\mathbb{C}^n \times \mathbb{C}$. Figure 7.3 shows the level sets for $r = 0.5$ and $r = 1.25$, respectively, of the offset surface of a parabola. The real locus of $\mathrm{Off}_{X,r}$ always contains the boundary of the tubular neighborhood $\mathrm{Tube}(X, r)$, which is $\partial\mathrm{Tube}(X, r) = \{\mathbf{u} \in \mathbb{R}^n \mid d(\mathbf{u}, X) = r\}$. Indeed, for any $\mathbf{u} \in \partial\mathrm{Tube}(X, r)$, there is a closest point $\mathbf{x} \in X$ with $\|\mathbf{x} - \mathbf{u}\| = r$. In particular, $\mathbf{x}$ is an ED critical point for $\mathbf{u}$ and $(\mathbf{x}, \mathbf{u}, r) \in \mathrm{OC}_X$, i.e., $\mathbf{u} \in \mathrm{Off}_{X, r}$. For the parabola in Figure 7.3, the real locus of $\mathrm{Off}_{X, 0.5}$ is equal to $\partial\mathrm{Tube}(X, 0.5)$. On the other hand, $\partial\mathrm{Tube}(X, 1.25)$ is strictly contained in the real part of $\mathrm{Off}_{X, 1.25}$. In this example, that change of qualitative behavior occurs when the radius r crosses the reach $\tau(X) = 0.5$.

It follows from Lemma 7.12 that Off_X is the zero set of a polynomial $g_X(\mathbf{u}, \varepsilon)$; i.e.,

$$\mathrm{Off}_X = V(g_X) \subset \mathbb{C}^n \times \mathbb{C}.$$

We call g_X the *offset polynomial*. It is also known as the *ED polynomial*; see [141].

Fig. 7.2: The offset hypersurface Off_X of the parabola is a surface in $\mathbb{R}^3$. The surface is symmetric along the ε-axis, because only even powers of ε appear in the offset polynomial $g(\mathbf{u}, \varepsilon)$. The parabola itself is visible at level $\varepsilon = 0$.

Example 7.14 We compute the offset polynomial of a parabola in Macaulay2 [73]:

```
R = QQ[x1, x2, u1, u2, eps];
f = x2 - x1^2;   d = (x1-u1)^2 + (x2-u2)^2 - eps^2;
J = matrix {{x1-u1, x2-u2}, {diff(x1, f), diff(x2, f)}};
OC = ideal {f, det(J), d};
O = eliminate({x1, x2}, OC);
g = sub((gens O)_0_0, QQ[x1, x2, u1, u2][eps])
```

Setting $X = V(x_2 - x_1^2)$, this gives $g_X(\mathbf{u}, \varepsilon) = g_0(\mathbf{u}) + g_1(\mathbf{u})\varepsilon^2 + g_2(\mathbf{u})\varepsilon^4 + g_3(\mathbf{u})\varepsilon^6$, where

$$g_0(\mathbf{u}) = (u_1^2 - u_2)^2(16u_1^2 + 16u_2^2 - 8u_2 + 1),$$
$$g_2(\mathbf{u}) = 48u_1^2 + 16u_2^2 + 32u_2 - 8,$$
$$g_1(\mathbf{u}) = -48u_1^4 - 32u_1^2u_2^2 + 8u_1^2u_2 - 32u_2^3 - 20u_1^2 - 8u_2^2 + 8u_2 - 1,$$
$$g_3(\mathbf{u}) = -16.$$

Figure 7.2 shows the real part of the offset surface $g(\mathbf{u}, \varepsilon) = 0$, which lives in the ambient space $\mathbb{R}^3 = \mathbb{R}^2 \times \mathbb{R} \subset \mathbb{C}^2 \times \mathbb{C}$. ◇

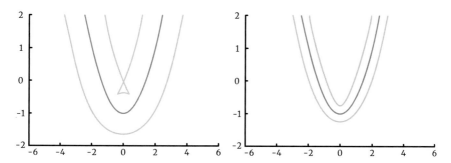

Fig. 7.3: The offset surface Off$_X$ of the parabola intersected with the planes $\varepsilon = 1.25$ and $\varepsilon = 0.5$.

We next examine some properties of the offset polynomial.

Proposition 7.15 (a) *For a general point $\mathbf{u} \in \mathbb{C}^n$, the zeros of the univariate polynomial function $\varepsilon \mapsto g_X(\mathbf{u}, \varepsilon)$ are the complex numbers $\varepsilon = \pm\sqrt{\|\mathbf{u} - \mathbf{x}\|^2}$, where $\mathbf{x}$ ranges over all ED critical points for $\mathbf{u}$ on X.*

(b) *The degree of the offset polynomial $g_X(\mathbf{u}, \varepsilon)$ in the unknown ε is two times the Euclidean distance degree of X.*

Proof We observe that the projection Off$_X \to \mathbb{C}^n$, $(\mathbf{u}, \varepsilon) \mapsto \mathbf{u}$ is dominant, because general points in $\mathbb{C}^n$ have ED critical points on X. Take a general $\mathbf{u} \in \mathbb{C}^n$. Then, $g_X(\mathbf{u}, \varepsilon) = 0$ if and only if there exists $\mathbf{x} \in X$ with $(\mathbf{x}, \mathbf{u}) \in \mathcal{E}_X$ and $\varepsilon^2 = \|\mathbf{x} - \mathbf{u}\|^2$. Therefore, $\varepsilon = \pm\sqrt{\|\mathbf{u} - \mathbf{x}\|^2}$, where $\mathbf{x}$ ranges over all ED critical points for $\mathbf{u}$. In particular, this shows that the univariate polynomial function $\varepsilon \mapsto g_X(\mathbf{u}, \varepsilon)$ has $2 \cdot \text{EDdegree}(X)$ many zeros for general $\mathbf{u}$. □

Example 7.16 In Example 7.14, we see that the offset polynomial $g_X(\mathbf{u}, \varepsilon)$ has degree six in ε. This reflects the fact that the ED degree of the parabola is three. For instance, any point $\mathbf{u} \in \mathbb{R}^2$ in Figure 7.4 that is both above the blue curve (the parabola) and above the yellow curve (the evolute) has three real ED critical points on the parabola. ◇

The coefficients of the offset polynomial were studied by Ottaviani and Sodomaco in [141]. They prove the following theorem (see [141, Proposition 4.4]).

Theorem 7.17 *If X is general enough and $g_X(\mathbf{u}, \varepsilon) = c_0(\mathbf{u}) + c_1(\mathbf{u})\varepsilon^2 + \cdots + c_k(\mathbf{u})\varepsilon^{2k}$ is its offset polynomial, then the leading coefficient $c_k(\mathbf{u})$ is a real constant.*

7.3 Offset Discriminant

In the previous section, we introduced the offset polynomial $g_X(\mathbf{u}, \varepsilon)$ of a variety $X \subset \mathbb{R}^n$. For any fixed $\mathbf{u} \in \mathbb{R}^n$, this encodes the distances from $\mathbf{u}$ to its ED critical points on X. If $\mathbf{u}$

is on the medial axis, then $g_X(\mathbf{u}, \varepsilon)$ must have a double root in ε. This motivates us to study the discriminant of the offset polynomial with respect to the distinguished variable ε. That discriminant is a polynomial in $\mathbf{u} = (u_1, \ldots, u_n)$.

Definition 7.18 The *offset discriminant* is the polynomial

$$\delta_X(\mathbf{u}) := \mathrm{Disc}_\varepsilon \, g_X(\mathbf{u}, \varepsilon).$$

Its zero set is the hypersurface

$$\Delta_X^{\mathrm{Off}} := V(\delta_X) \subset \mathbb{C}^n.$$

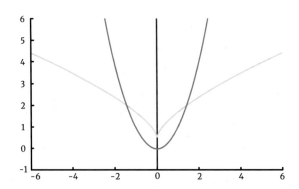

Fig. 7.4: The offset discriminant of the parabola $X = V(x_2 - x_1^2)$ has three real components: the parabola itself (blue), the algebraic medial axis M_X (the red vertical line) and the ED discriminant or evolute Σ_X (the yellow cubic curve).

Example 7.19 The discriminant of the offset polynomial of the parabola in Example 7.14 is

$$\delta_X(\mathbf{u}) = u_1^4 \cdot (u_1^2 - u_2) \cdot \delta_1(\mathbf{u})^6 \cdot \delta_2(\mathbf{u})$$
$$\text{where} \quad \delta_1(\mathbf{u}) = -27u_1^2 + 2(2u_2 - 1)^3 \quad \text{and} \quad \delta_2(\mathbf{u}) = 16u_1^2 + (4u_2 - 1)^2.$$

The factor $\delta_2(\mathbf{u})$ has no real zeros, and $u_1^2 - u_2$ is the polynomial of X. The real zero locus of the other factors u_1 and $\delta_1(\mathbf{u})$ are shown in Figure 7.4. The medial axis of the parabola is $\mathrm{Med}(X) = \{(0, u_2) \mid u_2 > \frac{1}{2}\}$ and $u_1 = 0$ is the Zariski closure of $\mathrm{Med}(X)$. The variety $\delta_1(\mathbf{u}) = 0$ is the ED discriminant or evolute of the parabola, also known as the *semicubical parabola*. Above the evolute, every point $\mathbf{u} \in \mathbb{R}^2$ has three real ED critical points on X, and below the evolute it has one real and two complex ED critical points.   ◇

The discriminant Δ_X^{Off} in the previous example has three irreducible components: the variety X, its algebraic medial axis, and its evolute. We show that this is a general fact, following [92]. For that, we define the *bisector hypersurface* Bis_X of the variety X. If we write $Bl_X \subset \mathbb{C}^n \times \mathbb{C}$ for the branch locus of the projection $\pi_2 : OC_X \to \mathbb{C}^n \times \mathbb{C}$, the bisector hypersurface is the set of all points $\mathbf{u}$ that arise from that branch locus:

$$Bis_X := \{ \mathbf{u} \mid (\mathbf{u}, \varepsilon) \in Bl_X \}.$$

Theorem 7.20 *The offset discriminant has the following decomposition:*

$$\Delta_X^{Off} = Bis_X \cup \Sigma_X \supseteq X \cup M_X \cup \Sigma_X.$$

The real locus of Δ_X^{Off} consists of the real points of the varieties X, M_X, and Σ_X.

Proof By Proposition 7.15, the offset discriminant is the locus of those $\mathbf{u}$ such that $g_X(\mathbf{u}, \varepsilon) \in \mathbb{C}[\varepsilon]$ has fewer than $2 \cdot EDdegree(X)$ distinct complex zeros. This can happen for two reasons: either $\mathbf{u}$ has fewer than $EDdegree(X)$ distinct ED critical points on X, or $\mathbf{u}$ has two distinct ED critical points

$$\mathbf{x}_1 \neq \mathbf{x}_2 \text{ with } \|\mathbf{x}_1 - \mathbf{u}\|^2 = \|\mathbf{x}_2 - \mathbf{u}\|^2. \tag{7.2}$$

The first case is accounted for by the ED discriminant Σ_X and the second case is the bisector hypersurface Bis_X. By definition, the medial axis $Med(X)$ is contained in Bis_X, and thus we also have the inclusion $M_X \subseteq Bis_X$. Since the ε that come from zeros of $g_X(\mathbf{u}, \varepsilon)$ come in signed pairs (cf. Proposition 7.15), we see that $X \times \{0\}$ is doubly covered by the projection π_2. This implies $X \subseteq Bis_X$. All other components of Bis_X that are not in $X \cup M_X$ consist of non-real points $\mathbf{u}$ that have ED critical points as in (7.2). $\square$

Suppose now that $\mathbf{u}_0 \in \mathbb{R}^n \backslash \Delta_X^{Off}$ is a real point outside the offset discriminant. Using Remark 7.9, we can compute a unit normal field from the offset polynomial by working locally near $\mathbf{u}_0$. Namely, we first choose ε_0 such that $(\mathbf{u}_0, \varepsilon_0) \in \mathbb{R}^n \times \mathbb{R}$ is a real zero of g_X. Then, by varying $\mathbf{u}$ in a small neighborhood of $\mathbf{u}_0$, we obtain a function $\mathbf{u} \mapsto \varepsilon(\mathbf{u})$. This function is defined implicitly by the equation $g_X(\mathbf{u}, \varepsilon) = 0$.

We consider the gradient $\frac{d\varepsilon}{d\mathbf{u}}(\mathbf{u}, \varepsilon)$ of the function $\mathbf{u} \mapsto \varepsilon(\mathbf{u})$. It follows from Remark 7.9 that this gradient is a unit normal vector which points from a real ED critical point on X towards $\mathbf{u}$. We compute that gradient by implicit differentiation. Namely, by differentiating the equation $g_X(\mathbf{u}, \varepsilon) = 0$, we obtain

$$\nabla \varepsilon := \frac{d\varepsilon}{d\mathbf{u}}(\mathbf{u}, \varepsilon) = -\left(\frac{\partial g_X}{\partial \varepsilon} \right)^{-1} \cdot \frac{\partial g_X}{\partial \mathbf{u}} \in \mathbb{R}^n. \tag{7.3}$$

Furthermore, by differentiating (7.3), we obtain the following formula for the Hessian matrix of $\mathbf{u} \mapsto \varepsilon(\mathbf{u})$:

$$\frac{d^2 \varepsilon}{d\mathbf{u}^2} = -\left(\frac{\partial g_X}{\partial \varepsilon} \right)^{-1} \cdot \left(\frac{\partial^2 g_X}{\partial \mathbf{u}^2} + \frac{\partial^2 g_X}{\partial \varepsilon^2} \cdot \nabla \varepsilon (\nabla \varepsilon)^\top + 2\nabla \varepsilon \left(\frac{\partial(\partial g_X / \partial \varepsilon)}{\partial \mathbf{u}} \right)^\top \right). \tag{7.4}$$

The expression in (7.4) is an $n \times n$ matrix. The entries of the gradient vector and the Hessian matrix are rational functions in $\mathbf{u}$, whose denominator vanishes when $\frac{\partial g_X}{\partial \varepsilon}(\mathbf{u}, \varepsilon) = 0$. This is why we assumed $\mathbf{u}_0 \notin \Delta_X^{\text{Off}}$. The next theorem shows that we can compute the second fundamental form of X from the Hessian of g_X.

Theorem 7.21 *Let $\mathbf{u} \in \mathbb{R}^n \setminus \Delta_X^{\text{Off}}$ and $\varepsilon > 0$ with $g_X(\mathbf{u}, \varepsilon) = 0$. Let $\mathbf{x} \in X$ be the (uniquely determined) ED critical point corresponding to $(\mathbf{u}, \varepsilon)$. The following two statements hold:*

(a) The gradient $\frac{d\varepsilon}{d\mathbf{u}}(\mathbf{u}, \varepsilon)$ is a unit normal vector at $\mathbf{x}$ pointing towards $\mathbf{u}$.

(b) Let $\mathbf{t} \in T_\mathbf{x} X$ be a tangent vector. The second fundamental form of X evaluated at $\mathbf{t}$ is

$$\mathrm{II}_{\mathbf{u}-\mathbf{x}}(\mathbf{t}) = \lim_{\substack{s \to 0 \\ s > 0}} \mathbf{t}^\top \left(\frac{d^2\varepsilon}{d\mathbf{u}^2}(\mathbf{x} + s(\mathbf{u} - \mathbf{x}), s\varepsilon) \right) \mathbf{t}.$$

Proof Item (a) follows from Remark 7.9 and we have discussed this above. We need to prove item (b). Since $\mathbf{u} \notin \Delta_X^{\text{Off}}$, there can only be finitely many points on the line through $\mathbf{x}$ and $\mathbf{u}$ that belong to Δ_X^{Off}. This implies that there exists $s' > 0$ such that $\mathbf{x} + s(\mathbf{u} - \mathbf{x}) \notin \Delta_X^{\text{Off}}$ for all $0 < s < s'$. Item (a) implies that, for all $0 < s < s'$, the vector $\frac{d\varepsilon}{d\mathbf{u}}(\mathbf{x} + s(\mathbf{u} - \mathbf{x}), s\varepsilon)$ is a unit normal vector at $\mathbf{x}$ pointing towards $\mathbf{u}$.

Fix $0 < s < s'$. By definition of the second fundamental form in (6.9), we know that $\mathrm{II}_{\mathbf{u}-\mathbf{x}}(\mathbf{t})$ is the directional derivative of the unit normal field above in direction $\mathbf{t}$, i.e.,

$$\mathrm{II}_{\mathbf{u}-\mathbf{x}}(\mathbf{t}) = \mathbf{t}^\top \left(\frac{d}{d\mathbf{x}} \frac{d\varepsilon}{d\mathbf{u}}(\mathbf{x} + s(\mathbf{u} - \mathbf{x}), s\varepsilon) \right) \mathbf{t} \quad \text{for all } 0 < s < s'. \tag{7.5}$$

Let $L_\mathbf{v}$ denote the Weingarten map at $\mathbf{x}$ in normal direction $\mathbf{v} = (\mathbf{u} - \mathbf{x})/\|\mathbf{u} - \mathbf{x}\|$. By (6.15),

$$\frac{d\mathbf{u}}{d\mathbf{x}} = I_n + \varepsilon \cdot \begin{bmatrix} L_\mathbf{v} & 0 \\ 0 & 0 \end{bmatrix}.$$

By applying the chain rule, we find that

$$\frac{d}{d\mathbf{x}} \frac{d\varepsilon}{d\mathbf{u}} = \frac{d^2\varepsilon}{d\mathbf{u}^2} \cdot \frac{d\mathbf{u}}{d\mathbf{x}}$$

$$= \frac{d^2\varepsilon}{d\mathbf{u}^2} \left(I_n + \varepsilon \cdot \begin{bmatrix} L_\mathbf{v} & 0 \\ 0 & 0 \end{bmatrix} \right).$$

Evaluating this equation at $(\mathbf{x} + s(\mathbf{u} - \mathbf{x}), s\varepsilon)$, we obtain

$$\frac{d^2\varepsilon}{d\mathbf{u}^2}(\mathbf{x} + s(\mathbf{u} - \mathbf{x}), s\varepsilon) = \frac{d}{d\mathbf{x}} \frac{d\varepsilon}{d\mathbf{u}}(\mathbf{x} + s(\mathbf{u} - \mathbf{x}), s\varepsilon)$$
$$- s\varepsilon \cdot \left(\frac{d^2\varepsilon}{d\mathbf{u}^2} \cdot \begin{bmatrix} L_{\nabla\varepsilon} & 0 \\ 0 & 0 \end{bmatrix} \right)(\mathbf{x} + s(\mathbf{u} - \mathbf{x}), s\varepsilon).$$

If we multiply both sides of this equation from the left by $\mathbf{t}^\top$ and from the right by $\mathbf{t}$ and then take the limit $s \to 0$, the formula (7.5) implies that we obtain $\mathrm{II}_{\mathbf{u}-\mathbf{x}}(\mathbf{t})$ on the right-hand side. This proves item (b).   □

Example 7.22 We compute the expression (7.3) for the parabola $X = V(x_2 - x_1^2)$. Using the offset polynomial $g_X(\mathbf{u}, \varepsilon)$ in Example 7.14, we find $\frac{d\varepsilon}{d\mathbf{u}}(\mathbf{u}, \varepsilon) = \frac{1}{p}(h_1, h_2)$, where

$$h_1 = -96u_1\,\varepsilon^4 + \left(192u_1^3 + 64u_1u_2^2 - 16u_1u_2 + 40u_1\right)\varepsilon^2 - 4u_1\left(u_1^2 - u_2\right)\left(24\,u_1^2 + 16\,u_2^2 - 16\,u_2 + 1\right)$$

$$h_2 = (-32u_2 - 32)\,\varepsilon^4 + \left(64u_1^2u_2 - 8u_1^2 + 96u_2^2 + 16u_2 - 8\right)\varepsilon^2$$
$$- 2\left(u_1^2 - u_2\right)\left(16\,u_1^2u_22 - 20\,u_1^2 - 32\,u_2^2 + 12\,u_2 - 1\right)$$

$$p = -96\varepsilon^5 + \left(192u_1^2 + 64u_2^2 + 128u_2 - 32\right)\varepsilon^3$$
$$+ \left(-96u_1^4 - 64u_1^2u_2^2 + 16u_1^2u_2 - 64u_2^3 - 40u_1^2 - 16u_2^2 + 16u_2 - 2\right)\varepsilon.$$

For instance, if we plug in $(u_1, u_2, \varepsilon) = (0, \frac{1}{4}, \frac{1}{4})$, we obtain $(h_1, h_2) = (0, 1)$, which is the unit normal vector on the parabola at $\mathbf{x} = (0, 0)$ pointing towards $\mathbf{u} = (0, \frac{1}{4})$.

Now we compute the Hessian matrix of $\varepsilon(\mathbf{u})$ using the formula in (7.4). This expression is very large and this is why we chose not to display it. Instead, we evaluate it directly at $(u_1, u_2, \varepsilon) = (0, s, s)$, where $s > 0$, and thereafter we let $s \to 0$. This yields the matrix

$$A = \begin{bmatrix} -2 & 0 \\ 0 & 0 \end{bmatrix}.$$

We see from Theorem 7.21 (b) that the (signed) curvature of the parabola at $\mathbf{x} = (0, 0)$ is $\mathbf{t} A \mathbf{t}^\top = -2$, where $\mathbf{t} = (1, 0)$ is the tangent direction of X at $\mathbf{x}$. Here, the negative sign arises because the normal field points "inwards", which causes the derivative of the normal field to point in the opposite direction; this can be seen in Figure 6.2.

It was shown in Example 7.3 that the reach of X is $\frac{1}{2}$. This confirms Theorem 7.8, which here states that the reach of the parabola is the inverse of its maximal curvature $C(X)$. Indeed, a parabola has no real bottlenecks, and its curvature is maximal at the apex. ◇

Chapter 8
Voronoi Cells

Every real algebraic variety X determines a Voronoi decomposition of its ambient Euclidean space $\mathbb{R}^n$. This is a partition of $\mathbb{R}^n$ into *Voronoi cells*, one for each point in X. The Voronoi cell of $\mathbf{y} \in X$ is the set of points in $\mathbb{R}^n$ whose nearest point on X is $\mathbf{y}$. Hence, $\mathbf{u}$ is in the Voronoi cell of $\mathbf{y}$ if $d(\mathbf{u}, X) = d(\mathbf{u}, \mathbf{y})$. The Voronoi cell is a convex, semialgebraic set in the normal space of X at $\mathbf{y}$. Most readers are familiar with the case when X is a finite set. In this chapter, we study Voronoi cells of varieties, with a primary focus on their algebraic boundaries. We consider them only for the Euclidean metric, but it also makes much sense to study Voronoi cells under the Wasserstein distance from Chapter 5 [16] or Kullback–Leibler divergence from Chapter 11 [5]. As before, $\langle \cdot, \cdot \rangle$ denotes the Euclidean inner product.

8.1 Voronoi Basics

Let X be a finite subset of $\mathbb{R}^n$. The *Voronoi cell* of $\mathbf{y} \in X$ collects all points whose closest point in X is $\mathbf{y}$. Writing $d(\mathbf{u}, \mathbf{x}) = \|\mathbf{u} - \mathbf{x}\|$ for the Euclidean norm, the Voronoi cell equals

$$\text{Vor}_X(\mathbf{y}) \; := \; \left\{ \mathbf{u} \in \mathbb{R}^n \mid \mathbf{y} \in \arg\min_{\mathbf{x} \in X} d(\mathbf{u}, \mathbf{x}) \right\}. \tag{8.1}$$

The study of these cells, and how they depend on the configuration X, is ubiquitous in computational geometry and its numerous applications. We begin with the fact that the Voronoi cell is a (possibly unbounded) convex polyhedron with at most $|X| - 1$ facets.

Proposition 8.1 *The Voronoi cell of a point $\mathbf{y}$ in the finite set $X \subset \mathbb{R}^n$ is the polyhedron*

$$\text{Vor}_X(\mathbf{y}) \; = \; \left\{ \mathbf{u} \in \mathbb{R}^n \mid \langle \mathbf{u}, \mathbf{x} - \mathbf{y} \rangle \leq \tfrac{1}{2}(\|\mathbf{x}\|^2 - \|\mathbf{y}\|^2) \; \text{for all} \; \mathbf{x} \in X \backslash \{\mathbf{y}\} \right\}. \tag{8.2}$$

Proof By definition, the Voronoi cells $\text{Vor}_X(\mathbf{y})$ consists of all points $\mathbf{u}$ such that $d(\mathbf{u}, \mathbf{x}) \geq d(\mathbf{u}, \mathbf{y})$ for all $\mathbf{x} \in X \backslash \{\mathbf{y}\}$. This is equivalent to $\|\mathbf{u} - \mathbf{x}\|^2 - \|\mathbf{u} - \mathbf{y}\|^2$ being nonnegative. But, this expression is equal to $\|\mathbf{x}\|^2 - \|\mathbf{y}\|^2 - 2\langle \mathbf{u}, \mathbf{x} - \mathbf{y} \rangle$. The main point is that the quadratic term drops out, so the expression is linear in $\mathbf{u}$. □

The collection of Voronoi cells, as $\mathbf{y}$ ranges over the set X, is also known as the *Voronoi diagram* of X. This is a polyhedral subdivision of $\mathbb{R}^n$ into finitely many convex cells.

© The Author(s) 2024
P. Breiding et al., *Metric Algebraic Geometry*, Oberwolfach Seminars 53,
https://doi.org/10.1007/978-3-031-51462-3_8

We now shift gears, and we replace the finite set X by a real algebraic variety of positive dimension. Thus, let X be a real algebraic variety of codimension c in $\mathbb{R}^n$, and consider a point $\mathbf{y} \in X$. The Voronoi cell $\mathrm{Vor}_X(\mathbf{y})$ is defined as before. It consists of all points $\mathbf{u}$ in $\mathbb{R}^n$ such that $\mathbf{y}$ is closer or equal to $\mathbf{u}$ than any other point $\mathbf{x} \in X$. Equation (8.2) still holds, and we conclude that $\mathrm{Vor}_X(\mathbf{y})$ is a convex set. We seek the Voronoi diagram $\{\mathrm{Vor}_X(\mathbf{y})\}_{\mathbf{y} \in X}$ in $\mathbb{R}^n$ where $\mathbf{y}$ runs over all (infinitely many) points in the variety X.

Proposition 8.2 *If $\mathbf{y}$ is a smooth point of the variety X, then its Voronoi cell $\mathrm{Vor}_X(\mathbf{y})$ is a convex semialgebraic full-dimensional subset of the c-dimensional affine normal space*

$$N_X(\mathbf{y}) := \mathbf{y} + N_{\mathbf{y}}X = \{\mathbf{u} \in \mathbb{R}^n \mid \mathbf{u} - \mathbf{y} \text{ is perpendicular to the tangent space of } X \text{ at } \mathbf{y}\}.$$

Proof Fix $\mathbf{u} \in \mathrm{Vor}_X(\mathbf{y})$. Consider any point $\mathbf{x}$ in X that is close to $\mathbf{y}$, and set $\mathbf{v} = \mathbf{x} - \mathbf{y}$. The inequality in (8.2) implies $\langle \mathbf{u}, \mathbf{v} \rangle \leq \frac{1}{2}(||\mathbf{y} + \mathbf{v}||^2 - ||\mathbf{y}||^2) = \langle \mathbf{y}, \mathbf{v} \rangle + \frac{1}{2}||\mathbf{v}||^2$. For any $\mathbf{w}$ in the tangent space of X at $\mathbf{y}$, there exists $\mathbf{v} = \varepsilon \mathbf{w} + O(\varepsilon^2)$ such that $\mathbf{x} = \mathbf{y} + \mathbf{v}$ is in X.

The inequality above yields $\langle \mathbf{u}, \mathbf{w} \rangle \leq \langle \mathbf{y}, \mathbf{w} \rangle$, and the same with $-\mathbf{w}$ instead of $\mathbf{w}$. Then $\langle \mathbf{u} - \mathbf{y}, \mathbf{w} \rangle = 0$, and hence $\mathbf{u}$ is in the normal space $N_X(\mathbf{y})$. We already argued that $\mathrm{Vor}_X(\mathbf{y})$ is convex. It is semialgebraic, by Tarski's Theorem on Quantifier Elimination. This allows us to eliminate $\mathbf{x}$ from the formula (8.2). Finally, the Voronoi cell $\mathrm{Vor}_X(\mathbf{y})$ is full-dimensional in the c-dimensional space $N_X(\mathbf{y})$ because every point $\mathbf{u}$ in an ε-neighborhood of $\mathbf{y}$ has a unique closest point in X. Moreover, if $\mathbf{u} \in N_X(\mathbf{y})$ and $|\mathbf{u} - \mathbf{y}| < \epsilon$, then $\mathbf{y}$ must be the point in X that is closest to $\mathbf{u}$.   □

One approach to understanding Voronoi cells of a variety X is to take a large but finite sample from X and consider the Voronoi diagram of that sample. This is a finite approximation to the desired limit object. By taking finer and finer samples, the Voronoi diagram should converge nicely to a subdivision with infinitely many lower-dimensional regions, namely the Voronoi cells $\mathrm{Vor}_X(\mathbf{y})$. This process was studied by Brandt and Weinstein in [25] for the case when $n = 2$ and X is a curve. An illustration, similar to [25, Figure 1], is shown in Figure 8.1. Note that, for $n \geq 3$, the Voronoi cells are generally not polyhedra.

Theorem 1 in [25] states that under some mild hypothesis, the limit of the Voronoi cells in this process converges indeed to the Voronoi cells $\mathrm{Vor}_X(\mathbf{y})$. The authors posted a delightful YouTube video, called *Mathemaddies' Ice Cream Map*. Please do watch that movie! Their curve X is the shoreline that separates the city of Berkeley from the San Francisco Bay. One hopes to find many ice cream shops at the shore.

The topological boundary of the Voronoi cell $\mathrm{Vor}_X(\mathbf{y})$ in the normal space $N_X(\mathbf{y})$ is denoted by $\partial \mathrm{Vor}_X(\mathbf{y})$. It consists of all points in $N_X(\mathbf{y})$ that have at least two closest points in X, including $\mathbf{y}$. Note that $\partial \mathrm{Vor}_X(\mathbf{y})$ is contained in the intersection of $N_X(\mathbf{y})$ with the medial axis $\mathrm{Med}(X)$ that was introduced in Chapter 7. We are interested in the *algebraic boundary* $\partial_{\mathrm{alg}} \mathrm{Vor}_X(\mathbf{y})$. This is a hypersurface in the complex affine space $N_X(\mathbf{y})_{\mathbb{C}} \simeq \mathbb{C}^c$, defined as the Zariski closure of $\partial \mathrm{Vor}_X(\mathbf{y})$. For the sake of algebraic consistency, here the Zariski closure should be computed over the field of definition of X (cf. Example 8.5).

Definition 8.3 The degree of the algebraic boundary of the Voronoi cell at $\mathbf{y} \in X$ is denoted

$$\nu_X(\mathbf{y}) := \deg(\partial_{\mathrm{alg}} \mathrm{Vor}_X(\mathbf{y})).$$

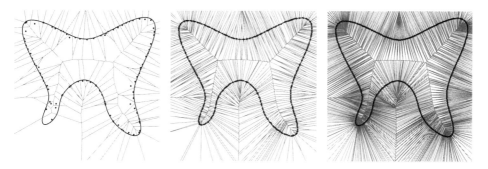

Fig. 8.1: We see the quartic curve $X = \{x^4 - x^2y^2 + y^4 - 4x^2 - 2y^2 - x - 4y + 1 = 0\}$, called *butterfly curve* or *tooth curve*. In each picture, the grey lines are the boundaries of the Voronoi diagram of a finite sample on or near the curve X. From left to right, the number of points in the sample increases. The Voronoi diagram becomes finer and approaches the Voronoi diagram of the curve. Note how the medial axis $\mathrm{Med}(X)$ becomes visible here. For the reach $\tau(X)$ see [25, Example 6.1].

The positive integer $\nu_X(\mathbf{y})$ is called the *Voronoi degree* of X at $\mathbf{y}$. If X is irreducible and $\mathbf{y}$ is a general point on X, then the Voronoi degree $\nu_X(\mathbf{y})$ does not depend on the choice of $\mathbf{y}$. We call that integer the Voronoi degree of the variety X.

Example 8.4 (Curves in 3-space) Let X be a general algebraic curve in $\mathbb{R}^3$. For $\mathbf{y} \in X$, the Voronoi cell $\mathrm{Vor}_X(\mathbf{y})$ is a convex set in the normal plane $N_X(\mathbf{y})$. Its algebraic boundary $\partial_{\mathrm{alg}}\mathrm{Vor}_X(\mathbf{y})$ is a plane curve of degree $\nu_X(\mathbf{y})$. This Voronoi degree can be expressed in terms of the degree and genus of X. Specifically, this degree is 12 when X is the intersection of two general quadrics in $\mathbb{R}^3$. Figure 8.2 shows one such curve X together with the normal plane at a point $\mathbf{y} \in X$. The Voronoi cell $\mathrm{Vor}_X(\mathbf{y})$ is the planar convex region highlighted on the right. Its algebraic boundary $\partial_{\mathrm{alg}}\mathrm{Vor}_X(\mathbf{y})$ is a curve of degree $\nu_X(\mathbf{y}) = 12$. The topological boundary $\partial\mathrm{Vor}_X(\mathbf{y})$ is only a small subset of that algebraic boundary. ◇

Example 8.5 (Surfaces in 3-space) Fix a general polynomial $f \in \mathbb{Q}[x_1, x_2, x_3]$ of degree $d \geq 2$ and let $X = V(f)$ be its surface in $\mathbb{R}^3$. The normal space at a general point $\mathbf{y} \in X$ is the affine line $N_X(\mathbf{y}) = \{\mathbf{y} + \lambda \cdot \nabla f(\mathbf{y}) \mid \lambda \in \mathbb{R}\}$. The Voronoi cell $\mathrm{Vor}_X(\mathbf{y})$ is a (possibly unbounded) line segment in $N_X(\mathbf{y})$ that contains $\mathbf{y}$. The topological boundary $\partial\mathrm{Vor}_X(\mathbf{y})$ consists of at most two points from among the zeros of an irreducible polynomial in $\mathbb{Q}[\lambda]$. We shall see that this univariate polynomial has degree $d^3 + d - 7$. Its complex zeros form the algebraic boundary $\partial_{\mathrm{alg}}\mathrm{Vor}_X(\mathbf{y})$. Thus, the Voronoi degree of the surface X is $d^3 + d - 7$. Note that, in this example, our hypothesis "over the field of definition" becomes important. Here, the field of definition is $\mathbb{Q}$. The $\mathbb{Q}$-Zariski closure of one boundary point is the collection of all $d^3 + d - 7$ points in $\partial_{\mathrm{alg}}\mathrm{Vor}_X(\mathbf{y})$. ◇

Example 8.6 (Quadrics in 3-space) We illustrate Example 8.5 in the case $\mathbf{y} = (0, 0, 0)$ and $d = 2$. We consider $f = x_1^2 + x_2^2 + x_3^2 - 3x_1x_2 - 5x_1x_3 - 7x_2x_3 + x_1 + x_2 + x_3$. Let $r_0 \approx -0.209$, $r_1 \approx -0.107$, $r_2 \approx 0.122$ be the roots of the cubic polynomial $368\lambda^3 + 71\lambda^2 - 6\lambda - 1$. The

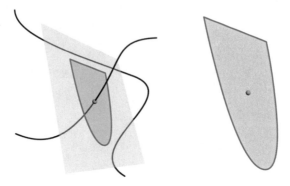

Fig. 8.2: A quartic space curve, shown with the Voronoi cell in one of its normal planes.

Voronoi cell $\mathrm{Vor}_X(\mathbf{y})$ is the line segment connecting the points (r_1, r_1, r_1) and (r_2, r_2, r_2). The topological boundary $\partial \mathrm{Vor}_X(\mathbf{y})$ consists of these two points, whereas the algebraic boundary $\partial_{\mathrm{alg}} \mathrm{Vor}_X(\mathbf{y})$ also contains (r_0, r_0, r_0). The cubic polynomial in the unknown λ was found with the algebraic method that is described in the next section. Namely, the Voronoi ideal in (8.4) equals $\mathrm{Vor}_I(0) = \langle u_1 - u_3,\ u_2 - u_3,\ 368u_3^3 + 71u_3^2 - 6u_3 - 1 \rangle$. This is a maximal ideal in $\mathbb{Q}[u_1, u_2, u_3]$, and it defines a field extension of degree 3 over $\mathbb{Q}$. ◇

8.2 Algebraic Boundaries

For any point in the ambient space, the ED problem asks the question "What point on the variety X am I closest to?" Another question one might ask is "How far do we have to get away from X before there is more than one answer to the closest point question?" The union of the boundaries of the Voronoi cells is the locus of points in $\mathbb{R}^n$ that have more than one closest point on X. This is the medial axis $\mathrm{Med}(X)$.

The distance from the variety to its medial axis, which is the answer to the "how far" question, is the reach of X. We have proved this in Proposition 7.2. The distance from a point $\mathbf{y}$ on X to the variety's medial axis is called the *local reach* of X. We formally define the local reach in (15.1), where we use it in a theorem that tells us how to compute the homology of X from finite samples. This is relevant for topological data analysis.

The material that follows is based on the article [50] by Cifuentes, Ranestad, Sturmfels, and Weinstein. We begin with the exact symbolic computation of the Voronoi boundary at $\mathbf{y}$ from the equations that define X. This uses a Gröbner-based algorithm whose input is $\mathbf{y}$ and the ideal of X and whose output is the ideal defining $\partial_{\mathrm{alg}} \mathrm{Vor}_X(\mathbf{y})$. In the next section, we present formulas for the Voronoi degree $v_X(\mathbf{y})$ when X and $\mathbf{y}$ are sufficiently general and $\dim(X) \leq 2$. The proofs of these formulas require some intersection theory. Thereafter, in Section 8.4, we study the case when $\mathbf{y}$ is a low-rank matrix and X is the variety of all matrices of bounded rank. This relies on the *Eckart–Young Theorem* (Theorem 2.9).

We now describe Gröbner basis methods for finding the Voronoi boundaries of a given variety. We start with an ideal $I = \langle f_1, f_2, \ldots, f_m \rangle$ in $\mathbb{Q}[x_1, \ldots, x_n]$ whose real variety $X = V(I) \subset \mathbb{R}^n$ is assumed to be non-empty. We assume that I is real radical and prime, so that the zero set of I in $\mathbb{C}^n$ is an irreducible variety whose real points are Zariski dense. Our aim is to compute the Voronoi boundary of a given point $\mathbf{y} \in X$. In our examples, the coordinates of $\mathbf{y}$ and the coefficients of the polynomials f_i are rational numbers. Under these assumptions, the following computations can be done in polynomial rings over $\mathbb{Q}$.

Fix the polynomial ring $R = \mathbb{Q}[x_1, \ldots, x_n, u_1, \ldots, u_n]$ where $\mathbf{u} = (u_1, \ldots, u_n)$ is an auxiliary point with unknown coordinates. As in Chapter 2, the *augmented Jacobian* $\mathcal{AJ}(\mathbf{x}, \mathbf{u})$ of X at $\mathbf{x}$ is the $(m + 1) \times n$ matrix with entries in R that is obtained by adding the row vector $\mathbf{u} - \mathbf{x}$ to the Jacobian matrix $(\partial f_i / \partial x_j)$.

Let N_I denote the ideal in R generated by I and the $(c + 1) \times (c + 1)$ minors of the augmented Jacobian matrix $\mathcal{AJ}(\mathbf{x}, \mathbf{u})$, where c is the codimension of the variety $X \subset \mathbb{R}^n$. If X is smooth, then the ideal N_I in R defines a subvariety of dimension n in $\mathbb{R}^{2n}$, namely the *Euclidean normal bundle* of X. Its points are pairs $(\mathbf{x}, \mathbf{u})$ where $\mathbf{x}$ is a point in the given variety X and $\mathbf{u}$ lies in the normal space of X at $\mathbf{x}$.

In Chapter 2, we have worked with the critical ideal. The ideal N_I is similar to the critical ideal, but the key difference is that now $\mathbf{u}$ is a vector of variables. In both cases, when X is singular, we may wish to saturate with respect to the ideal of the singular locus of X. In what follows, for any $\mathbf{y} \in X$, let $N_I(\mathbf{y})$ denote the linear ideal of the normal space. This is obtained from N_I by replacing the unknown point $\mathbf{x}$ by the specific point $\mathbf{y}$.

Example 8.7 (Cuspidal cubic) Let $n = 2$ and $I = \langle x_1^3 - x_2^2 \rangle$, so $X = V(I) \subset \mathbb{R}^2$ is a cubic curve with a cusp at the origin. The ideal of the Euclidean normal bundle of X is generated by two polynomials:

$$N_I = \left\langle x_1^3 - x_2^2, \det\begin{bmatrix} u_1 - x_1 & u_2 - x_2 \\ 3x_1^2 & -2x_2 \end{bmatrix} \right\rangle \subset R = \mathbb{Q}[x_1, x_2, u_1, u_2]. \tag{8.3}$$

Consider the normal line of X at $\mathbf{y} = (4, 8)$. Its ideal is $N_I(\mathbf{y}) = \langle u_1 + 3u_2 - 28 \rangle$. ◇

Returning to the general setting, we define for a fixed point $\mathbf{y}$ the following ideal:

$$C_I(\mathbf{y}) = N_I + N_I(\mathbf{y}) + \langle \|\mathbf{x} - \mathbf{u}\|^2 - \|\mathbf{y} - \mathbf{u}\|^2 \rangle \subset R.$$

The real variety of $C_I(\mathbf{y})$ lives in $\mathbb{R}^{2n}$. It consists of all pairs $(\mathbf{u}, \mathbf{x})$ such that $\mathbf{x}$ and $\mathbf{y}$ are equidistant from $\mathbf{u}$ and both are critical points of the distance function from $\mathbf{u}$ to X.

Definition 8.8 The *Voronoi ideal* in $\mathbb{Q}[u_1, \ldots, u_n]$ is obtained from the ideal $C_I(\mathbf{y})$ defined above by saturation and elimination, as follows:

$$\mathrm{Vor}_I(\mathbf{y}) = \left(C_I(\mathbf{y}) : \langle \mathbf{x} - \mathbf{y} \rangle^\infty \right) \cap \mathbb{Q}[u_1, \ldots, u_n]. \tag{8.4}$$

The geometric interpretation of each step in our construction implies the following result:

Proposition 8.9 *The affine variety in $\mathbb{C}^n$ defined by the Voronoi ideal $\mathrm{Vor}_I(\mathbf{y})$ contains the algebraic Voronoi boundary $\partial_{\mathrm{alg}}\mathrm{Vor}_X(\mathbf{y})$ of the given real variety X at its point $\mathbf{y}$.*

Remark 8.10 The verb "contains" sounds weak, but it is much stronger than it may seem. Indeed, in generic situations, the ideal $\text{Vor}_I(\mathbf{y})$ is prime and defines an irreducible hypersurface in the normal space of X at $\mathbf{y}$. This hypersurface equals the algebraic Voronoi boundary, so containment is an equality. We saw this in Examples 8.5 and 8.6. The ideal $\text{Vor}_I(\mathbf{y})$ usually defines a hypersurface in the normal space. For special data, it can have extraneous components, but these are easy to identify and remove when the dimension is low.

Example 8.11 We consider the cuspidal cubic $X = V(I) \subset \mathbb{R}^2$ in Example 8.7, where $I = \langle x_1^3 - x_2^2 \rangle$, and we fix the point $\mathbf{y} = (4, 8) \in X$. Going through the steps above, we find that the Voronoi ideal equals

$$\text{Vor}_I(\mathbf{y}) \;=\; \langle u_1 - 28, u_2 \rangle \cap \langle u_1 + 26, u_2 - 18 \rangle \cap \langle u_1 + 3u_2 - 28, \; 27u_2^2 - 486u_2 + 2197 \rangle. \quad (8.5)$$

The third component has no real roots and is hence extraneous. The Voronoi boundary consists of two points. Namely, we have $\partial \text{Vor}_X(\mathbf{y}) = \{(28, 0), (-26, 18)\}$. The Voronoi cell $\text{Vor}_X(\mathbf{y})$ is the line segment connecting these points. This segment is shown in green in Figure 8.3. Its right endpoint $(28, 0)$ is equidistant from $\mathbf{y}$ and the point $(4, -8)$. Its left endpoint $(-26, 18)$ is equidistant from $\mathbf{y}$ and the origin $(0, 0)$, which is the singular point of the curve X. Its Voronoi cell will be discussed in Remark 8.12.

The issue of saturation is subtle and interesting in this example. In (8.3), we did not saturate the ideal N_I, and this led to the three components in (8.5). By contrast, suppose we saturate by the singular point of our curve, i.e. we replace N_I by $N_I : \langle x_1, x_2 \rangle^\infty$. Then (8.4) yields a stricly larger ideal than in (8.5). The second prime ideal is gone, and $\text{Vor}_I(\mathbf{y})$ is only the intersection of the first and third prime ideals. Geometrically, we are losing the point $(-26, 18)$. This makes sense because that point has the same distance to $(4, 8)$ and to $(0, 0)$. The saturation step has removed the singular point $(0, 0)$ from our algebraic representation. The resulting Voronoi ideal only sees pairs of smooth points that are equidistant.

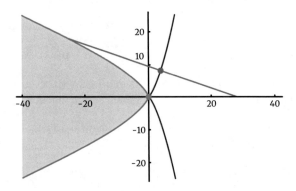

Fig. 8.3: The cuspidal cubic is shown in red. The Voronoi cell of a smooth point is a green line segment. The Voronoi cell of the cusp is the blue convex region bounded by the blue curve.

The cuspidal cubic X is very special. If we replace X by a general cubic (defined over $\mathbb{Q}$) in $\mathbb{R}^2$, then $\mathrm{Vor}_I(\mathbf{y})$ is generated modulo $N_I(\mathbf{y})$ by an irreducible polynomial of degree eight in $\mathbb{Q}[u_1, u_2]$. Thus, the expected Voronoi degree for general plane cubics is $v_X(\mathbf{y}) = 8$. ◇

Remark 8.12 (Singularities) Voronoi cells at singular points can be computed with the same tools as above. However, these Voronoi cells can have higher dimensions. For an illustration, consider the cuspidal cubic, and let $\mathbf{y} = (0,0)$ be the cusp. A Gröbner basis computation yields the Voronoi boundary $27u_2^4 + 128u_1^3 + 72u_1u_2^2 + 32u_1^2 + u_2^2 + 2u_1$. The Voronoi cell is the two-dimensional convex region bounded by this quartic, shown in blue in Figure 8.3. The Voronoi cell can also be empty at a singularity. This happens for instance for $V(x_1^3 + x_1^2 - x_2^2)$, which has an ordinary double point at $\mathbf{y} = (0,0)$. In general, the cell dimension depends on both the embedding dimension and the branches of the singularity.

Proposition 8.9 gives an algorithm for computing the Voronoi ideal $\mathrm{Vor}_I(\mathbf{y})$ when $\mathbf{y}$ is a smooth point in $X = V(I)$. Experiments with Macaulay2 [73] are reported in [50]. For small instances, the computation terminates and we obtain the defining polynomial of the Voronoi boundary $\partial_{\mathrm{alg}}\mathrm{Vor}_X(\mathbf{y})$. This polynomial is unique modulo the linear ideal of the normal space $N_I(\mathbf{y})$. For larger instances, we can only compute the degree of $\partial_{\mathrm{alg}}\mathrm{Vor}_X(\mathbf{y})$ but not its equation. This is done by working over a finite field and adding $c - 1$ random linear equations in $u_1, \ldots, u_n$ in order to get a zero-dimensional polynomial system.

Computations are easiest to set up for the case of hypersurfaces ($c = 1$). One can explore random polynomials f of degree d in $\mathbb{Q}[x_1, \ldots, x_n]$, both inhomogeneous and homogeneous. These are chosen among those that vanish at a preselected point $\mathbf{y}$ in $\mathbb{Q}^n$. In each iteration, the Voronoi ideal $\mathrm{Vor}_I(\mathbf{y})$ from (8.4) was found to be zero-dimensional. In fact, $\mathrm{Vor}_I(\mathbf{y})$ is a maximal ideal in $\mathbb{Q}[u_1, \ldots, u_n]$, and the Voronoi degree $v_X(\mathbf{y})$ is the degree of the field extension of $\mathbb{Q}$ that is defined by that maximal ideal.

We summarize our results in Tables 8.1 and 8.2, and we extract conjectural formulas.

$n\backslash d$	2	3	4	5	6	7	8	$v_X(y) = \mathrm{degree}(\partial_{\mathrm{alg}}\mathrm{Vor}_{(f)}(y))$
1	1	2	3	4	5	6	7	$d-1$
2	2	8	16	26	38	52	68	d^2+d-4
3	3	23	61	123	215	343		d^3+d-7
4	4	56	202	520	1112			$d^4-d^3+d^2+d-10$
5	5	125	631					$d^5-2d^4+2d^3+d-13$
6	6	266	1924					$d^6-3d^5+4d^4-2d^3+d^2+d-16$
7	7	551						$d^7-4d^6+7d^5-6d^4+3d^3+d-19$

Table 8.1: The Voronoi degree of an inhomogeneous polynomial f of degree d in $\mathbb{R}^n$.

Conjecture 8.13 The Voronoi degree of a generic hypersurface of degree d in $\mathbb{R}^n$ equals

$$(d-1)^n + 3(d-1)^{n-1} + \tfrac{4}{d-2}((d-1)^{n-1} - 1) - 3n.$$

Conjecture 8.14 The Voronoi degree of a generic homogeneous polynomial of degree d is

$$2(d-1)^{n-1} + \tfrac{4}{d-2}((d-1)^{n-1} - 1) - 3n + 2.$$

$n \backslash d$	2	3	4	5	6	7	8	$v_X(y) = \text{degree}(\partial_{\text{alg}}\text{Vor}_{\langle f \rangle}(y))$
2	2	4	6	8	10	12	14	$2d-2$
3	3	13	27	45	67	93	123	$2d^2-5$
4	4	34	96	202				$2d^3-2d^2+2d-8$
5	5	79	309					$2d^4-4d^3+4d^2-11$
6	6	172						$2d^5-6d^4+8d^3-4d^2+2d-14$
7	7	361						$2d^6-8d^5+14d^4-12d^3+6d^2-17$

Table 8.2: The Voronoi degree of a homogeneous polynomial f of degree d in $\mathbb{R}^n$.

Both conjectures are proved for $n \leq 3$ in [50, Section 4], where the geometric theory of Voronoi degrees of low-dimensional varieties is developed. The case $d = 2$ was analyzed in [49, Proposition 5.8]. For $n \geq 4$ and $d \geq 3$, the conjectures remain open.

8.3 Degree Formulas

To recap, the algebraic boundary of the Voronoi cell $\text{Vor}_X(y)$ is a hypersurface in the normal space to a variety $X \subset \mathbb{R}^n$ at a point $y \in X$. If y has rational coordinates, then that hypersurface is defined over $\mathbb{Q}$. We shall present formulas for the degree $v_X(y)$ of that hypersurface when X is a curve or a surface. All proofs appear in [50, Section 6]. We identify X and $\partial_{\text{alg}}\text{Vor}_X(y)$ with their Zariski closures in complex projective space $\mathbb{P}^n$, so there is a natural assigned hyperplane at infinity. We say that the variety X is in *general position* in $\mathbb{P}^n$ if the hyperplane at infinity intersects X transversally. The next result is [50, Theorem 5.1].

Theorem 8.15 Let $X \subset \mathbb{P}^n$ be a curve of degree d and geometric genus g with at most ordinary multiple points as singularities. If X is in general position, then the Voronoi degree at a general point $y \in X$ equals

$$v_X(y) = 4d + 2g - 6.$$

Example 8.16 If X is a smooth curve of degree d in the plane, then $2g - 2 = d(d - 3)$, so

$$v_X(y) = d^2 + d - 4.$$

This confirms our experimental results in the row $n = 2$ of Table 8.1.   ◇

Example 8.17 If X is a rational curve of degree d, then $g = 0$ and our formula gives $v_X(y) = 4d - 6$. If X is an elliptic curve ($g = 1$), then $v_X(y) = 4d - 4$. A space curve with $d = 4$ and $g = 1$ was studied in Example 8.4. Its Voronoi degree equals $v_X(y) = 12$.   ◇

The general position assumption in Theorem 8.15 is essential. For an example, let X be the twisted cubic curve in $\mathbb{P}^3$, with affine parameterization $t \mapsto (t, t^2, t^3)$. Here $g = 0$ and $d = 3$, so the expected Voronoi degree is 6. However, a computation shows that $v_X(y) = 4$. This drop arises because the plane at infinity in $\mathbb{P}^3$ intersects the curve X in a triple point.

After a general linear change of coordinates in $\mathbb{P}^3$, which amounts to a linear fractional transformation in $\mathbb{R}^3$, we correctly find $v_X(\mathbf{y}) = 6$.

We next present a formula for the Voronoi degree of a surface X, which is smooth and irreducible in $\mathbb{P}^n$. Our formula is in terms of its degree d and two further invariants. The first is the topological Euler characteristic $\chi(X)$. This equals the degree of the second Chern class of the tangent bundle; see Example 4.22 (b). The second invariant, denoted $g(X)$, is the genus of the curve obtained by intersecting X with a general quadratic hypersurface in $\mathbb{P}^n$. Thus, $g(X)$ is the quadratic analogue to the sectional genus of the surface X.

Theorem 8.18 (Theorem 5.4 in [50]) *Let $X \subset \mathbb{P}^n$ be a smooth surface of degree d. Then*

$$v_X(\mathbf{y}) = 3d + \chi(X) + 4g(X) - 11,$$

provided the surface X is in general position in $\mathbb{P}^n$ and $\mathbf{y}$ is a general point on X.

Example 8.19 If X is a smooth surface in $\mathbb{P}^3$ of degree d, then (4.10) and Example 2.21 yield $\chi(X) = \deg(c_2(X)) = d(d^2 - 4d + 6)$. A smooth quadratic hypersurface section of X is an irreducible curve of degree (d, d) in $\mathbb{P}^1 \times \mathbb{P}^1$ with genus $g(X) = (d-1)^2$. Thus,

$$v_X(\mathbf{y}) = 3d + d(d^2 - 4d + 6) + 4(d-1)^2 - 11 = d^3 + d - 7.$$

This confirms our experimental results in the row $n = 3$ of Table 8.1. ◇

Example 8.20 Let X be the Veronese surface of order e in $\mathbb{P}^{\binom{e+2}{2}-1}$ (see Definition 12.5), taken after a general linear change of coordinates in that ambient space. The degree of X equals $d = e^2$ (this is proved in Corollary 12.21). We have $\chi(X) = \chi(\mathbb{P}^2) = 3$, and the general quadratic hypersurface section of X is a curve of genus $g(X) = \binom{2e-1}{2}$. We conclude that the Voronoi degree of X at a general point $\mathbf{y}$ equals

$$v_X(\mathbf{y}) = 3e^2 + 3 + 2(2e-1)(2e-2) - 11 = 11e^2 - 12e - 4.$$

For instance, for the Veronese surface in $\mathbb{P}^5$ we have $e = 2$ and hence $v_X(\mathbf{y}) = 16$. This is smaller than the number 18 to be found later in Example 8.25. That example concerns the cone over the Veronese surface in $\mathbb{R}^6$, and not the Veronese surface in $\mathbb{R}^5 \subset \mathbb{P}^5$. ◇

We finally consider affine surfaces defined by homogeneous polynomials. Namely, let $X \subset \mathbb{R}^n$ be the affine cone over a general smooth curve of degree d and genus g in $\mathbb{P}^{n-1}$.

Theorem 8.21 (Theorem 5.7 in [50]) *If X is the cone over a smooth curve in $\mathbb{P}^{n-1}$ then*

$$v_X(\mathbf{y}) = 6d + 4g - 9,$$

provided that the curve is in general position and $\mathbf{y}$ is a general point.

Example 8.22 If $X \subset \mathbb{R}^3$ is the cone over a smooth curve of degree d in $\mathbb{P}^2$, then we have $2g - 2 = d(d-3)$, by the degree-genus formula for plane curves. We conclude that

$$v_X(\mathbf{y}) = 2d^2 - 5.$$

This confirms our experimental results on Voronoi degrees in the row $n = 3$ of Table 8.2. ◇

Let us comment on the assumption made in our theorems, namely that X is in general position in $\mathbb{P}^n$. If this is not satisfied, then the Voronoi degree may drop, assuming it remains zero-dimensional. Indeed, the Voronoi ideal $\mathrm{Vor}_I(\mathbf{y})$ depends polynomially on the description of X, and the degree of this ideal can only go down – and not up – when that description specializes. This follows from the Parameter Continuation Theorem 3.18.

8.4 Voronoi meets Eckart–Young

We now turn to a case of great interest in applications. Let X be the variety of real $m \times n$ matrices of rank at most r. We consider two norms on the space $\mathbb{R}^{m \times n}$ of real $m \times n$ matrices. Our first matrix norm is the *spectral norm* or *operator norm* $\|U\|_{\mathrm{op}} := \max_i \sigma_i(U)$ which extracts the largest singular value of the matrix U. Our second norm is the *Frobenius norm*
$$\|U\|_F := \sqrt{\sum_{ij} U_{ij}^2} = \sqrt{\mathrm{Trace}(U^\top U)}.$$ The Frobenius norm agrees with the Euclidean norm on $\mathbb{R}^{m \times n}$, so it fits into our setting. The case of symmetric matrices will be discussed later.

Fix a rank r matrix V in X. This is a nonsingular point in X. Consider any matrix U in the Voronoi cell $\mathrm{Vor}_X(V)$. This means that the closest point to U in the rank r variety X relative to the Frobenius norm is the matrix V. By the Eckart–Young Theorem (Theorem 2.9), the matrix V is derived from U by computing the singular value decomposition $U = \Sigma_1 D \Sigma_2$. Here Σ_1 and Σ_2 are orthogonal matrices of size $m \times m$ and $n \times n$, respectively, and D is a nonnegative diagonal matrix whose entries are the singular values. Let $D^{[r]}$ be the matrix that is obtained from D by replacing all singular values except for the r largest ones by zero. Then, according to Eckart–Young, we have $V = \Sigma_1 \cdot D^{[r]} \cdot \Sigma_2$.

Remark 8.23 The Eckart–Young Theorem works for both the Frobenius norm and the spectral norm. This means that $\mathrm{Vor}_X(V)$ is also the Voronoi cell for the spectral norm.

The following theorem describes the Voronoi cells for low-rank matrix approximation.

Theorem 8.24 *Let V be an $m \times n$-matrix of rank r. Let u be the r-th singular value of V. The Voronoi cell $\mathrm{Vor}_X(V)$ is the ball of radius u in the spectral norm on the space of $(m - r) \times (n - r)$-matrices.*

Proof Since the Frobenius norm is orthogonally invariant, we can assume that the matrix $V = (v_{ij}) \in X$ is a diagonal matrix whose entries are $v_{11} \geq v_{22} \geq \cdots \geq v_{rr} = u > 0$. These entries are the singular values of V. The normal space of the determinantal variety X is described in Lemma 9.12. In particular, the normal space of X at V consists of the matrices

$$A = \begin{bmatrix} 0 & 0 \\ 0 & U \end{bmatrix},$$

where U is an arbitrary $(m - r) \times (n - r)$ matrix. The matrix V is the rank r matrix closest to $V + A$ if and only if the spectral norm of U is less than u. This shows that the Voronoi cell $\mathrm{Vor}_X(V)$ is a full-dimensional convex body in the normal space, which we identified with the matrices U. Namely, $\mathrm{Vor}_X(V)$ equals u times the unit ball in $\mathbb{R}^{(m-r) \times (n-r)}$ under the spectral norm. $\square$

Fig. 8.4: The Voronoi cell of a symmetric 3×3 matrix of rank 1 is a convex body of dimension 3. It is shown for the Frobenius norm (left) and for the Euclidean norm (right). See Example 8.25 for a discussion.

The proof of Theorem 8.24 offers the following perspective on the Voronoi cells $\mathrm{Vor}_X(V)$ of the determinantal variety X. Suppose $m = n$. Consider the set Z of vectors in $\mathbb{R}^n$ that have at most r non-zero coordinates. This is a reducible variety with $\binom{n}{r}$ components, each a coordinate subspace. For a general point $\mathbf{y}$ in such a subspace, the Voronoi cell $\mathrm{Vor}_Z(\mathbf{y})$ is a convex polytope. It is congruent to a regular cube of dimension $n - r$, which is the unit ball in the L^∞-norm on $\mathbb{R}^{n-r}$. By Theorem 8.24, the Voronoi cell $\mathrm{Vor}_X(V)$ is the orbit of that cube under the group of orthogonal transformations. To make sense of this action, we here identify $\mathbb{R}^{n-r}$ with the space of diagonal matrices in $\mathbb{R}^{(n-r)\times(n-r)}$.

For example, consider the special case where $n = 3$ and $r = 1$. In this case, Z consists of the three coordinate axes in $\mathbb{R}^3$. The Voronoi decomposition of this reducible curve decomposes $\mathbb{R}^3$ into squares, each normal to a different point on the three lines. The image of this picture under orthogonal transformations is the Voronoi decomposition of $\mathbb{R}^{3\times3}$ associated with the affine variety of rank 1 matrices. That variety has dimension 5, and each Voronoi cell is a 4-dimensional convex body in the normal space.

If u is the smallest singular value of V, then $\lambda \cdot u$ is the smallest singular value of $\lambda \cdot V$, for every $\lambda \neq 0$. If we wish to compare the sizes of Voronoi cells, then we should work with some normalization and consider the relative size of Voronoi cells. Let us normalize rank-r matrices such that their largest singular value is equal to one. That is, we intersect the variety X with the unit sphere in the spectral norm. Let V be a matrix in this intersection. Then, by Theorem 8.24 the size of its Voronoi cell is $\lambda = \sigma_r(V)/\sigma_1(V)$. In Chapter 9 we will study this ratio of singular values. In particular, we will show in (9.4) that $\sigma_r(V)/\sigma_1(V)$ is the inverse of the *Turing condition number* of V, which is related to matrix inversion. Therefore, the relative size of the Voronoi cell of V is the inverse of the relative condition number of V. There is a close connection between condition numbers and errors in numerical computation. We explain this in Chapter 9.

Our problem is even more interesting when we restrict to matrices in a linear subspace. To see this, let now X denote the variety of *symmetric $n \times n$* matrices of rank at most r. We can regard X either as a variety in the matrix space $\mathbb{R}^{n\times n}$, or in the space $\mathbb{R}^{\binom{n+1}{2}}$ whose coordinates are the upper triangular entries. On the latter space we have the *Euclidean norm* and the *Frobenius norm* (aka *Bombieri–Weyl norm*). These are now different!

The Frobenius norm on $\mathbb{R}^{\binom{n+1}{2}}$ is the restriction of the Frobenius norm on $\mathbb{R}^{n \times n}$ to the subspace of symmetric matrices. For instance, if $n = 2$, we identify the vector (a, b, c) with the symmetric matrix $\begin{bmatrix} a & b \\ b & c \end{bmatrix}$. The Frobenius norm of this matrix is $\sqrt{a^2+2b^2+c^2}$ (cf. Example 9.15). On the other hand, the Euclidean norm is $\sqrt{a^2+b^2+c^2}$. The two norms have dramatically different properties with respect to low-rank approximation. The Eckart–Young Theorem remains valid for the Frobenius norm on $\mathbb{R}^{\binom{n+1}{2}}$, but it is not valid for the Euclidean norm. Some implications are explained in [60, Example 3.2]. In what follows, we elucidate this point by comparing the Voronoi cells with respect to the two norms.

Example 8.25 Let X be the variety of symmetric 3×3 matrices of rank ≤ 1. For the Euclidean metric, X lives in $\mathbb{R}^6$. For the Frobenius metric, X lives in a 6-dimensional subspace of $\mathbb{R}^{3 \times 3}$. Let V be a smooth point in X, i.e. a symmetric 3×3 matrix of rank 1. The normal space to X at V has dimension 3. Hence, in either norm, the Voronoi cell $\mathrm{Vor}_X(V)$ is a 3-dimensional convex body. Figure 8.4 shows these two bodies.

For the Frobenius metric, each Voronoi cell of the Veronese variety X is congruent to the set of matrices $\begin{bmatrix} a & b \\ b & c \end{bmatrix}$ whose eigenvalues are between -1 and 1. This semialgebraic set is bounded by the singular quadratic surfaces with defining polynomials $\det \begin{bmatrix} a+1 & b \\ b & c+1 \end{bmatrix}$ and $\det \begin{bmatrix} a-1 & b \\ b & c-1 \end{bmatrix}$. The Voronoi ideal is of degree 4, and it is generated by the product of these two determinants (modulo the normal space). The Voronoi cell is shown on the left in Figure 8.4. It is the intersection of two quadratic cones. The cell is the convex hull of the circle in which the two quadrics meet, together with the two vertices.

For the Euclidean metric, the Voronoi boundary at a generic point V in X is defined by an irreducible polynomial of degree 18 in a, b, c. This polynomial is found by the ideal computation in equation (8.4). In some cases, the Voronoi degree can drop. For instance, consider the special rank 1 matrix $V = \begin{bmatrix} 1 & 0 & 0 \\ 0 & 0 & 0 \\ 0 & 0 & 0 \end{bmatrix}$. For this point, the degree of the Voronoi boundary is only 12. This particular Voronoi cell is shown on the right in Figure 8.4. This cell is the convex hull of two ellipses, which are marked in red in the diagram.   ◇

Chapter 9
Condition Numbers

The concept of a condition number has its origin in numerical analysis. It measures how much the output value of a function we wish to evaluate can change for a small change in the input argument. In this chapter, we discuss condition numbers in the context of metric algebraic geometry. In the first section, we offer an introduction to the relevant notions for assessing errors in numerical computations.

Suppose the function to be evaluated is an algebraic function. For instance, we may wish to map the coefficients of a univariate polynomial to one of its roots. This function is well-defined locally, and it is well-behaved when the polynomial is far from the hypersurface defined by the discriminant. Indeed, the condition number stands in a reciprocal relationship to the distance to the variety of ill-posed instances. This variety is often a discriminantal hypersurface, and thus we are naturally led to the ED problem for discriminants. This topic will be studied in the third and last sections of this chapter. The classical discriminant is the dual variety to the Veronese variety, and other discriminants are dual to other toric varieties. We can apply ED duality (Theorem 2.23) to gain insight and computational speed.

A special case of a discriminant is the determinant of a square matrix. This arises when our function is matrix inversion. The relevant ED problem points us to the Eckart–Young Theorem (Theorem 2.9). This is why we include the proof of Eckart–Young in the second section, which is about the condition number of matrix inversion.

9.1 Errors in Numerical Computations

Input data for numerical algorithms can have errors. These may be caused by measurement errors. Hence, the output of the computation also has errors. We wish to compare the output error with the input error. This is a fundamental issue for numerical computations.

Example 9.1 (Exact algorithm) Given a matrix $A \in \mathbb{R}^{2\times 2}$ with $\det(A) \neq 0$, we want to compute the inverse matrix A^{-1}. We consider two instances of this problem. The errors are measured by the Euclidean norm.

© The Author(s) 2024
P. Breiding et al., *Metric Algebraic Geometry*, Oberwolfach Seminars 53,
https://doi.org/10.1007/978-3-031-51462-3_9

(a) First, let $A = \begin{bmatrix} 1 & 1 \\ -1 & 1 \end{bmatrix}$. We consider this matrix to be the true input data. A small measurement error gives the new input data $\tilde{A} = \begin{bmatrix} 1 & 1 \\ -1+\varepsilon & 1 \end{bmatrix}$, where $0 < \varepsilon \ll 1$. The *exact* solutions A^{-1} and $\tilde{A}^{-1}$ are then

$$A^{-1} = \frac{1}{2}\begin{bmatrix} 1 & -1 \\ 1 & 1 \end{bmatrix}, \quad \text{and} \quad \tilde{A}^{-1} = \frac{1}{2-\varepsilon}\begin{bmatrix} 1 & -1 \\ 1-\varepsilon & 1 \end{bmatrix}$$

$$= A^{-1} + \frac{\varepsilon}{2(2-\varepsilon)}\begin{bmatrix} 1 & -1 \\ -1 & 1 \end{bmatrix}.$$

Comparing the errors, we find that $\|A^{-1} - \tilde{A}^{-1}\| \approx \|A - \tilde{A}\|$. In words, the error in the input $\|A - \tilde{A}\|$ and the error in the output $\|A^{-1} - \tilde{A}^{-1}\|$ are roughly the same for small values of ε. They both are of the order $O(\varepsilon)$.

(b) The true input in our second example is the matrix $B = \begin{bmatrix} 1 & 1 \\ 1 & 1+\delta \end{bmatrix}$, where $|\delta| \neq 0$ is small. We perturb the input by adding ε to the lower left entry. The perturbed input is $\tilde{B} = \begin{bmatrix} 1 & 1 \\ 1+\varepsilon & 1+\delta \end{bmatrix}$. The matrix inverses are

$$B^{-1} = \frac{1}{\delta}\begin{bmatrix} 1+\delta & -1 \\ -1 & 1 \end{bmatrix} \quad \text{and} \quad \tilde{B}^{-1} = \frac{1}{\delta-\varepsilon}\begin{bmatrix} 1+\delta & -1 \\ -1-\varepsilon & 1 \end{bmatrix}$$

$$= B^{-1} + \frac{\varepsilon}{\delta(\varepsilon-\delta)}\begin{bmatrix} -(1+\delta) & 1 \\ 1+\delta & -1 \end{bmatrix}.$$

This implies $\|B^{-1} - \tilde{B}^{-1}\| \approx \frac{1}{\delta(\varepsilon-\delta)} \cdot \|B - \tilde{B}\|$. If $\varepsilon < \delta$, then the error is amplified by a factor of roughly δ^{-2}, which is large. The behavior here is different from that before.

We applied an exact algorithm to the problem of matrix inversion, and we observed considerable differences in the output. In the first example, the output $\tilde{A}^{-1}$ for the perturbed data $\tilde{A}$ was close to the true output A^{-1}. On the other hand, in the second example, the output $\tilde{B}^{-1}$ for the perturbed data $\tilde{B}$ was far from the true output B^{-1}.   ◊

The previous example shows that, even if we can compute the *exact* solution of a problem, small errors in the data may be amplified tremendously in the output. The theory of *condition numbers* helps us to understand when and why this happens. In simple terms, a condition number is a quantity associated with a *computational problem*, and it measures the sensitivity of the output to small errors in the input data.

Definition 9.2 A *computational problem* is a function $\phi : M \to N$ from a space M of inputs to a space N of outputs. For us, each space is a subset of a Euclidean space, and it carries the induced Euclidean metric.

Example 9.3 In Example 9.1, the input space and the output space are both given by $M = N = \{A \in \mathbb{R}^{2\times 2} \mid \det(A) \neq 0\}$. The computational problem is matrix inversion, so the relevant function is $\phi(A) = A^{-1}$.   ◊

Let (M, d_M) and (N, d_N) be metric spaces. The following definition is due to Rice [152].

Definition 9.4 The *(absolute) condition number* of $\phi : M \rightarrow N$ at the input $\mathbf{p} \in M$ is

$$\kappa[\phi](\mathbf{p}) := \lim_{\varepsilon \rightarrow 0} \sup_{\substack{\mathbf{q} \in M: \\ d_M(\mathbf{p}, \mathbf{q}) \leq \varepsilon}} \frac{d_N(\phi(\mathbf{p}), \phi(\mathbf{q}))}{d_M(\mathbf{p}, \mathbf{q})}.$$

The motivation for this definition is as follows: For small $d_M(\mathbf{p}, \mathbf{q})$ we have

$$d_N(\phi(\mathbf{p}), \phi(\mathbf{q})) \leq \kappa[\phi](\mathbf{p}) \cdot d_M(\mathbf{p}, \mathbf{q}) + o(d_M(\mathbf{p}, \mathbf{q})).$$

In words, a small error $\varepsilon = d_M(\mathbf{p}, \mathbf{q})$ in the input data causes an error of roughly $\kappa[\phi](\mathbf{p}) \cdot \varepsilon$ in the output data. This relationship is entirely independent of the algorithm that is used to evaluate $\phi(\mathbf{p})$. At this stage, given the lim-sup definition, it is unclear how condition numbers can be computed. As we shall see, this is where the geometric perspective of Demmel [55] comes in. But, let us first discuss an important variant.

Remark 9.5 Fix the Euclidean spaces $M = \mathbb{R}^n$ and $N = \mathbb{R}^m$. What does "small error" mean in this case? If $\|\mathbf{p}\| = 10^4$, is an error of size $\|\mathbf{p} - \mathbf{q}\| = 10^2$ small or large? To address a question like this, there is a notion of relative error in numerical analysis. By definition, the *relative error* between $\mathbf{p}$ and $\mathbf{q}$ is

$$\text{RelError}(\mathbf{p}, \mathbf{q}) := \frac{\|\mathbf{p} - \mathbf{q}\|}{\|\mathbf{p}\|} \quad \text{for } \mathbf{p}, \mathbf{q} \in M.$$

The *relative condition number* of the function ϕ at the input datum $\mathbf{p} \in M$ is as follows:

$$\kappa_{\text{REL}}[\phi](\mathbf{p}) := \lim_{\varepsilon \rightarrow 0} \sup_{\text{RelError}(\mathbf{p}, \mathbf{q}) \leq \varepsilon} \frac{\text{RelError}(\phi(\mathbf{p}), \phi(\mathbf{q}))}{\text{RelError}(\mathbf{p}, \mathbf{q})} = \kappa[\phi](\mathbf{p}) \cdot \frac{\|\mathbf{p}\|_M}{\|\phi(\mathbf{p})\|_N}.$$

In numerical analysis, relative errors are more significant than absolute errors because *floating-point* arithmetic introduces relative errors (see, e.g., [89] or [173, p. 91]). Modern architectures are optimized for computing with floating-point numbers. A floating-point number system $\mathcal{F}$ is a subset of the real numbers $\mathbb{R}$ that is specified by four integers $\beta, t, e_{\min}, e_{\max}$, where β is called *base*, t is called *precision*, and $[e_{\min}, e_{\max}]$ is called *exponential range*. Then, $\mathcal{F} := \{\pm\beta^e \sum_{i=1}^t \frac{d_i}{\beta^i} \mid 0 \leq d_i \leq \beta - 1, \ e_{\min} \leq e \leq e_{\max}\}$. The quantity $u = \frac{1}{2}\beta^{1-t}$ is referred to as the *relative precision* of $\mathcal{F}$.

The *range* of $\mathcal{F}$ is the set $G := \{x \in \mathbb{R} \mid \beta^{e_{\min}-1} \leq |x| \leq \beta^{e_{\max}}(1 - \beta^{-1})\}$. Note that $\mathcal{F} \subset G$. Consider the *rounding function* $\text{fl} : \mathbb{R} \rightarrow \mathcal{F}, x \mapsto \text{argmin}_{y \in \mathcal{F}} |x - y|$. One can show that every $x \in G$ satisfies $\text{fl}(x) = x(1 + \delta) \in \mathcal{F}$ for some δ with $|\delta| \leq u$. This is a crucial property. It tells us that every number in G can be approximated by a number of $\mathcal{F}$ with relative precision u. Namely, the following inequality holds:

$$\text{RelError}(x, \text{fl}(x)) = \|\delta\| \leq u \quad \text{for all } x \in G. \tag{9.1}$$

Many hardware floating-point units use the IEEE 754 standard. This is a system $\mathcal{F}$ with the specifications

	β	t	$e_{\min}$	$e_{\max}$	u
half (16 bit)	2	11	−14	$16 = 2^4$	$\approx 5 \cdot 10^{-4}$
single (32 bit)	2	24	−125	$128 = 2^7$	$\approx 6 \cdot 10^{-8}$
double (64 bit)	2	53	−1021	$1024 = 2^{10}$	$\approx 10^{-16}$

This is designed so that the arithmetic operations $\circ \in \{+, -, \times, /, \sqrt{\cdot}\}$ satisfy the property $\mathrm{fl}(x \circ y) = (x \circ y)(1 + \delta)$ for some $|\delta| \leq u$. For instance, the 64-bit floating-point number system, specified in the third row of the table, can approximate any real number within its range with a relative error of at most $u \approx 10^{-16}$.

After this digression into the practical aspects of numerical computing, we now return to the mathematical theory. The next theorem is also due to Rice [152]. In his article, M and N are arbitrary Riemannian manifolds. We here specialize to the algebraic setting. For us, each of M and N is a submanifold in a Euclidean space, described by a finite Boolean combination of polynomial equations and inequalities.

Theorem 9.6 *Fix a differentiable function $\phi : M \to N$. The condition number of this computational problem at the input datum $\mathbf{p} \in M$ is the maximal norm of the derivative over the unit sphere in the tangent space at $\mathbf{p}$; i.e.,*

$$\kappa[\phi](\mathbf{p}) = \max_{\mathbf{t} \in T_{\mathbf{p}}M:\ \|\mathbf{t}\|=1} \|D_{\mathbf{p}}\phi(\mathbf{t})\|.$$

We obtain the relative condition number $\kappa_{\mathrm{REL}}[\phi](\mathbf{p})$ by multiplying this with $\|\mathbf{p}\|/\|\phi(\mathbf{p})\|$.

A key step in computing the condition number with Rice's formula is to find an expression for the Jacobian $D_{\mathbf{p}}\phi$. For problems of interest to us, this step uses implicit differentiation or geometric arguments.

Example 9.7 (Roots of univariate polynomials) Following [35, Section 14.1.1], we examine the computational problem of finding one real root of a polynomial f in one variable z of degree d. We write

$$f(z) = f_0 + f_1 z + f_2 z^2 + \cdots + f_d z^d.$$

The coefficient vector $f = (f_0, f_1, \ldots, f_d)$ now serves as the input $\mathbf{p}$, and the output is a particular real number a which satisfies $f(a) = 0$. The function $f \mapsto a(f)$ is well defined in a small open subset U of the coefficient space $\mathbb{R}^{d+1}$. The set U must be small enough so that one root can be identified for each polynomial in U.

To find the derivative $D_f a$ of our root-finding function $f \mapsto a(f)$, we assume that the coefficients are differentiable functions $f_i(t)$ of a parameter t, and we set $\dot{f}_i = f_i'(0)$ and $\dot{f} = \sum_{i=0}^{d} \dot{f}_i z^i$. By differentiating the identity $f(a(f)) = 0$ with respect to t, we find the following formula for the desired derivative:

$$D_f a = -\frac{\dot{f}(a)}{f'(a)}. \tag{9.2}$$

We now think of a as a fixed root of a fixed polynomial $f(z)$. Theorem 9.6 implies that the condition number $\kappa[a](f)$ equals $|f'(a)|^{-1}$ times the maximal value $|\dot{f}(a)|$, where $\dot{f}$

runs over all points $(\dot{f}_0, \dot{f}_1, \ldots, \dot{f}_d)$ on the unit d-sphere. The Cauchy–Schwarz inequality implies that this maximum equals $|\dot{f}(a)| = \sqrt{\sum_{i=0}^d a^{2i}}$, and therefore we have the following formula for the desired condition number:

$$\kappa[a](f) \;=\; \frac{\sqrt{\sum_{i=0}^d a^{2i}}}{|f'(a)|}. \tag{9.3}$$

This quantity is $+\infty$ when a is a double zero of f. The further away from being a double zero, the smaller the condition number. The relative condition number for the univariate root-finding problem equals

$$\kappa_{\mathrm{REL}}[a](f) \;=\; \frac{\sqrt{(\sum_{i=0}^d f_i^2)(\sum_{i=0}^d a^{2i})}}{|a|\,|f'(a)|}.$$

This formula involves not just the polynomial $f(z)$ but also the zero a we seek to find. This is due to the fact that $a(f)$ is a *local map*, defined locally around the zero a; see [93]. ◇

Example 9.8 (The condition number of the ED minimization problem) We revisit the ED minimization problem (2.1) for a smooth algebraic variety $X \subset \mathbb{R}^n$ from the point of view of condition numbers. Recall the definition of the medial axis $\mathrm{Med}(X)$ from Chapter 7. Every point outside the medial axis has a unique closest point on X. Formulating this as a computational problem, we have the input space $M = \mathbb{R}^n\backslash\mathrm{Med}(X)$ and the output space $N = X$. Here, $\phi : (\mathbb{R}^n\backslash\mathrm{Med}(X)) \to X$ is the closest-point function $\phi(\mathbf{u}) = \mathrm{argmin}_{\mathbf{x}\in X} \|\mathbf{x}-\mathbf{u}\|$. We can apply Theorem 9.6 to compute the associated condition number $\kappa[\phi](\mathbf{u})$. This was carried out in detail in [32]. Suppose $\mathbf{x} = \phi(\mathbf{u})$. We have

$$\kappa[\phi](\mathbf{u}) \;=\; \max_{\mathbf{t}\in T_{\mathbf{x}}X:\ \|\mathbf{t}\|=1} \left\| (I_m - \lambda\cdot L_\mathbf{v})^{-1}\,\mathbf{t} \right\|,$$

where $\lambda = \|\mathbf{u} - \mathbf{x}\|$ is the distance from $\mathbf{u}$ to X, $\mathbf{v} = \lambda^{-1}(\mathbf{u} - \mathbf{x})$ is the normal vector at $\mathbf{x}$ pointing towards $\mathbf{u}$, and $L_\mathbf{v}$ is the Weingarten map (6.10) of X at $\mathbf{x}$ in normal direction $\mathbf{v}$. ◇

9.2 Matrix Inversion and Eckart–Young

In this section, we study the condition number for matrix inversion. This will lead us back to the Eckart–Young Theorem, for which we here present a proof. We begin by reviewing some norms on the space of real $m\times n$ matrices. The Euclidean norm (or Frobenius norm) is

$$\|A\| \;=\; \sqrt{\sum_{i,j} a_{ij}^2} \;=\; \sqrt{\mathrm{trace}(AA^\top)} \quad \text{for } A \in \mathbb{R}^{m\times n}.$$

By contrast, in Theorem 9.6, we used the spectral norm, or operator norm. This is given by

$$\|A\|_{\mathrm{op}} \;:=\; \max_{\|\mathbf{x}\|=1} \|A\mathbf{x}\|.$$

We have already met the spectral norm in Section 8.4. There, we introduced it as the largest singular value of A. To see that this is an equivalent definition, we first observe that, if U and V are orthogonal matrices, then $\|U^T A V\|_{op} = \|A\|_{op}$. In words, the spectral norm is *orthogonally invariant*. Suppose $m \geq n$. Let $A = UDV^T$ be the singular value decomposition of A, where $U \in \mathbb{R}^{m \times m}$ and $V \in \mathbb{R}^{n \times n}$ are orthogonal matrices and D is the $m \times n$ diagonal matrix with singular values $\sigma_1 \geq \cdots \geq \sigma_n \geq 0$ on the main diagonal. Orthogonal invariance implies $\|A\|_{op} = \|D\|_{op} = \sigma_1$. Similarly, the Frobenius norm is orthogonally invariant and hence satisfies $\|A\| = \|D\| = (\sigma_1^2 + \cdots + \sigma_n^2)^{1/2}$. Furthermore, in the case where $n = m$ and A is invertible, we have $A^{-1} = VD^{-1}U^T$, so that

$$\|A^{-1}\|_{op} = \sigma_n^{-1}.$$

We now focus on the case of square matrices ($m = n$). Matrix inversion is the map

$$\text{inv} : \mathcal{D} \to \mathcal{D}, \ A \mapsto A^{-1},$$

where $\mathcal{D} = \{A \in \mathbb{R}^{n \times n} \mid \det(A) \neq 0\}$. We shall prove the following characterization of the condition number of matrix inversion. For any $A \in \mathcal{D}$, the smallest singular value σ_n is a positive real number.

Theorem 9.9 *The condition number of matrix inversion at $A \in \mathcal{D}$ is*

$$\kappa[\text{inv}](A) = \|A^{-1}\|_{op}^2 = \sigma_n^{-2}.$$

Before we prove this theorem, let us briefly bring it into context with Remark 9.5. In numerical analysis, a popular choice for measuring the relative error is using the spectral norm. In this case, by Theorem 9.9, the relative condition number can be expressed as the ratio of the largest and smallest singular value of A:

$$\kappa_{\text{REL}}[\text{inv}](A) = \kappa[\text{inv}](A) \cdot \frac{\|A\|_{op}}{\|A^{-1}\|_{op}} = \frac{\sigma_1}{\sigma_n}. \tag{9.4}$$

This ratio is *Turing's condition number* and goes back to the work of Turing [174]. The meaning of (9.4) for Voronoi cells of determinantal varieties was discussed in Section 8.4.

Proof (of Theorem 9.9) Let $\text{adj}(A)$ denote the adjoint matrix of A. Since

$$\text{inv}(A) = A^{-1} = \frac{1}{\det(A)} \cdot \text{adj}(A),$$

the map inv is a polynomial function on the open set $\mathcal{D}$. In particular, inv is differentiable on $\mathcal{D}$. By Theorem 9.6, the condition number is $\kappa[\text{inv}](A) = \|D_A \text{inv}\|_{op}$. We compute the derivative of inv. Taking the derivative of $AB = I_n$ we have $\dot{A}B + A\dot{B} = 0$. Since $B = A^{-1}$, we conclude $\dot{B} = -A^{-1}\dot{A}A^{-1}$. For a tangent vector $\dot{A} = R \in \mathbb{R}^{n \times n}$ we therefore have that $D_A \text{inv}(R) = A^{-1}RA^{-1}$. Let $A = UDV^T$ be the singular value decomposition of A. Then, $A^{-1} = VD^{-1}U^T$. By orthogonal invariance of the Euclidean norm, we find

$$\max_{\|R\|=1} \|A^{-1}RA^{-1}\|^2 = \max_{\|R\|=1} \|D^{-1}RD^{-1}\|^2 = \max_{\Sigma_{i,j} r_{i,j}^2 = 1} \sum_{i,j} \frac{r_{i,j}^2}{\sigma_i^2 \sigma_j^2}.$$

The last expression is maximized when $r_{n,n} = 1$ and all other $r_{i,j}$ are zero. We conclude that $\kappa[\text{inv}](A) = \max_{\|R\|=1} \|D_A \text{inv}(R)\| = \frac{1}{\sigma_n^2} = \|A^{-1}\|_{\text{op}}^2$ is the condition number of matrix inversion at A. $\qquad\square$

Example 9.10 We revisit Example 9.1. The matrices were $A = \begin{bmatrix} 1 & 1 \\ -1 & 1 \end{bmatrix}$ and $B = \begin{bmatrix} 1 & 1 \\ 1 & 1+\delta \end{bmatrix}$. To be concrete, we set $\delta = 10^{-8}$. Since $A^\top A = 2 \cdot I_2$, we have that $\|A^{-1}\mathbf{x}\| = \frac{1}{2}\|\mathbf{x}\|$ for all $\mathbf{x} \in \mathbb{R}^2$. By Theorem 9.9, $\kappa[\text{inv}](A) = \|A^{-1}\|_{\text{op}}^2 = \frac{1}{4}$. On the other hand, we have $\|B^{-1}\|_{\text{op}} \geq \|B^{-1}\mathbf{e}_0\| \geq 10^8$, and this implies $\kappa[\text{inv}](B) = \|B^{-1}\|_{\text{op}}^2 \geq 10^{16}$. This explains the different behaviors of the outputs with respect to errors in the input in Example 9.1. ◇

The Eckart–Young Theorem (Theorem 2.9) yields a metric interpretation of Turing's condition number $\kappa[\text{inv}](A)$ from Theorem 9.9. The smallest singular value σ_n of a square matrix A equals the Euclidean distance of A to the variety of singular matrices. This is the hypersurface defined by the determinant:

$$\Sigma := \{A \in \mathbb{R}^{n\times n} \mid \det(A) = 0\}.$$

Turing's condition number is the squared inverse distance from A to the hypersurface Σ:

$$\kappa[\text{inv}](A) = \frac{1}{\text{dist}(A,\Sigma)^2} \quad \text{and} \quad \kappa_{\text{REL}}[\text{inv}](A) = \frac{\|A\|_{\text{op}}}{\text{dist}(A,\Sigma)}. \tag{9.5}$$

Example 9.11 Fix the 2×2-determinant $\Sigma = V(\det) \subset \mathbb{R}^{2\times2}$ and consider the set of matrices $A \in \mathbb{R}^{2\times2}$ for which $\kappa[\text{inv}](A) = \varepsilon^{-1}$. This is the offset hypersurface of Σ at level $\sqrt{\varepsilon}$. We can compute this as in Example 7.14. In what follows we compute the hypersurface defined by $\kappa_{\text{REL}}[\text{inv}](A) = \varepsilon^{-1}$ for $\varepsilon > 0$. To make the formula in (9.5) more convenient, we replace $\|A\|_{\text{op}}$ by $\|A\|$. We proceed as in Section 7.2 to compute a polynomial equation for $\text{dist}(A,\Sigma) = \varepsilon \cdot \|A\|$ in terms of A and ε, but for $B \in \Sigma$ we replace the affine sphere $\|A - B\| = \varepsilon$ by the homogeneous sphere $\|A - B\| = \varepsilon \cdot \|A\|$. We use Macaulay2 [73]:

```
R = QQ[a_0..a_3, b_0..b_3, eps];
f = b_0*b_3 - b_1*b_2;
nAsq = a_0^2 + a_1^2 + a_2^2 + a_3^2;
d = (a_0-b_0)^2+(a_1-b_1)^2+(a_2-b_2)^2+(a_3-b_3)^2-eps^2*nAsq;
J1 = {diff(b_0, f), diff(b_1, f), diff(b_2, f), diff(b_3, f)};
J2 = {diff(b_0, d), diff(b_1, d), diff(b_2, d), diff(b_3, d)};
J = matrix {J1, J2};
OC = ideal {f, minors(2, J), d};
O = eliminate({b_0, b_1, b_2, b_3}, OC);
g = sub((gens O)_0_0, QQ[a_0..a_3, b_0..b_3][eps])
```

The result of this computation is a polynomial of the form

$$g(\mathbf{a}, \varepsilon) = (a_0^2 + a_1^2 + a_2^2 + a_3^2)^2 \cdot (g_0(\mathbf{a}) + g_1(\mathbf{a})\varepsilon^2 + g_2(\mathbf{a})\varepsilon^4 + g_3(\mathbf{a})\varepsilon^6 + g_4(\mathbf{a})\varepsilon^8).$$

The coefficients of this *relative offset polynomial* are

$$g_4(\mathbf{a}) = (a_0^2 + a_1^2 + a_2^2 + a_3^2)^2$$

$$g_3(\mathbf{a}) = -3(a_0^2 + a_1^2 + a_2^2 + a_3^2)^2$$

$$g_2(\mathbf{a}) = 3a_0^4 + 6a_0^2a_1^2 + 3a_1^4 + 6a_0^2a_2^2 + 7a_1^2a_2^2 + 3a_2^4 - 2a_0a_1a_2a_3 + 7a_0^2a_3^2 + 6a_1^2a_3^2 + 6a_2^2a_3^2 + 3a_3^4$$

$$g_1(\mathbf{a}) = -(a_0^4 + 2a_0^2a_1^2 + a_1^4 + 2a_0^2a_2^2 + 4a_1^2a_2^2 + a_2^4 - 4a_0a_1a_2a_3 + 4a_0^2a_3^2 + 2a_1^2a_3^2 + 2a_2^2a_3^2 + a_3^4)$$

$$g_0(\mathbf{a}) = (a_1a_2 - a_0a_3)^2.$$

Figure 9.1 shows the zero set of $g(\mathbf{a}, \varepsilon)$ at level $\varepsilon = 0.5$ in the affine patch $a_0 = 1$. We remark that, if $\mathrm{dist}(A, \Sigma) = \varepsilon \cdot \|A\|$, then $\varepsilon = \sin \alpha$, where α is the minimal angle between the line $\mathbb{R} \cdot A$ and a line $\mathbb{R} \cdot B$ with $B \in \Sigma$. In this case, ε^{-1} is also called a *conic condition number* (see [35, Chapters 20 & 21]).   ◇

Fig. 9.1: This picture shows the determinantal hypersurface $\Sigma = \{\det(A) = 0\} \subset \mathbb{R}^{2\times 2}$ (the red-blue surface in the middle) together with the hypersurface defined by $\mathrm{dist}(A, \Sigma) = \varepsilon \cdot \|A\|$ with $\varepsilon = 0.5$ (the union of the two green-yellow surfaces on the outside). The picture is drawn in the affine patch where the upper left entry of A is fixed to be one.

In the remainder of this section, we give a proof of the Eckart–Young Theorem. For this, we return now to rectangular matrices and denote by $X_r := \{A \in \mathbb{R}^{m\times n} \mid \mathrm{rank}(A) \le r\}$ the variety of real matrices of rank at most r. Recall that $\mathrm{Sing}(X_r) = X_{r-1}$. We first compute the normal space of X_r at a smooth point. Lemma 9.12 can be viewed as a variant of Example 2.22, but presented in a more down-to-earth manner.

Lemma 9.12 *Fix $A \in X_r$ of rank r. Suppose that $A = RS^\top$, where $R \in \mathbb{R}^{m\times r}$ and $S \in \mathbb{R}^{n\times r}$ have rank r. The normal space of X_r at A has dimension $(m-r)(n-r)$ and it equals*

$$N_A X_r = \mathrm{span}\left\{ \mathbf{u}\mathbf{v}^\top \mid \mathbf{u}^\top R = 0 \text{ and } S^\top \mathbf{v} = 0 \right\}.$$

Proof Let $R(t) \in \mathbb{R}^{m\times r}$ and $S(t) \in \mathbb{R}^{n\times r}$ be smooth curves with $R(0) = R$ and $S(0) = S$. Then, $\gamma(t) := R(t)S(t)^\top$ is a smooth curve with $\gamma(0)=A$. By the product rule from calculus,

$$\left.\frac{\partial}{\partial t}\gamma(t)\right|_{t=0} = R\left(\left.\frac{\partial}{\partial t}S(t)\right|_{t=0}\right)^\top + \left(\left.\frac{\partial}{\partial t}R(t)\right|_{t=0}\right)S^\top. \tag{9.6}$$

Let $V := \{RP^\top \mid P \in \mathbb{R}^{n\times r}\}$ and $W := \{QS^\top \mid Q \in \mathbb{R}^{m\times r}\}$. The equation (9.6) shows that the tangent space of X_r at A is the sum $T_A X_r = V + W$. (Aside for students: what is the intersection $V \cap W$?) We note that V consists of all matrices $L \in \mathbb{R}^{m\times n}$ such that $\mathbf{u}^\top L\mathbf{x} = 0$ for all $\mathbf{u}$ with $\mathbf{u}^\top R = 0$ and $\mathbf{x} \in \mathbb{R}^n$ arbitrary. Since $\mathbf{u}^\top L\mathbf{x} = \mathrm{Trace}(L^\top \mathbf{u}\mathbf{x}^\top)$, this shows that the normal space of V is spanned by matrices of the form $\mathbf{u}\mathbf{x}^\top$. Similarly, the normal space of W is spanned by $\mathbf{y}\mathbf{v}^\top$, where $S^\top \mathbf{v} = 0$ and $\mathbf{y} \in \mathbb{R}^m$ arbitrary. Therefore, the normal space of $T_A X_r = V + W$ is spanned by all $\mathbf{u}\mathbf{v}^\top$ with $\mathbf{u}$ and $\mathbf{v}$ as above. $\square$

Corollary 9.13 $\dim X_r = nm - (m - r)(n - r) = r(m + n - r)$.

We use Lemma 9.12 to prove the Eckart–Young Theorem.

Proof (of Theorem 2.9) Let $B \in X_r$ be a matrix of rank r. We consider a singular value decomposition $B = UDV^\top$. The matrices $U \in \mathbb{R}^{m\times r}$ and $V \in \mathbb{R}^{n\times r}$ have orthonormal columns, and $D = \mathrm{diag}(\sigma_1, \ldots, \sigma_r)$ with $\sigma_1, \ldots, \sigma_r > 0$ (not necessarily ordered). To derive Theorem 2.9, we shall prove a claim that is reminiscent of Theorem 8.24. Namely, the singular value decomposition of all matrices $A \in \mathbb{R}^{m\times n}$ such that B is an ED critical point for A has the form $A = [U\ U']\begin{bmatrix} D & 0 \\ 0 & D' \end{bmatrix}[V\ V']^\top$, where $D' = \mathrm{diag}(\sigma_{r+1}, \ldots, \sigma_n)$.

By orthogonal invariance, we can assume that $B = \begin{bmatrix} D & 0 \\ 0 & 0 \end{bmatrix}$. Let $A \in B + N_B X_r$ be a matrix in the normal space of B. By Lemma 9.12, we have $A = B + \sum_{i=r+1}^{m} \sum_{j=r+1}^{n} a_{ij}\, \mathbf{e}_i \mathbf{e}_j^\top$ for some coefficients $a_{ij} \in \mathbb{R}$, i.e.

$$A = \begin{bmatrix} D & 0 \\ 0 & A' \end{bmatrix}, \quad \text{where } A' = (a_{ij}) \in \mathbb{R}^{(m-r)\times(n-r)}.$$

Let now $A' = U'D'V'^\top$ be the singular value decomposition of A'. Then,

$$A = [I_r\ U']\begin{bmatrix} D & 0 \\ 0 & D' \end{bmatrix}[I_r\ V']^\top. \tag{9.7}$$

This is the desired singular value decomposition of the $m \times n$ matrix A. $\square$

Remark 9.14 Our proof of the Eckart–Young Theorem also works for symmetric matrices. It implies that the rank r matrix B, which minimizes the distance to a symmetric matrix A, is also symmetric.

9.3 Condition Number Theorems

The formula in (9.5) states that the condition number of matrix inversion at the input $A \in \mathbb{R}^{n\times n}$ is the inverse squared distance to the variety of singular matrices Σ. To a numerical analyst, the elements of a set like Σ are the *ill-posed inputs* of the computational problem. A result that connects a condition number and an inverse distance to ill-posed inputs is called a *condition number theorem*. Equation (9.5) yields a condition number theorem for matrix inversion. Condition number theorems were derived in [176] for computing eigenvalues of matrices, and in [93] for computing zeros of polynomials, as in Example 9.7.

Condition number theorems connect metric algebraic geometry with numerical analysis. They relate the numerical difficulty of an input datum **p** to the distance of **p** to the locus of ill-posed inputs. This is explained in detail in Demmel's paper [55]. The determinant of a square matrix is a special case of a discriminant, and Σ will generally be a hypersurface. This will be made precise in the next section. In this section, we prove a condition number theorem for solving systems of polynomial equations. We consider homogeneous polynomials and their zeros in projective space. This differs from Example 9.7, where a was a zero on the affine line. While this difference may seem insignificant to an algebraic geometer, it becomes significant in our metric setting. Namely, affine space with its Euclidean metric and projective space with the metric described below are markedly different as metric spaces.

We fix the real projective space $\mathbb{P}^n_{\mathbb{R}}$ and we write $\mathcal{H}_d$ for the vector space of homogeneous polynomials of degree d in $\mathbf{x} = (x_0, \ldots, x_n)$. For $m \leq n$ and $\mathbf{d} = (d_1, \ldots, d_m)$ we abbreviate

$$\mathcal{H}_{\mathbf{d}} := \mathcal{H}_{d_1} \times \cdots \times \mathcal{H}_{d_m}.$$

Suppose $m = n$, let $F \in \mathcal{H}_{\mathbf{d}}$ and let $\mathbf{a} \in \mathbb{P}^n_{\mathbb{R}}$ a regular zero of F. By the Implicit Function Theorem, there is a locally defined solution function $a : U \to V$, where $U \subset \mathcal{H}_{\mathbf{d}}$ is a neighborhood of F and $V \subset \mathbb{P}^n_{\mathbb{R}}$ is a neighborhood of $\mathbf{a}$, such that $F(a(F)) = 0$. Our goal is to give a formula for the condition number of the solution function a.

For this, we first need to introduce metrics on the space of polynomials and the space of zeros. Since a is defined locally, we can replace projective space $\mathbb{P}^n_{\mathbb{R}}$ by the n-sphere $\mathbb{S}^n \subset \mathbb{R}^{n+1}$. The latter carries the natural metric inherited from its ambient space. In fact, we can use the Riemannian metric on the sphere to define a metric on $\mathbb{P}^n_{\mathbb{R}}$, so that it becomes a Riemannian manifold (this approach is worked out in Section 12.3 for complex projective space). In this way, $\mathbb{P}^n_{\mathbb{R}}$ becomes curved, while affine space is flat.

For $\mathcal{H}_d$, we use the Bombieri–Weyl metric. Let $A := \{\alpha \in \mathbb{N}^{n+1} \mid \alpha_0 + \cdots + \alpha_n = d\}$. The *Bombieri–Weyl inner product* between two homogeneous polynomials $f = \sum_{\alpha \in A} f_\alpha \mathbf{x}^\alpha$ and $g = \sum_{\alpha \in A} g_\alpha \mathbf{x}^\alpha$ in $\mathcal{H}_d$ is defined by the formula

$$\langle f, g \rangle_{\text{BW}} := \sum_{\alpha \in A} \frac{\alpha_0! \cdots \alpha_n!}{d!} f_\alpha \cdot g_\alpha. \tag{9.8}$$

For $d = 1$, this is the usual Euclidean inner product in $\mathbb{R}^{n+1}$.

The reason for the multinomial coefficients in (9.8) is that the Bombieri-Weyl inner product is orthogonally invariant: If $U \in O(n+1)$ is an orthogonal matrix, then

$$\langle f \circ U, g \circ U \rangle_{\text{BW}} = \langle f, g \rangle_{\text{BW}}.$$

Kostlan [110, 111] showed that (9.8) is the unique (up to scaling) orthogonally invariant inner product on $\mathcal{H}_d$ such that monomials are pairwise orthogonal. This inner product extends to $\mathcal{H}_{\mathbf{d}}$. Namely, for tuples $F = (f_1, \ldots, f_m)$ and $G = (f_1, \ldots, f_m)$, we define

$$\langle F, G \rangle_{\text{BW}} := \langle f_1, g_1 \rangle_{\text{BW}} + \cdots + \langle f_m, g_m \rangle_{\text{BW}}.$$

The Bombieri–Weyl norm is $\|F\|_{BW} := \sqrt{\langle F, F\rangle_{BW}}$, and the corresponding distance is

$$\text{dist}_{BW}(F, G) := \|F - G\|_{BW} \quad \text{for} \quad F, G \in \mathcal{H}_d.$$

Example 9.15 ($n = 1, d = 2$) Consider two quadrics $f(x_0, x_1) = ax_0^2 + bx_0x_1 + cx_1^2 = \mathbf{x}^\top A\mathbf{x}$ and $g(x_0, x_1) = \alpha x_0^2 + \beta x_0x_1 + \gamma x_1^2 = \mathbf{x}^\top B\mathbf{x}$, where $\mathbf{x} = (x_0, x_1)^\top$, with symmetric matrices

$$A = \begin{bmatrix} a & b/2 \\ b/2 & c \end{bmatrix} \quad \text{and} \quad B = \begin{bmatrix} \alpha & \beta/2 \\ \beta/2 & \gamma \end{bmatrix}.$$

The inner product of the two quadratic forms is the trace inner product of the two matrices:

$$\langle f, g\rangle_{BW} = \frac{2! \cdot 0!}{2!} a\alpha + \frac{1! \cdot 1!}{2!} b\beta + \frac{2! \cdot 0!}{2!} c\gamma = a\alpha + \frac{1}{2} b\beta + c\gamma = \text{Trace}(A^\top B).$$

An analogous formula holds for quadrics in more than two variables. Thus, the Bombieri–Weyl product generalizes the trace inner product for symmetric matrices to homogeneous polynomials of any degree. ◇

We now turn to the condition number of the solution function $a : U \to V$. The zero $\mathbf{a} = a(F)$ is represented by a unit vector in $\mathbb{S}^n$. Proceeding as in Example 9.7, we differentiate the equation $F(a(F)) = 0$. This gives $JF(\mathbf{a})\dot{\mathbf{a}} + \dot{F}(\mathbf{a})$. Here, $JF(\mathbf{a}) = \left(\frac{\partial f_i}{\partial x_j}(\mathbf{a})\right) \in \mathbb{R}^{n\times(n+1)}$ is the Jacobian matrix of F at $\mathbf{a}$, and $\dot{\mathbf{a}} \in T_{\mathbf{a}}\mathbb{S}^n$. The tangent space $T_{\mathbf{a}}\mathbb{S}^n$ is the image of the projection matrix $P_{\mathbf{a}} := I_{n+1} - \mathbf{a}\mathbf{a}^\top$. Note that $\det(P_{\mathbf{a}}) = 1 - \|\mathbf{a}\|^2 = 0$. When $\mathbf{a}$ is a regular zero of F, the rank of $JF(\mathbf{a})$ is n, and we obtain

$$\dot{\mathbf{a}} = (JF(\mathbf{a})|_{T_{\mathbf{a}}\mathbb{S}^n})^{-1}\dot{F}(\mathbf{a}).$$

This formula is a multivariate version of (9.2). By Theorem 9.6, the condition number for a is $\kappa[a](F) = \max_{\|\dot{F}\|_{BW}=1} \|(JF(\mathbf{a})|_{T_{\mathbf{a}}\mathbb{S}^n})^{-1}\dot{F}(\mathbf{a})\|$. We use orthogonal invariance of the Bombieri–Weyl metric to compute this maximum. Fix $U \in O(n+1)$ such that $\mathbf{a} = U\mathbf{e}_0$, where $\mathbf{e}_0 = (1, 0, \ldots, 0) \in \mathbb{S}^n$. We have $\|\dot{F}\|_{BW} = \|\dot{F} \circ U\|_{BW}$ for all $\dot{F} \in \mathcal{H}_d$. This implies that $\kappa[a](F)$ is also obtained by maximizing $\|(JF(\mathbf{a})|_{T_{\mathbf{a}}\mathbb{S}^n})^{-1}\dot{F}(\mathbf{e}_0)\|$ over all $\dot{F} \in \mathcal{H}_d$ with $\|\dot{F}\|_{BW} = 1$. Let us write $\dot{F}(\mathbf{x}) = x_0^d \cdot \mathbf{b} + G(\mathbf{x})$, where $\mathbf{b} \in \mathbb{R}^n$ and G involves only powers of x_0 of degree less than d. Then, $\dot{F}(\mathbf{e}_0) = \mathbf{b}$ and so $\|\dot{F}(\mathbf{e}_0)\| = \|\mathbf{b}\| = \sqrt{1 - \|G\|_{BW}^2}$.

This uses the fact that $x_0^d \cdot \mathbf{b}$ and G are orthogonal for the Bombieri–Weyl inner product. We have thus shown that $\{\dot{F}(\mathbf{a}) \mid \|\dot{F}\|_{BW} = 1\}$ is the unit ball in $\mathbb{R}^{n+1}$. Consequently,

$$\kappa[a](F) = \|(JF(\mathbf{a})|_{T_{\mathbf{a}}\mathbb{S}^n})^{-1}\|_{op}. \tag{9.9}$$

Corollary 9.16 *The condition number for solving n polynomial equations on $\mathbb{P}^n$ equals*

$$\kappa[a](F) = \frac{1}{\sigma_n(JF(\mathbf{a})P_{\mathbf{a}})}. \tag{9.10}$$

Proof The matrix $JF(\mathbf{a})P_{\mathbf{a}}$ represents the linear map $JF(\mathbf{a})|_{T_{\mathbf{a}}\mathbb{S}^n}$. Hence, the two have the same singular values. □

Example 9.17 (Univariate polynomials revisited) We examine our formula for $n = 1$. The computation takes a binary form $f(x_0, x_1)$ to one of its zeros $\mathbf{a} = (a_0, a_1)$ in $\mathbb{S}^1$. This is the projective version of Example 9.7. The Jacobian equals $Jf(\mathbf{a}) = [\partial f/\partial x_0(\mathbf{a}) \ \partial f/\partial x_1(\mathbf{a})]$. Equation (9.10) yields the following analogue to Equation (9.3):

$$\kappa[a](F) = \frac{1}{\sqrt{(\partial f/\partial x_0(\mathbf{a}))^2 + (\partial f/\partial x_1(\mathbf{a}))^2}}.$$

Here, $\mathbf{a}$ is a point on the circle $\mathbb{S}^1$, representing a zero of f in $\mathbb{P}^1$.   ◇

We now return to the general case $m \le n$, i.e. we allow systems with fewer equations.

Proposition 9.18 *For $\mathbf{x} \in \mathbb{P}^n_\mathbb{R}$, denote $\Sigma(\mathbf{x}) := \{F \in \mathcal{H}_\mathbf{d} \mid F(\mathbf{x}) = 0 \text{ and } \operatorname{rank} JF(\mathbf{x}) < m\}$ and set $D = \operatorname{diag}(d_1, \dots, d_m)$. The distance from F to the discriminant $\Sigma(\mathbf{x})$ equals*

$$\operatorname{dist}_{\mathrm{BW}}(F, \Sigma(\mathbf{x})) = \sqrt{\|F(\mathbf{x})\|^2 + \sigma_m(D^{-1/2} JF(\mathbf{x}) P_\mathbf{x})^2}.$$

Hence, if $\mathbf{a}$ is a zero of the system F, then we have $\operatorname{dist}_{\mathrm{BW}}(F, \Sigma(\mathbf{a})) = \sigma_m(D^{-1/2} JF(\mathbf{a}) P_\mathbf{a})$.

Proof Let $U \in O(n + 1)$ with $\mathbf{x} = U\mathbf{e}_0$ and set $F_0 := F \circ U$. By orthogonal invariance, $\operatorname{dist}_{\mathrm{BW}}(F, \Sigma(\mathbf{x})) = \operatorname{dist}_{\mathrm{BW}}(F_0, \Sigma(\mathbf{e}_0))$. We shall compute the latter. We write

$$F_0(\mathbf{x}) = x_0^d \cdot \mathbf{b} + x_0^{d-1} \cdot B(x_1, \dots, x_n)^\top + H(\mathbf{x}),$$

where $\mathbf{b} \in \mathbb{R}^m$, $B \in \mathbb{R}^{m \times n}$ and H has degree $\le d - 2$ in x_0. Note that $\mathbf{b} = F_0(\mathbf{e}_0) = F(\mathbf{x})$ and $B = JF_0(\mathbf{e}_0) P_{\mathbf{e}_0} = JF(\mathbf{x}) P_\mathbf{x} U^\top$. Recall that $G \in \Sigma(\mathbf{e}_0)$ if and only if $G(\mathbf{e}_0) = 0$ and the Jacobian $JG(\mathbf{e}_0)$ has rank at most $m - 1$. This means that the distance from $\Sigma(\mathbf{e}_0)$ to F_0 is minimized at a polynomial system of the form $G_0(\mathbf{x}) = x_0^{d-1} \cdot A(x_1, \dots, x_n)^\top \in \Sigma(\mathbf{e}_0)$, where A is a matrix of rank at most $m - 1$. We have

$$\|F_0 - G_0\|_{\mathrm{BW}}^2 = \|\mathbf{b}\|^2 + \|D^{-1/2}(A - B)\|^2.$$

The Eckart–Young Theorem implies that the distance from $D^{-1/2}B$ to the variety of matrices of rank at most $m - 1$ equals $\sigma_m(D^{-1/2}B)$. From this, we infer the desired formula for the distance to the variety $\Sigma(\mathbf{x})$:

$$\operatorname{dist}_{\mathrm{BW}}(F, \Sigma(\mathbf{x})) = \sqrt{\|\mathbf{b}\|^2 + \sigma_m(D^{-1/2}B)^2} = \sqrt{\|F(\mathbf{x})\|^2 + \sigma_m(D^{-1/2} JF(\mathbf{x}) P_\mathbf{x})^2}.$$

Here, we use the fact that the singular values are invariant under multiplication with the orthogonal matrix $U^\top$.   □

Proposition 9.18 can be viewed as a condition number theorem. If $m = n$ and $\mathbf{a}$ is a zero of F, then the formula for the condition number in (9.10) is equal to the reciprocal distance to $\Sigma(\mathbf{a})$, up to the scaling factor by the diagonal matrix D. The variety $\Sigma(\mathbf{a})$ plays the role of a discriminant, but it still depends on the point $\mathbf{a}$. The actual discriminant is defined as

$$\Sigma := \{F \in \mathcal{H}_\mathbf{d} \mid \text{there is } \mathbf{x} \in \mathbb{P}^n_\mathbb{R} \text{ s.t. } F(\mathbf{x}) = 0 \text{ and } \operatorname{rank} JF(\mathbf{x}) < m\}. \tag{9.11}$$

This is the union of the local discriminants $\Sigma(\mathbf{a})$ where $\mathbf{a}$ runs over $\mathbb{P}^n$. Note that Σ is the discriminant since we are asking for $\mathbf{x}$ to be a real point. In the next section, we replace Σ by its Zariski closure. This is the variety that is obtained by allowing $\mathbf{x}$ to be complex.

The next theorem gives a formula for the distance from a polynomial system F to the discriminant Σ. This was proved by Raffalli [151] for the case $m = 1$. Bürgisser and Cucker [35] cover the case $m = n$.

Theorem 9.19 Let $m \leq n$ and $\mathbf{d} = (d_1, \ldots, d_m)$ be a tuple of degrees. Let $F \in \mathcal{H}_{\mathbf{d}}$. Then,

$$\mathrm{dist}_{\mathrm{BW}}(F, \Sigma) \;=\; \min_{\mathbf{x} \in \mathbb{S}^n} \sqrt{\|F(\mathbf{x})\|^2 + \sigma_m(D^{-1/2}\, JF(\mathbf{x})\, P_{\mathbf{x}})^2}\,,$$

where, as before, $D = \mathrm{diag}(d_1, \ldots, d_m)$ and $P_{\mathbf{x}} := I_{n+1} - \mathbf{x}\mathbf{x}^\top$ is the projection onto $T_{\mathbf{x}}\mathbb{S}^n$.

Proof By definition, the discriminant equals $\Sigma = \bigcup_{\mathbf{x} \in \mathbb{S}^n} \Sigma(\mathbf{x})$. This implies

$$\mathrm{dist}_{\mathrm{BW}}(F, \Sigma) \;=\; \min_{\mathbf{x} \in \mathbb{S}^n} \mathrm{dist}_{\mathrm{BW}}(F, \Sigma(\mathbf{x})).$$

The minimum is attained since $\mathbb{S}^n$ is compact. Proposition 9.18 now yields the claim.   □

Remark 9.20 Theorem 9.19 can be generalized as follows. Fix $1 \leq k < m$. Consider the distance from $F \in \mathcal{H}_{\mathbf{d}}$ to the space of polynomial systems $G \in \mathcal{H}_{\mathbf{d}}$ such that there exists $\mathbf{x} \in \mathbb{P}^n$ with $G(\mathbf{x}) = 0$ and rank $JG(\mathbf{x}) < k$. This distance equals

$$\min_{\mathbf{x} \in \mathbb{S}^n} \sqrt{\|F(\mathbf{x})\|^2 + \sum_{i=k}^{m} \sigma_i(D^{-1/2}\, JF(\mathbf{x})\, P_{\mathbf{x}})^2}\,,$$

where $\sigma_k(\cdot), \ldots, \sigma_m(\cdot)$ are the $m - k + 1$ smallest singular values. This is because the distance from a matrix $A \in \mathbb{R}^{m \times n}$ to the nearest matrix of rank at most $k - 1$ is equal to $\sqrt{\sum_{i=k}^{m} \sigma_i(A)^2}$, by the Eckart–Young Theorem.

Remark 9.21 Let $m = 1$. We have only one polynomial $f \in \mathcal{H}_d$, and Theorem 9.19 yields $\mathrm{dist}_{\mathrm{BW}}(f, \Sigma) = \min_{\mathbf{x} \in \mathbb{S}^n} \sqrt{f(\mathbf{x})^2 + \frac{1}{d}\|P_{\mathbf{x}} \nabla f(\mathbf{x})\|^2}$, where the column vector $\nabla f(\mathbf{x})$ is the gradient of the polynomial f at $\mathbf{x}$. By Euler's formula for homogeneous functions, we have $\mathbf{x}^\top \cdot \nabla f(\mathbf{x}) = d \cdot f(\mathbf{x})$, and hence $P_{\mathbf{x}} \nabla f(\mathbf{x}) = \nabla f(\mathbf{x}) - (d \cdot f(\mathbf{x}))\, \mathbf{x}$. We obtain

$$\mathrm{dist}_{\mathrm{BW}}(f, \Sigma) \;=\; \min_{\mathbf{x} \in \mathbb{S}^n} \sqrt{f(\mathbf{x})^2 + \frac{1}{d}\|\nabla f(\mathbf{x}) - (d \cdot f(\mathbf{x}))\, \mathbf{x}\|^2}\,. \tag{9.12}$$

Example 9.22 ($m = 1, n = 2, d = 3$) We compute the discriminant Σ for ternary cubics:

```
R = QQ[x,y,z,c0,c1,c2,c3,c4,c5,c6,c7,c8,c9];
c0*x^3+c1*y^3+c2*z^3+c3*x^2*y+c4*x^2*z+c5*x*y^2
    +c6*y^2*z+c7*x*z^2+c8*y*z^2+c9*x*y*z;
I = ideal(diff(x,oo),diff(y,oo),diff(z,oo), z-1);
disc = first first entries gens eliminate({x,y,z},I);
toString disc
degree disc, # terms disc
```

We see that Σ is given by a polynomial with 2040 terms of degree 12 in 10 unknowns.

We now compute the distance from the Fermat cubic $f(\mathbf{x}) = x_0^3 + x_1^3 + x_2^3$ to the discriminant Σ. This is easy if we use (9.12). We have $\text{dist}_{\text{BW}}(f, \Sigma) = \min_{\mathbf{x} \in \mathbb{S}^n} \sqrt{h(\mathbf{x})}$, where $h(\mathbf{x}) = (x_0^3 + x_1^3 + x_2^3)^2 + 3 \sum_{i=0}^{2} \left(x_i^2 - (x_0^3 + x_1^3 + x_2^3)x_i\right)^2$. The polynomial function $h : \mathbb{S}^2 \to \mathbb{R}$ is minimized at the point $\mathbf{x}_0 = \frac{1}{\sqrt{3}}(1, 1, 1)$, and we conclude that

$$\text{dist}_{\text{BW}}(f, \Sigma) = \sqrt{h(\mathbf{x}_0)} = \frac{1}{\sqrt{3}}.$$

Since Σ is a cone in $\mathcal{H}_3$, we can compute the minimal angle (measured in the Bombieri–Weyl metric) between f and any polynomial in Σ as in Example 9.11. Since the norm of the Fermat cubic is $\|f\|_{\text{BW}} = \sqrt{3}$, the minimal angle is $\arcsin(1/3) \approx 0.21635 \cdot \frac{\pi}{2}$.   ◇

9.4 Distance to the Discriminant

Theorem 9.19 expresses the distance from a polynomial system F to its discriminant Σ as the optimal value of an optimization problem over the unit sphere $\mathbb{S}^n$. Thus, we are solving the ED problem from Chapter 2 for discriminants. In this section, we examine this problem through the lens of algebraic geometry. This uses the ED duality in Theorem 2.23 and the fact that discriminants are projectively dual to toric varieties.

In the previous section, we considered dense systems, where the polynomials involve all monomials of a fixed degree. In that setting, we used the orthogonal invariance of the Bombieri–Weyl metric to prove Theorem 9.19. In many applications, however, polynomial systems are not dense but *sparse*, and there is no such invariant metric. To address this, we now work in the setting of n sparse Laurent polynomials in n variables. Our equations are not homogeneous. The given data is a collection of finite support sets $\mathcal{A}_1, \mathcal{A}_2, \ldots, \mathcal{A}_n \subset \mathbb{Z}^n$.

Our system of equations takes the form $f_1(\mathbf{x}) = f_2(\mathbf{x}) = \cdots = f_n(\mathbf{x}) = 0$, where

$$f_i(\mathbf{x}) = \sum_{\mathbf{a} \in \mathcal{A}_i} c_{i,\mathbf{a}} \mathbf{x}^{\mathbf{a}} \quad \text{for } i = 1, 2, \ldots, n. \tag{9.13}$$

The BKK Theorem [19] tells us that the number of solutions in $(\mathbb{C}^*)^n$ equals the mixed volume $\text{MV}(P_1, P_2, \ldots, P_n)$. Here, P_i is the convex hull of $\mathcal{A}_i$ and the coefficients $c_{i,\mathbf{a}}$ are assumed to be generic. Thus, $f_i(\mathbf{x})$ is a polynomial with Newton polytope P_i. The BKK theorem was discussed in Example 3.31. We assume that the mixed volume is at least 2.

The *discriminant* of the system (9.13) is the irreducible polynomial $\Delta(\mathbf{c})$ which vanishes whenever our equations have a double root in $(\mathbb{C}^*)^n$. Here $\mathbf{c} = (c_{i,\mathbf{a}})$ denotes the vector of all coefficients. The discriminant Δ is unique up to scaling. We refer to [42] for many details about Δ, including a more precise definition. Our goal here is to solve the Euclidean distance problem for the discriminant hypersurface

$$\Sigma = \{\mathbf{c} \in \mathbb{R}^N \mid \Delta(\mathbf{c}) = 0\}. \tag{9.14}$$

Remark 9.23 The real discriminant equals the real locus in the discriminant Σ. This means that the real hypersurface defined (for the dense case) in (9.11) agrees with the definition above. Indeed, let $\mathbf{c}$ be a generic real point $\mathbf{c}$ in the hypersurface Σ. Then the corresponding polynomial system $F = (f_1, \ldots, f_n)$ has a complex double zero $\mathbf{x}$. That double zero $\mathbf{x}$ is unique. Since F has real coefficients, $\mathbf{x}$ is invariant under complex conjugation. Therefore $\mathbf{x}$ has real coordinates, and $\mathbf{c}$ is also in the set (9.11). In fact, there is a formula for expressing $\mathbf{x}$ by rational operations in $\mathbf{c}$. See (9.15) and [71, Equation (1.28) in Section 12.1.B].

From now on, we work over the field $\mathbb{C}$ and we view the discriminant Σ as a complex projective hypersurface, defined by an irreducible homogeneous polynomial with integer coefficients. This is consistent with [71] and with our earlier chapters.

Example 9.24 We consider the case $n = 1$ and $\mathcal{A} = \{0, 2, 5, 6\}$. The discriminant of the polynomial $f(x) = c_1 + c_2 x^2 + c_3 x^5 + c_4 x^6$ equals

$$\Delta = 46656 c_1^4 c_4^5 + 32400 c_1^3 c_2 c_3^2 c_4^3 - 3125 c_1^3 c_3^6 + 13824 c_1^2 c_2^3 c_4^4 - 1500 c_1^2 c_2^2 c_3^3 c_4$$
$$+ 192 c_1 c_2^4 c_3^2 c_4^2 + 1024 c_2^6 c_4^3 - 108 c_2^5 c_3^4.$$

This defines a surface Σ of degree 9 in $\mathbb{P}^3$. For any point $\mathbf{c} \in \Sigma$, we find the double root x of f by evaluating the gradient of the discriminant. This yields

$$\left(\frac{\partial \Delta}{\partial c_1} : \frac{\partial \Delta}{\partial c_2} : \frac{\partial \Delta}{\partial c_3} : \frac{\partial \Delta}{\partial c_4} \right)(\mathbf{c}) = \left(1 : x^2 : x^5 : x^6\right). \tag{9.15}$$

This identity follows from the duality between discriminants and toric varieties. In particular, if $\mathbf{c}$ is a real vector, then x is a real number. $\diamond$

Given any sparse system F, we seek the distance from its coefficient vector $\mathbf{u}$ to the discriminant Σ. We will use the formulation of the ED problem given in Theorem 2.23. Let X be the toric variety naturally associated with the tuple $(\mathcal{A}_1, \mathcal{A}_2, \ldots, \mathcal{A}_n)$. Then, the dual variety $X^\vee$ is precisely the discriminant hypersurface Σ. This observation is known as the *Cayley trick*; see [71] and [42, Section 2] for expositions.

We write $\mathbf{x}$ for a point in X and $\mathbf{c}$ for a point in Σ. The conormal variety N_X consists of pairs $(\mathbf{x}, \mathbf{c})$ such that the hyperplane $\mathbf{c}$ is tangent to X at the point $\mathbf{x}$. Equivalently, $\mathbf{c}$ is a polynomial system whose variety is singular at $\mathbf{x}$. Given any system $\mathbf{u}$, our task is to solve the equations $\mathbf{x} + \mathbf{c} = \mathbf{u}$ for $(\mathbf{x}, \mathbf{c}) = (\mathbf{x}, \mathbf{u} - \mathbf{x}) \in N_X$. The desired distance from $\mathbf{u}$ to Σ is the minimum of $||\mathbf{x}||$ over all solutions to these equations. In symbols, we have

$$\mathrm{dist}(\mathbf{u}, \Sigma) = \min_{\mathbf{c} \in \Sigma} ||\mathbf{u} - \mathbf{c}|| = \min_{(\mathbf{x}, \mathbf{u} - \mathbf{x}) \in N_X} ||\mathbf{x}||. \tag{9.16}$$

Here, the minimum is taken over all real points. Of course, the minimum value depends on the metric that we choose. In the previous section, we took the Bombieri–Weyl metric.

Example 9.25 ($n = 1, d = 3$) Fix a polynomial $\mathbf{u} = u_0 + u_1 t + u_2 t^2 + u_3 t^3$ with three distinct complex roots. The coefficients u_0, u_1, u_2, u_3 are fixed numbers. Here $\mathcal{A} = \{0, 1, 2, 3\}$, so $X = X_\mathcal{A}$ is the twisted cubic curve in $\mathbb{P}^3$. The discriminant $X^\vee = \Sigma$ is given by

$$\Delta = c_1^2 c_2^2 - 4 c_0 c_2^3 - 4 c_1^3 c_3 + 18 c_0 c_1 c_2 c_3 - 27 c_0^2 c_3^2,$$

as computed in Example 4.13. We consider the following problem:

minimize $||\mathbf{u} - \mathbf{c}||^2 = (u_0 - c_0)^2 + (u_1 - c_1)^2 + (u_2 - c_2)^2 + (u_3 - c_3)^2$ subject to $\mathbf{c} \in \Sigma$.

The conormal variety N_X is a 4-dimensional affine variety in $\mathbb{C}^8$. Its prime ideal equals

$$\langle x_1^2 - x_0 x_2, x_1 x_2 - x_0 x_3, x_2^2 - x_1 x_3, \Delta,$$
$$c_0 x_0 - c_2 x_2 - 2c_3 x_3, c_1 x_1 + 2c_2 x_2 + 3c_3 x_3, 3c_0 x_1 + 2c_1 x_2 + c_2 x_3, c_1 x_0 + 2c_2 x_1 + 3c_3 x_2 \rangle.$$

The equation $\mathbf{x} + \mathbf{c} = \mathbf{u}$ has 7 complex solutions in N_X, since X has ED degree 7. We seek the one that minimizes $||\mathbf{x}||^2$. For a numerical example, consider the polynomial $3t^3 - 26t^2 + 61t - 30$, which is represented by the vector $\mathbf{u} = (-30, 61, -26, 3)$. We solve the equations $\mathbf{x} + \mathbf{c} = \mathbf{u}$ and $(\mathbf{x}, \mathbf{c}) \in N_X$. Among the seven complex solutions, three are real. The optimal value is $||\mathbf{x}||^2 = 0.0251866$, and this is attained by the point

$$\mathbf{c} = \mathbf{u} - \mathbf{x} = (-29.997, 61.011, -25.959, 3.153).$$

In larger cases, where computing the ideal of N_X is infeasible, we can still solve the ED problem with the monomial parametrization of X. In our example, the problem is:

$$\text{minimize } \sum_{i=0}^{3} (st^i)^2 \text{ over all critical points of } (s, t) \mapsto \sum_{i=0}^{3} (u_i - st^i)^2. \tag{9.17}$$

The optimal point is $(s, t) = (-0.002824, 3.78442)$, and the optimal value is 0.0251866. $\diamond$

We now turn to the general case $n > 1$. The toric variety X has dimension $2n - 1$, and it is associated with the Cayley configuration of $\mathcal{A}_1, \mathcal{A}_2, \ldots, \mathcal{A}_n$. To define this, we introduce new variables $y_1, y_2, \ldots, y_n$. We encode our polynomial system $F = (f_1, \ldots, f_n)$ into the single polynomial $\psi(\mathbf{x}, \mathbf{y}) = \sum_{i=1}^{n} \sum_{\mathbf{a} \in \mathcal{A}_i} c_{i,\mathbf{a}} y_i \mathbf{x}^{\mathbf{a}}$. The toric variety $X = X_{\mathcal{A}}$ is parametrized by all monomials $y_i \mathbf{x}^{\mathbf{a}}$, where $i = 1, \ldots, n$ and $\mathbf{a} \in \mathcal{A}_i$.

Consider the hypersurface $\Omega := \{(\mathbf{x}, \mathbf{y}) \in (\mathbb{C}^*)^{2n} \mid \psi(\mathbf{x}, \mathbf{y}) = 0\}$. The discriminant Σ is a hypersurface in the space of coefficient vectors $\mathbf{c} = (c_{i,\mathbf{a}})$. It comprises all $\mathbf{c}$ such that Ω has a singular point. Therefore, the toric variety X is dual to the discriminant Σ. Points $\mathbf{u}$ in the common ambient space of X and Σ are identified with polynomial systems (9.13) that have concrete numerical coefficients. The distance from such a system to the discriminant Σ can be computed as described in (9.16). We summarize our discussion as follows:

Theorem 9.26 *The Euclidean distance from a polynomial system $\mathbf{u}$ to the discriminant Σ equals the smallest norm $||\mathbf{x}||$ among all points $\mathbf{x}$ in the toric variety X that are critical for the distance to $\mathbf{u}$.*

We emphasize again that the distance crucially depends on the norm that we choose. This was Bombieri–Weyl in Theorem 9.19. The points on X that are critical for the distance to $\mathbf{u}$ are best computed with the monomial parametrization. For $n = 1$, we saw this in (9.17).

Example 9.27 (Hyperdeterminant) Fix $n = 2$ and $\mathcal{A}_1 = \mathcal{A}_2 = \{(0, 0), (0, 1), (1, 0), (1, 1)\}$. The Cayley configuration is the 3-cube, and X is a 4-dimensional affine toric variety in $\mathbb{C}^8$, namely the cone over the Segre variety $\mathbb{P}^1 \times \mathbb{P}^1 \times \mathbb{P}^1 \subset \mathbb{P}^7$. The points $\mathbf{c}$ in $\mathbb{C}^8$ correspond to pairs of bilinear equations in two unknowns:

$$c_{11} + c_{12}t_1 + c_{13}t_2 + c_{14}t_1t_2 \;\; = \;\; c_{21} + c_{22}t_1 + c_{23}t_2 + c_{24}t_1t_2 \;\; = \;\; 0. \qquad (9.18)$$

For generic $\mathbf{c}$, this polynomial system has 2 solutions, and the two solutions coincide when the discriminant Δ vanishes. The discriminant for $(\mathcal{A}_1, \mathcal{A}_2)$ is the $2 \times 2 \times 2$ *hyperdeterminant* computed in Example 3.16:

$$\Delta = c_{11}^2 c_{24}^2 + c_{12}^2 c_{23}^2 + c_{13}^2 c_{22}^2 + c_{14}^2 c_{21}^2 + 4c_{11}c_{14}c_{22}c_{23} + 4c_{12}c_{13}c_{21}c_{24} - 2c_{11}c_{12}c_{23}c_{24}$$
$$-2c_{11}c_{13}c_{22}c_{24} - 2c_{11}c_{14}c_{21}c_{24} - 2c_{12}c_{13}c_{22}c_{23} - 2c_{12}c_{14}c_{21}c_{23} - 2c_{13}c_{14}c_{21}c_{22}.$$

Given any polynomial system $\mathbf{u}$ of the form (9.18), we seek its squared distance to the hyperdeterminantal hypersurface $\Sigma = \{\Delta = 0\}$ in $\mathbb{R}^8$. This number is the optimal value of:

minimize $\;\; \sum_{i=1}^2 y_i^2(1 + t_1^2 + t_2^2 + t_1^2 t_2^2) \;$ over all critical points of the function

$$(y_1, y_2, t_1, t_2) \;\mapsto\; \sum_{i=1}^2 \big((u_{i1} - y_i)^2 + (u_{i2} - y_i t_1)^2 + (u_{i3} - y_i t_2)^2 + (u_{i4} - y_i t_1 t_2)^2\big).$$

This objective function has 6 critical points over $\mathbb{C}$. The ED degree is 6, by [60, Example 8.2]. The ED degree jumps to 34 if the Euclidean norm is replaced by any nearby generic quadratic form. Indeed, the generic ED degree of the hyperdeterminant Δ equals 34. This can be seen by summing the polar degrees in the column labeled $k = 3$ in Table 5.2. ◇

Chapter 10
Machine Learning

One of the principal goals of machine learning is to learn in an automated way functions that represent the relationship between data points. Suppose we are given the data set $\mathcal{D} = \{(\mathbf{x}_1, \mathbf{y}_1), \ldots, (\mathbf{x}_d, \mathbf{y}_d)\} \subset \mathbb{R}^n \times \mathbb{R}^m$. The vectors $\mathbf{x}_i$ are the input data. The vectors $\mathbf{y}_i$ are the output data. For instance, in image classification, the $\mathbf{x}_i$ might encode images of certain objects and the $\mathbf{y}_i$ are the respective classifiers. Another popular example from generative AI is the scenario where $\mathbf{x}_i$ and $\mathbf{y}_i$ encode word tokens that co-occur in text data.

The goal is to find a function $f : \mathbb{R}^n \to \mathbb{R}^m$ such that $f(\mathbf{x}_i) \approx \mathbf{y}_i$. The meaning of $\approx$ rests on a *loss function* $l(\mathbf{y}', \mathbf{y})$. We seek to minimize the mean loss $\ell_{\mathcal{D}}(f) := \frac{1}{d} \sum_{i=1}^{d} l(f(\mathbf{x}_i), \mathbf{y}_i)$ over a class of functions f. Often, the loss function comes from a metric. For instance, the *squared-error loss* is the squared Euclidean distance $l(\mathbf{y}', \mathbf{y}) = \|\mathbf{y}' - \mathbf{y}\|^2$ in output space $\mathbb{R}^m$, but also the Wasserstein distance from Chapter 5 and Kullback–Leibler divergence from Chapter 11 are used. One usually restricts to a model, consisting of functions that are specified by N real *parameters* $\theta = (\theta_1, \ldots, \theta_N) \in \mathbb{R}^N$. To learn a function means to compute θ that minimizes the loss for the data $\mathcal{D}$.

Problems of the type described above fall under the header *supervised learning*. Here, every input data point $\mathbf{x}_i$ has an associated output $\mathbf{y}_i$, also called *label*, and the goal is to find a relationship between them. By contrast, in *unsupervised learning*, we are given data points $\mathcal{D} = \{\mathbf{x}_1, \ldots, \mathbf{x}_d\} \subset \mathbb{R}^n$ without labels. The goal in unsupervised learning is to describe the data with as few parameters as possible. For instance, assuming the *variety hypothesis*, our goal could be to learn a variety $X \subset \mathbb{R}^n$ that represents the data $\mathcal{D}$. Here, the parameters define the polynomials that cut out X. We discuss this approach in Section 10.3.

The next sections offer a glimpse of machine learning from a geometry perspective. For a systematic introduction, we refer to the book *Mathematical Aspects of Deep Learning* [75].

10.1 Neural Networks

The model we focus on is the *(feedforward) neural network*. Such a network is a composition

$$f := f_{L,\theta} \circ \cdots \circ f_{2,\theta} \circ f_{1,\theta}.$$

© The Author(s) 2024
P. Breiding et al., *Metric Algebraic Geometry*, Oberwolfach Seminars 53,
https://doi.org/10.1007/978-3-031-51462-3_10

Here, $f_{i,\theta}$ is a function that depends on a parameter θ. The function f is called the *end-to-end function*. We call L the *number of layers*. The dimension of the domain of $f_{i,\theta}$ is the *width* of the i-th layer. The number of layers, their widths, and the type of each layer function $f_{i,\theta}$ constitute the *network's architecture*. The variables of the i-th layer $f_{i,\theta}$ are *neurons* or *nodes*. A neural network is often visualized by a graph as in Figure 10.1.

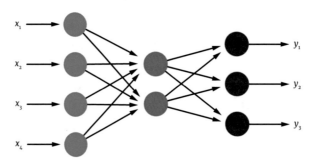

Fig. 10.1: A fully connected neural network with two layers $f_{1,\theta} : \mathbb{R}^4 \to \mathbb{R}^2$ and $f_{2,\theta} : \mathbb{R}^2 \to \mathbb{R}^3$.

We say that the network is *fully connected* if, for every index $i = 1, \dots, L$, every component of the i-th layer $f_{i,\theta}$ depends on every variable/neuron/node. Figure 10.1 shows a fully connected network. On the other hand, Figure 10.2 shows a network that is not fully connected. Some of the edges are missing.

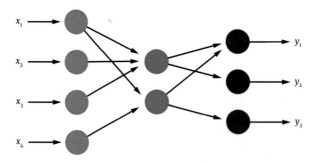

Fig. 10.2: A neural network with two layers that is not fully connected.

Most commonly, the layer functions are compositions of an affine linear map $\alpha_{i,\theta}$ with a (typically nonlinear) map $\sigma_i : \mathbb{R} \to \mathbb{R}$ that is applied componentwise:

$$f_{i,\theta} = (\sigma_i, \dots, \sigma_i) \circ \alpha_{i,\theta}. \tag{10.1}$$

The map σ_i is called the *activation function*. One example of an activation function is the *Rectified Linear Unit (ReLU)*, which is $\sigma(z) = \max\{0, z\}$. Methods from (metric or tropical) algebraic geometry can be used to study neural networks when the activation function has an algebraic structure (identity, polynomial, ReLU, etc.).

Here, we consider the simplest case where the activation function is the identity. The theoretical study of neural networks is much more challenging if that function is nonlinear. For instance, the end-to-end function of a ReLU neural network (i.e., all activation functions are ReLU) is piecewise linear. In fact, every piecewise linear function with finitely many pieces arises from a fully connected ReLU network [8]. ReLU end-to-end functions can be interpreted as tropical rational functions [180]. That perspective was further developed in [134] to provide sharp bounds on the number of linear regions of the end-to-end functions. Another geometric perspective on ReLU networks is described in [74]. A first algebro-geometric study for networks with polynomial activation functions appears in [105].

The term *expressivity* refers to which functions neural networks with fixed architecture can express. Thus, expressivity is the study of the space M of all functions that are given by our network. This function space M is often called the *neuromanifold* of the neural network architecture. Note that M is typically not a smooth manifold. The neuromanifold is the image of the network's parametrization map

$$\mu : \mathbb{R}^N \to M, \quad \theta \mapsto f_{L,\theta} \circ \cdots \circ f_{2,\theta} \circ f_{1,\theta}. \tag{10.2}$$

The loss function ℓ_D is defined on M. We can pull it back to the parameter space via μ, i.e. we consider the composition

$$\mathcal{L} : \mathbb{R}^N \xrightarrow{\mu} M \xrightarrow{\ell_D} \mathbb{R}. \tag{10.3}$$

Instead of minimizing the loss function ℓ_D over the neuromanifold M, training a neural network means minimizing $\mathcal{L}$ over the parameter space $\mathbb{R}^N$ of the neuromanifold.

Remark 10.1 The numerical uncertainty in minimizing the loss depends on the condition number of the function $\mathcal{L}$. Following Chapter 9, it would be interesting to compare the condition numbers of $\mathcal{L}$ and ℓ_D and how they are related to the network's architecture. There is lots of room for metric algebraic geometry to contribute.

One of the big mysteries in machine learning theory is why training neural networks results in "nice" minima. The concrete meaning of the adjective nice varies in the literature. See [130] for an algebro-geometric perspective on "flat" minima. Training a neural network minimizes the loss function $\mathcal{L} = \ell_D \circ \mu$ in (10.3) on the parameter space $\mathbb{R}^N$. The meaningful critical points are those that come from critical points of ℓ_D on the neuromanifold M. Formally, such *pure critical points* θ of $\mathcal{L}$ satisfy that $\mu(\theta)$ is a critical point of the functional ℓ_D restricted to the smooth locus of M. The network parametrization map μ can induce additional *spurious critical points* of the function $\mathcal{L}$.

The following two questions play a central role in the study of learning problems:

1. How does the network architecture affect the geometry of the neuromanifold M?
2. How does the geometry of the neuromanifold impact the training of the network?

We shall discuss these questions for two specific network architectures. In the remainder of this section, we start with the simplest class, namely networks that are linear and fully connected. Thereafter, in Section 10.2, we turn to *linear convolutional networks*. This will lead us to interesting algebraic varieties.

In a *linear fully connected neural network*, the activation function σ_i in layer i is the identity and the layer functions $f_{i,\theta}$ are arbitrary linear maps, for $i = 1, \ldots, L$. The parameters θ are the entries of the matrices representing the linear layer functions. The network parametrization map (10.2) specializes to

$$\mu : \mathbb{R}^{k_1 \times k_0} \times \mathbb{R}^{k_2 \times k_1} \times \cdots \times \mathbb{R}^{k_L \times k_{L-1}} \rightarrow \mathbb{R}^{k_L \times k_0},$$
$$(W_1, W_2, \ldots, W_L) \mapsto W_L \cdots W_2 W_1. \tag{10.4}$$

Let us first discuss the expressivity of this model. The image of the parametrization map μ consists of all matrices with rank at most $\min(k_0, k_1, \ldots, k_L)$. Hence, the neuromanifold

$$\mathcal{M} = \{W \in \mathbb{R}^{k_L \times k_0} \mid \mathrm{rank}(W) \leq \min(k_0, k_1, \ldots, k_L)\}$$

is a determinantal variety. If the input or output dimension is one of the minimal widths, then $\mathcal{M}$ is equal to the whole ambient space. Equivalently, in symbols, if $\min(k_0, k_1, \ldots, k_L) = \min(k_0, k_L)$, then $\mathcal{M} = \mathbb{R}^{k_L \times k_0}$. Otherwise, $\mathcal{M}$ is a lower-dimensional Zariski closed subset. The singular locus of $\mathcal{M}$ is the variety of matrices of rank at most $\min(k_0, k_1, \ldots, k_L) - 1$, and therefore, it is the neuromanifold for a smaller network architecture.

Next, we turn to questions about optimization. Theoretical studies on the optimization problem of training neural networks can be roughly grouped into *static* and *dynamic* studies. Static investigations concern the loss landscape [124] and the critical points of (10.3), while dynamic studies focus on a training algorithm, e.g., by investigating its convergence.

We first discuss static properties. For linear fully connected networks, the pure and spurious critical points were characterized in [33]. The critical points θ of $\mathcal{L}$ such that $\mu(\theta)$ has maximal rank are pure [33, Proposition 6]. All spurious critical points θ correspond to lower-rank matrices $\mu(\theta)$. These are essentially always saddles [33, Proposition 9]. If $\ell_{\mathcal{D}}$ is smooth and convex, then all non-global local minima of $\mathcal{L}$ (often called "bad" minima) are pure critical points. It is a common, but false, belief that linear fully connected networks do not have bad minima. Our next result follows from [33, Proposition 10]:

Theorem 10.2 *Consider a linear fully connected network and a smooth and convex function $\ell_{\mathcal{D}}$. The function $\mathcal{L} = \ell_{\mathcal{D}} \circ \mu$ has non-global local minima if and only if $\ell_{\mathcal{D}}|_{\mathrm{Reg}(\mathcal{M})}$ has non-global local minima.*

This result immediately implies that linear fully connected networks *in two special settings* do not have bad minima. First, if $\mathcal{M}$ equals the ambient space (i.e., the input or output dimension of the network is its minimal width), then any convex function $\ell_{\mathcal{D}}$ has exactly one minimum on $\mathcal{M}$, and so $\mathcal{L}$ does not have any bad minima. This is the main result from [119]. The second setting concerns the squared-error loss

$$\ell_{\mathcal{D}} : \mathbb{R}^{k_L \times k_0} \rightarrow \mathbb{R}, \quad W \mapsto \frac{1}{d} \sum_{i=1}^{d} \|W\mathbf{x}_i - \mathbf{y}_i\|^2, \tag{10.5}$$

for the training data $\mathcal{D} = \{(\mathbf{x}_1, \mathbf{y}_1), \ldots, (\mathbf{x}_d, \mathbf{y}_d)\} \subset \mathbb{R}^{k_0} \times \mathbb{R}^{k_L}$. Writing $X \in \mathbb{R}^{k_0 \times d}$ and $Y \in \mathbb{R}^{k_L \times d}$ for the data matrices whose columns are x_i and y_j, the squared-error loss becomes the squared Frobenius norm

$$\ell_{\mathcal{D}}(W) = \frac{1}{d}\|WX - Y\|^2. \tag{10.6}$$

If $XX^\top$ has full rank, then we record the data in the matrix $U = YX^\top((XX^\top)^{\frac{1}{2}})^{-1}$. Minimizing (10.6) is equivalent to minimizing the squared Euclidean distance $\|W - U\|^2$ over all $W \in \mathrm{Reg}(\mathcal{M})$; see [33, Section 3.3]. According to the Eckart–Young Theorem, this optimization problem has a unique local and global minimum, provided the singular values of U are distinct and positive. Hence, given many generic data pairs $(\mathbf{x}_i, \mathbf{y}_i)$, the squared-error loss $\ell_{\mathcal{D}}$ has no non-global minima on $\mathrm{Reg}(\mathcal{M})$. Theorem 10.2 now implies that the squared-error loss $\mathcal{L}$ on the parameter space has no bad minima. This is a prominent result in machine learning, often attributed to [11] or [103]. However, the two settings where either $\mathcal{M}$ is a vector space or we consider the squared-error loss are rather special. Non-global local minima are expected for other loss functions and architectures where $\mathcal{M}$ is a proper determinantal variety; see [33, Example 13].

Now, we turn to dynamical properties. Optimization algorithms used in the training of neural networks are variations of gradient descent. After picking initial parameters θ, gradient descent adapts the parameters successively with the goal of minimizing the loss $\mathcal{L}(\theta)$. When training linear fully connected networks with the squared-error loss, where the data matrix $XX^\top$ has full rank, gradient descent converges for almost all initializations (under reasonable assumptions on its step sizes) to a critical point θ of the loss $\mathcal{L}$ [136, Theorem 2.4]. Moreover, the matrix $\mu(\theta) \in \mathcal{M}$ is a global minimum of $\ell_{\mathcal{D}}$ restricted to the smooth manifold of all matrices of the same format and rank [136, Theorem 2.6]. The authors of [136] conjecture that the matrix $\mu(\theta)$ has maximal possible rank, so it is a smooth point on the determinantal variety $\mathcal{M}$.

An essential ingredient in the convergence analysis of [136] are the *algebraic invariants* of gradient flow. The curve in parameter space traced by gradient flow is typically transcendental, but it does satisfy some algebraic relations. In other words, its Zariski closure is not the whole ambient parameter space.

Proposition 10.3 ([9]) *Consider a linear fully connected network with parametrization* (10.4). *Let* $\theta(t) = (W_1(t), W_2(t), \ldots, W_L(t))$ *be the curve traced out by gradient flow, starting at* $t = 0$. *Then, the* $L - 1$ *matrices*

$$W_i^\top(t) W_i(t) - W_{i-1}(t) W_{i-1}^\top(t), \quad where \ i = 2, 3, \ldots, L,$$

remain constant for all times $t \geq 0$.

We discuss some practical consequences of this result. One calls a parameter tuple $\theta = (W_1, \ldots, W_L)$ *balanced* if $W_i^\top W_i = W_{i-1}W_{i-1}^\top$ for $i \in \{2, \ldots, L\}$. Note that all matrices in a balanced tuple have the same Frobenius norm: $\|W_1\| = \|W_2\| = \cdots = \|W_L\|$. If a linear network is initialized at a balanced tuple $\theta(0)$, then every tuple $\theta(t)$ along the gradient flow curve is balanced, by Proposition 10.3. Since all matrices in $\theta(t)$ have the same Frobenius norm, it cannot happen that one matrix converges to zero while another matrix has entries that diverge to infinity. In fact, if one matrix converges to zero, then so does the whole tuple $\theta(t)$. Algebraic invariants of gradient flow were also studied in [108, Proposition 5.13] for linear convolutional networks (cf. Section 10.2). For ReLU networks and other networks, see [177, Lemma 3] and [62, Theorems 2.1–2.2].

Remark 10.4 (Nonlinear Autoencoders) Invariants also play a crucial role in the study of attractors of autoencoders. An *autoencoder* is a composition of two neural networks: an encoder and a decoder network, such that the input dimension of the encoder equals the output dimension of the decoder. Given training data $\mathbf{x}_1, \ldots, \mathbf{x}_d$, the composed network is typically trained by minimizing the *autoencoding loss*

$$ f \;\longmapsto\; \sum_{i=1}^{d} \| f(\mathbf{x}_i) - \mathbf{x}_i \|^2. $$

Note that $\mathbf{x}_i$ serves both as input and as output data. It is shown in [150] that an autoencoder trained with gradient descent on a single training example $\mathbf{x}$ (i.e., $d = 1$) memorizes $\mathbf{x}$ as an attractor (under suitable assumptions on the activation function and initialization). Namely, $\mathbf{x}$ is a fixed point of the learned function f such that the sequence $(f^i(\mathbf{y}))_{i \in \mathbb{N}}$ converges to $\mathbf{x}$ for any $\mathbf{y}$ in an open neighborhood of $\mathbf{x}$.

10.2 Convolutional Networks

A linear neural network parametrizes matrices that admit the structured product decomposition (10.4). When each layer is a one-dimensional convolution, the matrices W_i are generalized Toeplitz matrices. An example of such a matrix is

$$ \begin{bmatrix} w_0 & w_1 & w_2 & 0 & 0 & 0 & 0 \\ 0 & 0 & w_0 & w_1 & w_2 & 0 & 0 \\ 0 & 0 & 0 & 0 & w_0 & w_1 & w_2 \end{bmatrix}. \tag{10.7} $$

A *convolution* on one-dimensional signals is a linear map that depends on a *filter* $\mathbf{w} = (w_0, \ldots, w_{k-1}) \in \mathbb{R}^k$ and a *stride* $s \in \mathbb{N}$. It computes the inner product of the filter $\mathbf{w}$ with parts of a given input vector $\mathbf{x}$, and traverses the whole vector $\mathbf{x}$ by moving $\mathbf{w}$ through it with stride s. The formula for this linear map is

$$ \mathbb{R}^{s(m-1)+k} \;\rightarrow\; \mathbb{R}^m, \quad \mathbf{x} \longmapsto \left(\sum_{j=0}^{k-1} w_j \cdot \mathbf{x}_{is+j} \right)_{i=0}^{m-1}. \tag{10.8} $$

For instance, for $m = 3$, the convolution in (10.7) has filter size $k = 3$ and stride $s = 2$.

A *linear convolutional neural network* is the composition of L convolutions with filter sizes $\mathbf{k} = (k_1, k_2, \ldots, k_L)$ and strides $\mathbf{s} = (s_1, s_2, \ldots, s_L)$. All activation functions in this network are the identity. The resulting end-to-end function is also a convolution with filter size $k = k_1 + \sum_{l=2}^{L}(k_l - 1)\prod_{i=1}^{l-1} s_i$ and stride $s = s_1 s_2 \cdots s_L$. This was shown in [108, Proposition 2.2]; see also [109, Section 2]. The point is this: any product of generalized Toeplitz matrices like (10.7) is again a generalized Toeplitz matrix. The parametrization (10.4) sends the filters of each layer to the filter of the end-to-end convolution:

$$\mu : \mathbb{R}^{k_1} \times \mathbb{R}^{k_2} \times \cdots \times \mathbb{R}^{k_L} \to \mathbb{R}^k. \qquad (10.9)$$

The map μ can be evaluated via polynomial multiplication. This is done as follows. For any positive integers k and t, we write $\mathbb{R}[x^t]_{\leq k-1}$ for the space of univariate polynomials of degree at most $k - 1$. The variable in these polynomials is the power x^t. We identify any filter of size k with the coefficient vector of such a polynomial via

$$\pi_{t,k} : \mathbb{R}^k \to \mathbb{R}[x^t]_{\leq k-1}, \quad \mathbf{w} \mapsto w_0 x^{t(k-1)} + w_1 x^{t(k-2)} + \cdots + w_{k-2} x^t + w_{k-1}.$$

The filter $\mathbf{w}$ of the end-to-end convolution corresponds to the polynomial $\pi_{1,k}(\mathbf{w})$ that admits a sparse factorization into polynomials corresponding to the filters $\mathbf{w}_1, \mathbf{w}_2, \ldots, \mathbf{w}_L$:

$$\pi_{1,k}(\mu(\mathbf{w}_1, \ldots, \mathbf{w}_L)) = \pi_{t_L, k_L}(\mathbf{w}_L) \cdots \pi_{t_2, k_2}(\mathbf{w}_2) \cdot \pi_{t_1, k_1}(\mathbf{w}_1), \quad t_l := s_1 s_2 \cdots s_{l-1}.$$

The powers $\mathbf{t} = (t_1, \ldots, t_L)$ of the factors in the above formula uniquely encode the strides $\mathbf{s} = (s_1, \ldots, s_{L-1})$ of the convolutions. The last stride s_L does not influence the filter $\mu(\mathbf{w}_1, \ldots, \mathbf{w}_L)$ of the end-to-end convolution. We summarize our discussion in the following result, which is found in [109, Proposition 2.2].

Proposition 10.5 *The neuromanifold $\mathcal{M}_{\mathbf{k}, \mathbf{t}}$ for a linear convolutional neural network with filter sizes $\mathbf{k} = (k_1, k_2, \ldots, k_L)$ and powers $\mathbf{t} = (t_1, t_2, \ldots, t_L)$ is parametrized by polynomial multiplication, namely*

$$\mu : \mathbb{R}[x^{t_1}]_{\leq k_1-1} \times \mathbb{R}[x^{t_2}]_{\leq k_2-1} \times \cdots \times \mathbb{R}[x^{t_L}]_{\leq k_L-1} \to \mathbb{R}[x]_{\leq k-1}, \qquad (10.10)$$
$$(p_1, p_2, \ldots, p_L) \mapsto p_1 p_2 \cdots p_L.$$

Example 10.6 ($L = 2, \mathbf{t} = (1, 2)$) For $\mathbf{k} = (3, 2)$ and $\mathbf{k} = (2, 3)$, the matrix products are

$$\begin{bmatrix} v_0 & v_1 \end{bmatrix} \begin{bmatrix} u_0 & u_1 & u_2 & 0 & 0 \\ 0 & 0 & u_0 & u_1 & u_2 \end{bmatrix} = \begin{bmatrix} u_0 v_0 & u_1 v_0 & u_0 v_1 + u_2 v_0 & u_1 v_1 & u_2 v_1 \end{bmatrix},$$

$$\begin{bmatrix} v_0 & v_1 & v_2 \end{bmatrix} \begin{bmatrix} u_0 & u_1 & 0 & 0 & 0 & 0 \\ 0 & 0 & u_0 & u_1 & 0 & 0 \\ 0 & 0 & 0 & 0 & u_0 & u_1 \end{bmatrix} = \begin{bmatrix} u_0 v_0 & u_1 v_0 & u_0 v_1 & u_1 v_1 & u_0 v_2 & u_1 v_2 \end{bmatrix}.$$

These correspond to multiplying pairs of polynomials, as specified by Proposition 10.5:

$$(u_0 x^2 + u_1 x + u_2)(v_0 x^2 + v_1) \quad \text{and} \quad (u_0 x + u_1)(v_0 x^4 + v_1 x^2 + v_2). \qquad (10.11)$$

The neuromanifold $\mathcal{M}_{\mathbf{k}, \mathbf{t}}$ consists of polynomials that admit such a factorization. ◇

For linear convolutional networks, the powers $\mathbf{t} = (t_1, t_2, \ldots, t_L)$ satisfy that $t_1 = 1$ and t_{l-1} divides t_l for each $l = 2, \ldots, L$. From the algebraic perspective, it is interesting to study the polynomial multiplication map (10.10) for arbitrary positive integers t_l. Its image $\mathcal{M}_{\mathbf{k},\mathbf{t}}$ lives in the vector space $\mathbb{R}[x]_{\leq k-1} \simeq \mathbb{R}^k$, where $k - 1 = \sum_{l=1}^{L} t_l(k_l - 1)$. By Tarski's Theorem on Quantifier Elimination, $\mathcal{M}_{\mathbf{k},\mathbf{t}}$ is a semialgebraic set, i.e., it can be described by a Boolean combination of polynomial inequalities.

As is customary in applied algebraic geometry, we simplify our problem by replacing $\mathcal{M}_{\mathbf{k},\mathbf{t}}$ with its Zariski closure $\mathcal{V}_{\mathbf{k},\mathbf{t}}$ in the complex projective space $\mathbb{P}^{k-1}$. Thus, by definition, the variety $\mathcal{V}_{\mathbf{k},\mathbf{t}}$ is the image of the map

$$\mathbb{P}^{k_1-1} \times \mathbb{P}^{k_2-1} \times \cdots \times \mathbb{P}^{k_L-1} \to \mathbb{P}^{k-1}, \quad (p_1, p_2, \ldots, p_L) \mapsto p_1 p_2 \cdots p_L. \quad (10.12)$$

Proposition 10.7 *The map in (10.12) has no base points, so it is a morphism. Moreover, the map (10.12) is finite-to-one. If the integers $t_1, t_2, \ldots, t_L$ are pairwise distinct, then the map is generically one-to-one.*

Proof A product of polynomials can only be zero if one of its factors is zero. Hence, the projective multiplication map (10.12) is well-defined at all points in its domain. The map is finite-to-one because the irreducible factorization of a polynomial is unique, up to reordering the factors. If all factors have distinct powers, then the factors of a generic polynomial cannot be swapped, and the map is generically one-to-one.   □

Corollary 10.8 *The variety $\mathcal{V}_{\mathbf{k},\mathbf{t}}$ has dimension $k_1 + k_2 + \cdots + k_L - L$, and its degree is*

$$\frac{(k_1+k_2+\cdots+k_L - L)!}{(k_1 - 1)! \, (k_2 - 1)! \, \cdots \, (k_L - 1)!} \quad \text{divided by the size of the general fiber of (10.12).}$$

Proof We consider the variety $\mathbb{P}^{k_1-1} \times \mathbb{P}^{k_2-1} \times \cdots \times \mathbb{P}^{k_L-1}$ in its Segre embedding (see Definition 12.4). The map (10.12) is the composition of this Segre embedding followed by a linear projection. That projection has no base points by Proposition 10.7 and thus preserves dimension. Moreover, the linear projection is finite-to-one and of the same degree δ as (10.12). Hence, the degree of its image $\mathcal{V}_{\mathbf{k},\mathbf{t}}$ is the degree of the Segre variety $\mathbb{P}^{k_1-1} \times \cdots \times \mathbb{P}^{k_L-1}$ divided by δ. The degree of the Segre variety is computed in Corollary 12.21 below. It is the multinomial coefficient in the stated formula.   □

Example 10.9 If $\mathbf{k} = (2, 2, \ldots, 2)$ and $\mathbf{t} = (1, 1, \ldots, 1)$, we simply multiply linear polynomials. Here $k - 1 = L$. The map (10.12) is onto, by the Fundamental Theorem of Algebra, and hence $\mathcal{V}_{\mathbf{k},\mathbf{t}} = \mathbb{P}^L$. The fiber has cardinality $L!$ since the linear factors can be reordered arbitrarily. The multinomial coefficient in Corollary 10.8 is also $L!$, so the degree of the image is 1. The neuromanifold $\mathcal{M}_{\mathbf{k},\mathbf{t}}$ is Zariski dense in $\mathbb{R}[x]_{\leq L}$. It consists of all real-rooted polynomials of degree $\leq L$, so it is an interesting semialgebraic set.   ◇

The size of the general fiber in Corollary 10.8 can be read off from the filter sizes $\mathbf{k}$ and the powers $\mathbf{t}$, namely, it is the order of the symmetry group of the factorization (10.10). Such a symmetry arises whenever $t_i = t_j$ for $i \neq j$. In this case, we can merge the factors $\mathbb{R}[x^{t_i}]_{\leq k_i-1}$ and $\mathbb{R}[x^{t_j}]_{\leq k_j-1}$ into a single factor $\mathbb{R}[x^{t_i}]_{\leq k_i+k_j-2}$ without changing the image variety $\mathcal{V}_{\mathbf{k},\mathbf{t}}$. Therefore, for computing the degree or other properties of $\mathcal{V}_{\mathbf{k},\mathbf{t}}$, we can always assume that $\mathbf{t}$ has distinct coordinates. For example, suppose that $L = 6$, $\mathbf{k} = (5, 4, 3, 7, 6, 3)$ and $\mathbf{t} = (1, 5, 5, 5, 3, 3)$. Then $\mathcal{V}_{\mathbf{k},\mathbf{t}} \subset \mathbb{P}^{80}$ has dimension $22 = \sum_{i=1}^{6}(k_i - 1)$. After merging as described above, our new parameter vectors are $\mathbf{k} = (5, 12, 8)$ and $\mathbf{t} = (1, 5, 3)$. Using Proposition 10.7, we now see that the degree of $\mathcal{V}_{\mathbf{k},\mathbf{t}}$ equals $\frac{22!}{4!\,11!\,7!} = 232792560$.

Equipped with Corollary 10.8, we explore the prime ideal of $\mathcal{V}_{\mathbf{k},\mathbf{t}}$ in some small cases. This ideal lives in the polynomial ring $\mathbb{Q}[c_0, c_1, \ldots, c_{k-1}]$ where $p(x) = \sum_{i=0}^{k-1} c_i x^i$ is in the codomain of (10.10). We start with Example 10.6. The two varieties are cubic threefolds, in $\mathbb{P}^4$ resp. $\mathbb{P}^5$. Their prime ideals are

$$\langle\, c_0 c_3^2 + c_1^2 c_4 - c_1 c_2 c_3 \,\rangle \quad \text{and} \quad \langle\, c_0 c_3 - c_1 c_2,\ c_0 c_5 - c_1 c_4,\ c_2 c_5 - c_3 c_4 \,\rangle. \qquad (10.13)$$

The second ideal shows that $\mathcal{V}_{(2,3),(1,2)} = \mathbb{P}^1 \times \mathbb{P}^2 \subset \mathbb{P}^5$. The first ideal defines the variety $\mathcal{V}_{(3,2),(1,2)}$, which is a projection of that Segre threefold from $\mathbb{P}^5$ into $\mathbb{P}^4$.

Example 10.10 ($L=3, \mathbf{t} = (1, 2, 2)$) We extend the polynomials from (10.11) by multiplying them with a sparse quadratic factor:

$$p(x) = (u_2 + u_1 x + u_0 x^2)(v_1 + v_0 x^2)(w_1 + w_0 x^2) \quad \text{for} \quad \mathbf{k} = (3, 2, 2),$$
$$\text{and} \quad p(x) = (u_1 + u_0 x)(v_2 + v_1 x^2 + v_0 x^4)(w_1 + w_0 x^2) \quad \text{for} \quad \mathbf{k} = (2, 3, 2).$$

The resulting varieties $\mathcal{V}_{\mathbf{k},\mathbf{t}}$ live in $\mathbb{P}^6$ and $\mathbb{P}^7$ respectively. We can understand them by examining the factors of $p(x) + p(-x)$ and $p(x) - p(-x)$. Both varieties are 4-dimensional and they admit nice determinantal representations. The ideal of $\mathcal{V}_{(3,2,2),(1,2,2)}$ is generated by the four 3×3 minors of the 3×4 matrix

$$\begin{bmatrix} c_0 & c_2 & c_4 & c_6 \\ c_1 & c_3 & c_5 & 0 \\ 0 & c_1 & c_3 & c_5 \end{bmatrix}.$$

While the parametrization for $\mathbf{k} = (2, 3, 2)$ is not a monomial map, the variety $\mathcal{V}_{(2,3,2),(1,2,2)}$ is still toric. Its prime ideal is generated by the six 2×2 minors of the 2×4 matrix

$$\begin{bmatrix} c_0 & c_2 & c_4 & c_6 \\ c_1 & c_3 & c_5 & c_7 \end{bmatrix}.$$

We conclude that $\mathcal{V}_{(2,3,2),(1,2,2)}$ is the Segre embedding of $\mathbb{P}^1 \times \mathbb{P}^3$ into $\mathbb{P}^7$. $\diamond$

These examples suggest that our varieties $\mathcal{V}_{\mathbf{k},\mathbf{t}}$ are interesting objects for further study in combinatorial commutative algebra. From the perspective of metric algebraic geometry, one should pursue the concepts introduced in the previous chapters. One of these is Euclidean distance optimization. Computations show that the two varieties in Example 10.6 have ED degrees 10 and 2, while the varieties in Example 10.10 have ED degrees 23 and 2. Can we find a general formula for the ED degree of $\mathcal{V}_{\mathbf{k},\mathbf{t}}$ in terms of $\mathbf{k}$ and $\mathbf{t}$?

We turn our attention to describing the singular locus of the variety $\mathcal{V}_{\mathbf{k},\mathbf{t}}$ that represents convolutional networks. As argued above, we can restrict ourselves to the case when the powers $\mathbf{t} = (t_1, \ldots, t_L)$ are pairwise distinct integers. The singular locus was determined for those varieties $\mathcal{V}_{\mathbf{k},\mathbf{t}}$ that come from linear convolutional networks (i.e., under the assumption that $t_{l-1} | t_l$) in [109, Theorem 2.8]:

Theorem 10.11 *Let* $1 = t_1 < t_2 < \ldots < t_L$ *be integers such that* $t_{l-1}|t_l$ *for all* $l = 2, \ldots, L$. *The singular locus of* $\mathcal{V}_{\mathbf{k},\mathbf{t}}$ *consists of all its proper subvarieties* $\mathcal{V}_{\mathbf{k}',\mathbf{t}}$.

This result says that the neuromanifold is singular along smaller neuromanifolds of convolutional networks with the same strides. Hence, as in the case of linear fully connected networks, discussed in Section 10.1, the singular locus of the neuromanifold is parametrized by smaller network architectures.

In the remainder of this section, we discuss critical points of training linear convolutional networks. For that, we have to return to the semialgebraic neuromanifold $\mathcal{M}_{\mathbf{k},\mathbf{t}}$. This set is closed in the Euclidean topology. Describing its Euclidean relative boundary is a challenging problem, studied in [109, Section 6].

When all strides are one (i.e., $\mathbf{t} = (1, \ldots, 1)$), the neuromanifold $\mathcal{M}_{\mathbf{k},\mathbf{t}}$ is a full-dimensional semialgebraic subset of the ambient space $\mathbb{R}^k$. Here, critical points of the loss function often correspond to points on the Euclidean boundary of $\mathcal{M}_{\mathbf{k},\mathbf{t}}$. These are critical points of the network parametrization map μ. This stands in sharp contrast to convolutional networks where all strides are larger than one (i.e, $t_1 < t_2 < \ldots$) and $\mathcal{M}_{\mathbf{k},\mathbf{t}}$ is a lower-dimensional subset of $\mathbb{R}^k$. The following results from [109, Theorem 2.11].

Theorem 10.12 *Consider the map* μ *in* (10.10) *with integers* $1 = t_1 < t_2 < \ldots < t_L$ *such that* $t_{l-1}|t_l$ *for all* $l = 2, \ldots, L$. *Let* $d \geq k$, *and let* $\ell_{\mathcal{D}}$ *be the squared-error loss* (10.5) *for the end-to-end convolution* W. *For almost all* d-*tuples* $\mathcal{D}$ *of training data, every critical point* θ *of* $\mathcal{L} = \ell_{\mathcal{D}} \circ \mu$ *satisfies:*

(a) *either* $\mu(\theta) = 0$, *or*
(b) θ *is a regular point of* μ *and* $\mu(\theta)$ *lies in the smooth locus of* $\mathcal{V}_{\mathbf{k},\mathbf{t}}$ *and the Euclidean interior of* $\mathcal{M}_{\mathbf{k},\mathbf{t}}$.

We summarize our discussion on the training of linear networks with the squared-error loss. For fully connected networks as in Section 10.1, the neuromanifold $\mathcal{M}$ is a classical determinantal variety. Its Euclidean relative boundary is empty. Nevertheless, spurious critical points commonly appear; they correspond to singular points of $\mathcal{M}$. For convolutional networks of stride one, $\mathcal{M}$ is semialgebraic, Euclidean closed, and full-dimensional. Here, the singular locus (of its Zariski closure) is empty, but often critical points are on its Euclidean boundary and are thus critical points of μ. Finally, for convolutional networks with all strides larger than one, $\mathcal{M}$ is a semialgebraic, Euclidean closed, lower-dimensional subset. Its Euclidean relative boundary and its singular locus are usually nontrivial. Nevertheless, these loci are not relevant for training the network when using a sufficient amount of generic data: All critical points – except when a filter in one of the layers is zero – are pure and correspond to interior smooth points of $\mathcal{M}$.

Remark 10.13 In this section, we studied convolutions on one-dimensional signals (i.e., vectors). Many practical neural networks, notably in image processing, use convolutions on two-dimensional signals. Higher-dimensional convolutions move a *filter tensor* through an *input tensor*. The composition of such convolutions corresponds to the multiplication of multivariate polynomials [108, Section 4.3].

10.3 Learning Varieties

Suppose that we are given d data points $\mathcal{D} = \{\mathbf{x}_1, \ldots, \mathbf{x}_d\} \subset \mathbb{R}^n$. We assume the *variety hypothesis*. This means we assume that the points in $\mathcal{D}$ are (possibly noisy) samples from a real algebraic variety $X \subset \mathbb{R}^n$. The goal is to learn the variety X from the data $\mathcal{D}$. The unknown variety X is represented by the ideal $I(X)$ of all polynomials that vanish on X. Assuming X to be irreducible, the ideal $I(X)$ is prime. However, $I(X)$ is unknown. All we are given is the finite set $\mathcal{D}$. In this section, we discuss strategies to learn an ideal $I_{\mathcal{D}}$ which is meant to approximate $I(X)$. Our discussion is an invitation to the article [26].

Remark 10.14 In some situations, one is interested in specific features of X, such as its dimension or homology. In this case, learning the ideal means asking for more information than what is actually needed. Other strategies can be more efficient: Theorem 15.2 in Chapter 15 gives conditions under which one can recover the homology of X from the finite sample $\mathcal{D}$ by computing the Čech complex of a union of balls. For estimating the dimension of X, see [26, Section 3] and the vast literature on data dimensionality.

We now focus on learning the approximate ideal $I_{\mathcal{D}}$. First, we need to find an appropriate loss function. One immediate option is to use the mean squared loss function $\ell_{\mathcal{D}}(f) = \frac{1}{d} \sum_{i=1}^{d} f(\mathbf{x}_i)^2$. Next, we must define a threshold $\varepsilon \geq 0$ so that we keep only those polynomials for which $\ell_{\mathcal{D}}(f) \leq \varepsilon$. Moreover, since $f \in I(X)$ if and only if $\lambda \cdot f \in I(X)$ for every $\lambda \in \mathbb{R}$, we must work with some kind of normalization for f. The next example illustrates one approach to computing the approximate ideal $I_{\mathcal{D}}$ using linear algebra.

Example 10.15 Suppose $n = 2$ and that we have $d = 3$ data points $\mathcal{D} = \{\mathbf{x}, \mathbf{y}, \mathbf{z}\} \subset \mathbb{R}^2$. Given $\varepsilon > 0$ we want to compute polynomials $f \in \mathbb{R}[x_1, x_2]$ of degree at most two that

(a) have mean squared loss at most $\varepsilon \geq 0$, and
(b) satisfy the normalization
$$f_1^2 + \cdots + f_6^2 = 1,$$
  where $f = f_1 x_1^2 + f_2 x_1 x_2 + f_3 x_2^2 + f_4 x_1 + f_5 x_2 + f_6$.

For this, we set up the bivariate Vandermonde matrix of degree ≤ 2 for our data points:

$$M_{\mathcal{D}} := \begin{bmatrix} x_1^2 & x_1 x_2 & x_2^2 & x_1 & x_2 & 1 \\ y_1^2 & y_1 y_2 & y_2^2 & y_1 & y_2 & 1 \\ z_1^2 & z_1 z_2 & z_2^2 & z_1 & z_2 & 1 \end{bmatrix} \in \mathbb{R}^{3 \times 6}. \tag{10.14}$$

The mean squared loss equals $\frac{1}{d} \sum_{i=1}^{d} f(\mathbf{x}_i)^2 = \frac{1}{d} \|M_{\mathcal{D}} f\|^2$. Suppose that $f \in \mathbb{R}^6$ is a right-singular vector of $M_{\mathcal{D}}$ with singular value σ. Then, $\|M_{\mathcal{D}} f\|^2 = \sigma^2$. This means one can compute the singular value decomposition of $M_{\mathcal{D}}$ and keep those right singular vectors of $M_{\mathcal{D}}$ whose squared singular values are at most $d \cdot \varepsilon$. (Caveat: this does *not* yield all polynomials with mean squared loss at most ε). The polynomials corresponding to these singular vectors generate an ideal, which is a candidate for $I_{\mathcal{D}}$.   ◇

The approach from Example 10.15 can be generalized to any ambient dimension n and any number of points d. The matrix in (10.14) is called a *multivariate Vandermonde matrix*.

In the general setting, it is a $d \times m$-matrix, where m is the number of monomials we are considering for our polynomials in $I_{\mathcal{D}}$. In Example 10.15, all monomials of degree at most two are used. In general, the choice of monomials is a hyperparameter for this approach. This means that we have to specify the monomials before solving for f.

In Example 10.15, we used the singular value decomposition for computing polynomials that generate our ideal $I_{\mathcal{D}}$. Two alternative methods are discussed in [26, Section 5], namely the QR-decomposition and row-echelon, also for the Vandermonde matrix. All three methods have their advantages and disadvantages.

In other situations, one might have prior knowledge on the polynomials that generate $I(X)$. This might come from geometric considerations (as in [26, Section 2]). These might suggest determinantal representations, or torus actions on X, which ensure that $I(X)$ is homogeneous with respect to some multigrading. This amounts to information about sparsity. Sometimes, we have a priori bounds for the degrees of the generators of $I(X)$. The "variety of varieties" in [26, Section 2.2] suggests many options.

In order to search for sparse polynomials, we can utilize methods from *compressed sensing*. This field is concerned with computing sparse solutions to underdetermined systems of linear equations. Moreover, polynomials coming from specific applications often have rational or integer coefficients, so one might seek polynomials that also have this property. For this, finite field techniques can be very useful.

For the normalization of the polynomials in Example 10.15, we used the Euclidean norm of the vector of coefficients. An alternative approach is to fix a monomial ordering on $\mathbb{R}[x_1, \ldots, x_n]$ (see Chapter 3), and then require the leading coefficient of every f to be 1. This leads to the following definition.

Definition 10.16 Let $\mathcal{D} = \{\mathbf{x}_1, \ldots, \mathbf{x}_d\} \subset \mathbb{R}^n$ be a finite sample of points and $\varepsilon \geq 0$. The *ε-approximate vanishing ideal* over $\mathcal{D}$ is

$$I_{\mathcal{D}}^{\varepsilon} := \left\langle \left\{ f \mid \tfrac{1}{d} \sum_{i=1}^{d} f(\mathbf{x}_i)^2 \leq \varepsilon \text{ and the leading coefficient of } f \text{ is } 1 \right\} \right\rangle.$$

We always have the inclusion $I(X) \subseteq I_{\mathcal{D}}^{\varepsilon}$. In the limit, when the sample $\mathcal{D}$ consists of all points in X and $\varepsilon = 0$, we have $I_X^0 = I(X)$. However, it is not known under which conditions the dimension of the variety of $I_{\mathcal{D}}^{\varepsilon}$ is equal to the dimension of X, or how their degrees are related (if at all). There is a vast literature on the approximate vanishing ideal and how to compute it. We refer to [148] and the references therein.

Wirth and Pokutta [148] present a novel algorithm for computing the approximate vanishing ideal $I_{\mathcal{D}}^{\varepsilon}$, given any tolerance $\varepsilon \geq 0$. In contrast to the approach from Example 10.15, for their algorithm, they do not have to specify monomials; i.e., monomials are not hyperparameters. The algorithm of [148] returns sparse generators for the ideal $I_{\mathcal{D}}^{\varepsilon}$.

Both approaches presented above use the mean squared loss that comes from evaluating polynomials. A more geometric choice for the loss would be $\ell_{\mathcal{D}} = \tfrac{1}{d} \sum_{i=1}^{d} \|\mathbf{x}_i - X\|^2$, where $\|\mathbf{x} - X\|$ is the Euclidean distance from $\mathbf{x}$ to X. This idea has not been worked out, neither from a theoretical nor from an algorithmic perspective. This desirable aim underscores the relevance of the ED problem (2.1) for machine learning.

Remark 10.17 In Chapter 15, we present methods for sampling points from algebraic varieties given their equations. Thus, once we have learned equations, we can set up a *generative model* that produces synthetic data. For instance, Figure 15.4 shows points that have been generated near the Trott curve. For enlightening varieties in higher dimensions, see [26, Section 2.2]. One data set $\mathcal{D}$ to start with is [26, Example 2.13].

We now shift gears. In the remainder of this section, we discuss how machine learning can be helpful for theoretical research in algebraic geometry. Several articles have explored this theme, and we shall offer a brief guide to this emerging literature. However, it is not clear yet what type of information can be learned and which type of tasks can be solved by training neural networks. Despite its massive success in classification tasks (e.g., "Is there a cat in this picture?"), machine learning with neural networks has not improved explicitly geometric tasks such as solving systems of polynomial equations. For instance, the problem of reconstructing 3D scenes from images taken by unknown cameras (see Section 13.3) amounts to solving certain polynomial systems. There have been numerous attempts to solve that problem with machine learning methods. However, none of them have been as successful as traditional symbolic techniques with Gröbner bases or resultants [158, 181].

The works that have used machine learning techniques to answer questions in algebraic geometry come roughly in two flavors. On the one hand, machine learning has been used to directly compute geometric properties, e.g. of Hilbert series [12], of irreducible representations [47], or numerical Calabi–Yau (Ricci flat Kähler) metrics [59]. Those approaches trade off the reliability of the output with performance, and this can yield valuable insights into problem instances that lie outside of the scope of traditional techniques.

Many algebro-geometric algorithms depend on a heuristic that has to be chosen by the user and that might heavily influence their performance. For instance, to compute a Gröbner basis, a monomial ordering has to be chosen. Machine learning has been successful at predicting such a heuristic, which is then used to speed up the computation using traditional algorithms. In that way, the performance can be enhanced without compromising the reliability of the final output. This approach has been used to speed up Buchberger's Algorithm by learning S-pair selection strategies [144], and Cylindrical Algebraic Decomposition by learning a variable ordering and by exploring Gröbner basis preconditioning [97].

In a similar spirit, the computation of periods of hypersurfaces is enhanced in [84]. The authors study pencils of hypersurfaces, and they use neural networks to predict the complexity of the Gauss–Manin connection for the period matrix along the pencil. Based on that prediction, they explore all smooth quartic surfaces in $\mathbb{P}^3$ whose polynomials are sums of five monomials, and they guess when their periods are computable by Gauss–Manin. This leads them to determine the periods of 96% of those surfaces.

Although neural networks have not shown great potential for solving systems of polynomials, they can be trained to predict their number of real solutions [18, 27]. Such a prediction might be used by homotopy methods that track real solutions only. This assumes that one has pre-computed a starting system for each possible number of real solutions, and – ideally – one for each chamber of the real discriminant.

That would yield a reliable numerical computation that requires less computation time since it tracks fewer paths. A more drastic approach was taken in [95] where the authors learn a *single* starting solution for a real homotopy that has a good chance of reaching a good solution of the desired target system. That way of computing produces a less reliable solution, but since they propose their method as part of a random sample consensus (RANSAC) scheme, bad solutions can be detected and disregarded. Since tracking a single solution can be very fast, they can simply repeat their approach for each bad solution.

Chapter 11
Maximum Likelihood

In Chapter 2, we discussed the problem of minimizing the Euclidean distance from a data point $\mathbf{u}$ to a model X in $\mathbb{R}^n$ that is described by polynomial equations. In Chapter 5, we studied the analogous problem in the setting of algebraic statistics [167], where the model X represents a family of probability distributions, and we used the Wasserstein metric to measure the distance from $\mathbf{u}$ to X. Finally, in Chapter 9, we considered the distance from a polynomial system $\mathbf{u}$ to the discriminant in the context of numerical analysis.

In this chapter, we return to statistical models, but we now replace the Wasserstein distance with the Kullback–Leibler (KL) divergence. This will be defined in Section 11.1. We show that minimizing KL divergence is equivalent to maximum likelihood estimation. In Section 11.2, we introduce the maximum likelihood (ML) degree, which is an analogue to the ED degree from Chapter 2. An interesting connection to physics is featured in Section 11.3, and in Section 11.4, we study the ML degree for Gaussian models.

11.1 Kullback–Leibler Divergence

The two scenarios of most interest for statisticians are Gaussian models and discrete models. We start with discrete models, where we take as the state space the finite set $\{0, 1, \ldots, n\}$. In Chapter 5, we have worked with the closed probability simplex Δ_n. Here, we use the open simplex of probability distributions, and we denote it by

$$\Delta_n^o := \big\{ \mathbf{p} = (p_0, p_1, \ldots, p_n) \in \mathbb{R}^{n+1} \mid p_0 + p_1 + \cdots + p_n = 1 \text{ and } p_0, p_1, \ldots, p_n > 0 \big\}.$$

Definition 11.1 The *Kullback–Leibler (KL) divergence* of two distributions $\mathbf{q}, \mathbf{p} \in \Delta_n^o$ is

$$D_{\mathrm{KL}}(\mathbf{q} \,\|\, \mathbf{p}) = \sum_{i=0}^{n} q_i \cdot \log(q_i/p_i). \tag{11.1}$$

P. Breiding et al., *Metric Algebraic Geometry*, Oberwolfach Seminars 53, https://doi.org/10.1007/978-3-031-51462-3_11

The KL divergence is not symmetric, i.e. $D_{KL}(\mathbf{q} \| \mathbf{p}) \neq D_{KL}(\mathbf{p} \| \mathbf{q})$ holds in general. Nevertheless, we interpret KL divergence as a kind of metric on the open simplex Δ_n^o.

The definition of KL divergence has the following background. Fix a random variable Z on the set $\{0, 1, \ldots, n\}$ with probability distribution $\mathbf{q}$. In information theory, the quantity $H = -\sum_{i=0}^{n} q_i \cdot \log(q_i)$ is called the *entropy* of $\mathbf{q}$. The *information content* of the event $Z = i$ equals $\log(1/q_i) = -\log(q_i)$, so H is the expected value of the information content.

We explain the motivation for this definition. The probability of i and its information content are inversely proportional: the more likely i, the smaller is its information content. We anticipate events with high probability. Hence, we gain little information once we observe such an event. On the other hand, if we observe an event of small probability, then we should receive comparably more information. The entropy of $\mathbf{q}$ is the expected value of the information content, and the KL divergence $D_{KL}(\mathbf{q} \| \mathbf{p})$ is the expected loss of information content, when we approximate $\mathbf{q}$ by $\mathbf{p}$.

Lemma 11.2 *The KL divergence is nonnegative. It is zero if and only if the two distributions agree. In symbols, $D_{KL}(\mathbf{q} \| \mathbf{p}) \geq 0$ for all $\mathbf{p}, \mathbf{q} \in \Delta_n^o$, with equality if and only if $\mathbf{p} = \mathbf{q}$.*

Proof We use the calculus fact that the function $x \mapsto (x - 1) - \log(x)$ is nonnegative for $x \in \mathbb{R}_{>0}$. Its only zero occurs at $x = 1$. The sum in (11.1) is bounded below as follows:

$$D_{KL}(\mathbf{q} \| \mathbf{p}) = -\sum_{i=0}^{n} q_i \cdot \log(p_i/q_i) \geq -\sum_{i=0}^{n} q_i \cdot (p_i/q_i - 1) = \sum_{i=0}^{n} q_i - \sum_{i=0}^{n} p_i = 0.$$

Moreover, equality holds if and only if $p_i/q_i = 1$ for all indices i.   □

Our statistical model is a subset X of the probability simplex Δ_n^o. That subset is defined by homogeneous polynomial equations. As before, for venturing beyond linear algebra, we identify X with its Zariski closure in complex projective space $\mathbb{P}^n$. In this chapter, we present the algebraic approach to maximum likelihood estimation for X. Our sources include the articles [41, 63, 90, 100, 101, 167] and references therein.

Suppose we are given N i.i.d. samples. This data is summarized in the *data vector* $\mathbf{u} = (u_0, u_1, \ldots, u_n)$, whose ith coordinate u_i is the number of samples that were in state i. We assume that $u_i > 0$ for all i. The *sample size* is $N = |\mathbf{u}| := u_0 + u_1 + \cdots + u_n$. The associated log-likelihood function equals

$$\ell_{\mathbf{u}} : \Delta_n^o \to \mathbb{R}, \quad \mathbf{p} \mapsto u_0 \cdot \log(p_0) + u_1 \cdot \log(p_1) + \cdots + u_n \cdot \log(p_n).$$

Performing ML estimation for the model X means solving the optimization problem:

$$\text{Maximize } \ell_{\mathbf{u}}(\mathbf{p}) \text{ subject to } \mathbf{p} \in X. \tag{11.2}$$

If we write $\mathbf{q} = \frac{1}{N}\mathbf{u}$ for the empirical distribution in Δ_n^o corresponding to $\mathbf{u}$, then the maximum likelihood estimation problem (11.2) is equivalent to:

$$\text{Minimize } D_{KL}(\mathbf{q} \| \mathbf{p}) \text{ subject to } \mathbf{p} \in X, \quad \text{where } \mathbf{q} = \frac{1}{N}\mathbf{u}. \tag{11.3}$$

This holds because the KL divergence can be written as the negative entropy of $\mathbf{q}$ minus the log-likelihood function: $D_{KL}(\mathbf{q} \| \mathbf{p}) = -H - \frac{1}{N}\ell_{\mathbf{u}}(\mathbf{p}) = \sum_{i=0}^{n} q_i \log(q_i) - \frac{1}{N}\ell_{\mathbf{u}}(\mathbf{p})$.

Therefore, viewed through the lens of metric algebraic geometry, this problem amounts to minimizing a certain distance, namely KL divergence, to the variety X. The objective function in the optimization problem (11.2) involves logarithms and is not algebraic. However, each of its partial derivatives is a rational function, and therefore we can study this problem using algebraic geometry.

Example 11.3 The independence model for two binary random variables is a quadratic surface X in the tetrahedron Δ_3^o. We studied this model under the optimization of a Wasserstein distance in Theorem 5.14. The model X is described by the constraints

$$\det \begin{bmatrix} p_0 & p_1 \\ p_2 & p_3 \end{bmatrix} = 0 \quad \text{and} \quad p_0 + p_1 + p_2 + p_3 = 1 \quad \text{and} \quad p_0, p_1, p_2, p_3 > 0. \quad (11.4)$$

The data for this model takes the form of a 2×2 matrix

$$\mathbf{u} = \begin{bmatrix} u_0 & u_1 \\ u_2 & u_3 \end{bmatrix}.$$

The sample size of the data is $|\mathbf{u}| = u_0 + u_1 + u_2 + u_3$.

Minimizing the KL divergence from $\mathbf{u}$ to the quadratic surface X means solving the optimization problem (11.3). This is equivalent to solving the constrained optimization problem (11.2). To do this, we apply Lagrange multipliers to the constraints in (11.4), bearing in mind that the gradient of the objective function $\ell_\mathbf{u}(\mathbf{p})$ equals

$$\nabla \ell_\mathbf{u}(\mathbf{p}) = \left(u_0/p_0, \ u_1/p_1, \ u_2/p_2, \ u_3/p_3 \right)^\top.$$

We solve the Lagrange multiplier equations, and we find that it has a unique solution $\hat{\mathbf{p}}$. This unique critical point is the maximum likelihood estimate for the model X given the data $\mathbf{u}$. Its coordinates are

$$\begin{array}{ll} \hat{p}_0 = |\mathbf{u}|^{-2}(u_0+u_1)(u_0+u_2), & \hat{p}_1 = |\mathbf{u}|^{-2}(u_0+u_1)(u_1+u_3), \\ \hat{p}_2 = |\mathbf{u}|^{-2}(u_2+u_3)(u_0+u_2), & \hat{p}_3 = |\mathbf{u}|^{-2}(u_2+u_3)(u_1+u_3). \end{array} \quad (11.5)$$

In words, we multiply the row sums with the column sums of $\frac{1}{|\mathbf{u}|}\mathbf{u}$. ◇

Let $X \subset \mathbb{P}^n$ be any fixed real projective variety, viewed as a statistical model as above. For any given data vector $\mathbf{u}$, we write $\hat{\mathbf{p}}$ for the optimal solution to our optimization problem (11.2) – (11.3) in the statistical model $X \cap \Delta_n^o$. Note that $\hat{\mathbf{p}}$ is an algebraic function of $\mathbf{u}$. In the next section, we study the algebraic geometry of the function $\mathbf{u} \mapsto \hat{\mathbf{p}}$. A key player will be the very affine variety X^o; see Theorem 11.7.

11.2 Maximum Likelihood Degree

We fix a real projective variety X in $\mathbb{P}^n$, and we consider the problem in (11.2) or (11.3).

Definition 11.4 The *maximum likelihood degree* (ML degree) of the variety X is defined to be the number of complex critical points of the optimization problem (11.2) for generic data $\mathbf{u}$. We denote it by MLdegree(X).

For arbitrary data $\mathbf{u}$, the optimal solution in the statistical model $X \cap \Delta_n^o$ is called the *maximum likelihood estimate (MLE)* of the model X for the data $\mathbf{u}$. It is denoted by $\hat{\mathbf{p}}$. Thus, the ML degree is the analogue to the ED degree, when now KL divergence replaces the Euclidean distance. The ML degree measures the algebraic complexity of the MLE.

The critical equations for (11.2) are similar to those of the ED problem in Chapter 2. We shall now describe these equations. Let $I_X = \langle f_1, \ldots, f_k \rangle$ be the homogeneous ideal of the model X. In addition, we consider the inhomogeneous linear polynomial $f_0 := p_0 + p_1 + \cdots + p_n - 1$. Let $\mathcal{J} = (\partial f_i / \partial p_j)$ denote the Jacobian matrix of size $(k+1) \times (n+1)$ for these polynomials, and set $c = \mathrm{codim}(X)$. Following Chapters 2 and 5, the *augmented Jacobian* $\mathcal{AJ}$ is obtained from the Jacobian matrix $\mathcal{J}$ by prepending one more row, namely the gradient of the objective function $\nabla \ell_{\mathbf{u}} = (u_0/p_0, u_1/p_1, \ldots, u_n/p_n)^{\top}$. To obtain the critical ideal, we enlarge I_X by the $(c+2) \times (c+2)$ minors of the $(k+2) \times (n+1)$ matrix $\mathcal{AJ}$, then we clear denominators, and finally we remove extraneous components by saturation. Thus, MLdegree(X) is the degree of the critical ideal.

Example 11.5 (Space curves) Let $n = 3$ and X be the curve in Δ_3^o defined by two general polynomials f_1 and f_2 of degrees d_1 and d_2 in p_0, p_1, p_2, p_3. The augmented Jacobian is

$$\mathcal{AJ} = \begin{bmatrix} u_0/p_0 & u_1/p_1 & u_2/p_2 & u_3/p_3 \\ 1 & 1 & 1 & 1 \\ \partial f_1/\partial p_0 & \partial f_1/\partial p_1 & \partial f_1/\partial p_2 & \partial f_1/\partial p_3 \\ \partial f_2/\partial p_0 & \partial f_2/\partial p_1 & \partial f_2/\partial p_2 & \partial f_2/\partial p_3 \end{bmatrix}. \tag{11.6}$$

Since the codimension of X equals $c = 2$, we need to enlarge I_X by the determinant of the 4×4 matrix $\mathcal{AJ}$. Clearing denominators amounts to multiplying the ith column of $\mathcal{AJ}$ by p_i, so the determinant contributes a polynomial of degree $d_1 + d_2 + 1$ to the critical equations. The generators of I_X have degrees d_1 and d_2 respectively. We therefore conclude that the ML degree of X equals $d_1 d_2 (d_1 + d_2 + 1)$ by Bézout's Theorem. ◇

The following general upper bound on the ML degree is established in [90, Theorem 5].

Proposition 11.6 *Let X be a model of codimension c in the probability simplex Δ_n^o whose ideal I_X is generated by $k \geq c$ polynomials $f_1, \ldots, f_k$ of degrees $d_1 \geq \cdots \geq d_k$. Then*

$$\mathrm{MLdegree}(X) \leq d_1 d_2 \cdots d_c \cdot \sum_{i_1 + i_2 + \cdots + i_c \leq n-c} d_1^{i_1} d_2^{i_2} \cdots d_c^{i_c}. \tag{11.7}$$

Equality holds when X is a generic complete intersection of codimension c (hence $c = k$).

The formula in Theorem 11.7 below shows that, under some assumptions, the ML degree is equal to a signed Euler characteristic. This is the likelihood analogue to Theorem 2.13, which relates the ED degree to polar degrees, and hence to Chern classes (Chapter 4).

Given our variety X in the complex projective space $\mathbb{P}^n$, we let X^o be the open subset of X that is obtained by removing the hyperplane arrangement $\{p_0 p_1 \cdots p_n (\sum_{i=0}^n p_i) = 0\}$.

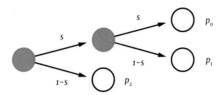

Fig. 11.1: Probability tree that describes the coin toss model in Example 11.8.

We recall from [99, 100] that a *very affine variety* is a closed subvariety of an algebraic torus $(\mathbb{C}^*)^r$. In our setting, the open set X^o is a very affine variety with $r = n + 2$. The following formula works for any very affine variety.

Theorem 11.7 *Suppose that the very affine variety X^o is nonsingular. Then,*

$$\text{MLdegree}(X) = (-1)^{\dim(X)} \cdot \chi(X^o),$$

where the latter is the topological Euler characteristic of the manifold X^o.

Proof (and Discussion) This was proved under additional assumptions by Catanese, Hoşten, Khetan, and Sturmfels in [41, Theorem 19], and then in full generality by Huh in [99, Theorem 1]. If the very affine variety X^o is singular, then the Euler characteristic can be replaced by the Chern–Schwartz–MacPherson class, as shown in [99, Theorem 2]. □

Varieties for which the ML degree is equal to one are of special interest, both in statistics and in geometry. For a model X to have ML degree one means that the MLE $\hat{\mathbf{p}}$ is a rational function of the data $\mathbf{u}$. This happens for the independence model in Example 11.3. The ML degree of the surface $X = V(p_0 p_3 - p_1 p_2)$ is one because the MLE in (11.5) is a rational function of the data. To be precise, $\hat{\mathbf{p}}$ is a homogeneous rational function of degree 0 in $\mathbf{u}$. Here is another model which has ML degree one.

Example 11.8 Given a biased coin, we perform the following experiment: *Flip the biased coin. If it shows heads, flip it again.* The outcome of this experiment is the number of heads: 0, 1, or 2. This describes a generative statistical model on three states, illustrated in Figure 11.1. If s denotes the bias of our coin, then the model is the parametric curve X given by $(0, 1) \to X \subset \Delta_2^o$, $s \mapsto (s^2, s(1-s), 1-s)$. The underlying variety is the conic $X = V(p_0 p_2 - (p_0 + p_1)p_1) \subset \mathbb{P}^2$. Its MLE is given by the formula

$$\hat{p} = (\hat{p}_0, \hat{p}_1, \hat{p}_2) = \left(\frac{(2u_0 + u_1)^2}{(2u_0 + 2u_1 + u_2)^2}, \frac{(2u_0 + u_1)(u_1 + u_2)}{(2u_0 + 2u_1 + u_2)^2}, \frac{u_1 + u_2}{2u_0 + 2u_1 + u_2} \right). \quad (11.8)$$

Since the coordinates of $\hat{\mathbf{p}}$ are rational functions, the ML degree of X is equal to one.   ◇

The following theorem explains what we saw in equations (11.5) and (11.8):

Theorem 11.9 *If $X \subset \Delta_n^o$ is a model of ML degree one, so that $\hat{\mathbf{p}}$ is a rational function of $\mathbf{u}$, then each coordinate $\hat{p}_i$ is an alternating product of linear forms with positive coefficients.*

Proof (and Discussion) This was proved by Huh [100] in the setting of arbitrary complex very affine varieties. It was adapted to real algebraic geometry, and hence to statistical models, by Duarte, Marigliano, and Sturmfels [63]. These two articles offer precise statements via Horn uniformization for A-discriminants [71]. A-discriminants are hypersurfaces dual to toric varieties, as we already saw at the end of Chapter 9. For additional information, we refer to the paper by Huh and Sturmfels [101, Corollary 3.12].   □

Models given by rank constraints on matrices and tensors are particularly important in applications since these represent conditional independence. Consider two random variables, having n_1 and n_2 states respectively, which are conditionally independent given a hidden random variable with r states. For us, this model is the determinantal variety X_r in $\mathbb{P}^{n_1 n_2 - 1}$ that is defined by the $(r+1) \times (r+1)$ minors of an $n_1 \times n_2$ matrix (p_{ij}). It appeared as the neuromanifold of linear fully connected neural networks in Section 10.1. The ML degree of this rank r model was first studied by Hauenstein, Rodriguez, and Sturmfels in [82], who obtained the following results using methods from numerical algebraic geometry.

Proposition 11.10 *For small values of n_1 and n_2, the ML degrees of low-rank models X_r are presented in the following table:*

$r \backslash (n_1, n_2)$	(3,3)	(3,4)	(3,5)	(4,4)	(4,5)	(4,6)	(5,5)	
1	1	1	1	1	1	1	1	
2	10	26	58	191	843	3119	6776	(11.9)
3	1	1	1	191	843	3119	61326	
4				1	1	1	6776	
5							1	

The table in (11.9) can be viewed as a Kullback–Leibler analogue to Corollary 5.20. Every entry in the $r = 1$ row equals 1 because the MLE for the independence model is a rational function in the data (u_{ij}). One finds $\hat{\mathbf{p}} = (\hat{p}_{ij})$ by multiplying the column vector of row sums of $\mathbf{u}$ with the row vector of column sums of $\mathbf{u}$, and then dividing by $|\mathbf{u}|^2$, as shown in (11.5). The other entries are more interesting, and they give precise information on the algebraic complexity of minimizing the Kullback–Leibler divergence from a given data matrix $\mathbf{u}$ to the conditional independence model X_r. Here is an example taken from [82].

Example 11.11 We fix $n_1 = n_2 = 5$. Following [82, Example 7], we consider the data

$$
\mathbf{u} \;=\; \begin{bmatrix}
2864 & 6 & 6 & 3 & 3 \\
2 & 7577 & 2 & 2 & 5 \\
4 & 1 & 7543 & 2 & 4 \\
5 & 1 & 2 & 3809 & 4 \\
6 & 2 & 6 & 3 & 5685
\end{bmatrix}.
$$

For $r = 2$ and $r = 4$, this instance of our ML estimation problem has the expected number of 6776 distinct complex critical points. In both cases, 1774 of these are real and 90 of these are real and positive. This illustrates the last statement in Theorem 11.12 below. The number of local maxima for $r = 2$ equals 15, and the number of local maxima for $r = 4$ equals 6. For $r = 3$, we have 61326 critical points, of which 15450 are real. Of these, 362 are positive and 25 are local maxima. We invite our readers to critically check these claims, by running software for solving polynomial equations, as explained in Chapter 3.   ◇

The columns of the table in (11.9) exhibit an obvious symmetry. This was conjectured in [82], and it was proved by Draisma and Rodriguez in their article [61] on maximum likelihood duality. We now state their result. Given an $n_1 \times n_2$ matrix $\mathbf{u}$, we write $\Omega_{\mathbf{u}}$ for the matrix whose (i, j) entry equals

$$\frac{u_{ij} u_{i+} u_{+j}}{(u_{++})^3}.$$

Here, we use the following notation for the row sums, the column sums, and the sample size:

$$u_{i+} = \sum_{j=1}^{n_2} u_{ij}, \quad u_{+j} = \sum_{i=1}^{n_1} u_{ij}, \quad \text{and} \quad u_{++} = \sum_{i=1}^{n_1} \sum_{j=1}^{n_2} u_{ij}.$$

In the following theorem, the symbol $\star$ denotes the Hadamard product (or entrywise product) of two matrices. That is, if $\mathbf{p} = (p_{ij})$ and $\mathbf{q} = (q_{ij})$, then $\mathbf{p} \star \mathbf{q} = (p_{ij} \cdot q_{ij})$.

Theorem 11.12 *Fix $n_1 \leq n_2$ and $\mathbf{u}$ an $n_1 \times n_2$-matrix with strictly positive integer entries. There exists a bijection between the complex critical points $\mathbf{p}_1, \mathbf{p}_2, \ldots, \mathbf{p}_s$ of the likelihood function for $\mathbf{u}$ on X_r and the complex critical points $\mathbf{q}_1, \mathbf{q}_2, \ldots, \mathbf{q}_s$ on $X_{n_1 - r + 1}$ such that*

$$\mathbf{p}_1 \star \mathbf{q}_1 = \mathbf{p}_2 \star \mathbf{q}_2 = \cdots = \mathbf{p}_s \star \mathbf{q}_s = \Omega_{\mathbf{u}}. \tag{11.10}$$

In particular, this bijection preserves reality, positivity, and rationality of the critical points.

This result is a multiplicative version of Theorem 2.23 on duality for ED degrees. By "multiplicative" we mean that $\mathbf{u}_i/\mathbf{p}_i$ instead of $\mathbf{u}_i - \mathbf{p}_i$ appears in the first row of the augmented Jacobian matrix. Theorem 11.12 concerns the case of determinantal varieties. In the ED theory, these led us to the Eckart–Young Theorem.

It is a challenge in intersection theory and singularity theory to find general formulas for the ML degrees in Proposition 11.10. This problem was solved for $r = 2$ by Rodriguez and Wang in [154]. They give a recursive formula in [154, Theorem 4.1], and they present impressive values in [154, Table 1]. They unravel the recursion, and they obtain the explicit formulas for the ML degree of conditional independence models in many cases. In particular, they obtain the following result, which had been stated as a conjecture in [82].

Theorem 11.13 (Rodriguez–Wang [154]) *Consider the variety $X_2 \subset \mathbb{P}^{3n-1}$ whose points are the $3 \times n$ matrices of rank at most two. The ML degree of this variety is*

$$\mathrm{MLdegree}(X_2) = 2^{n+1} - 6.$$

11.3 Scattering Equations

We now turn to a connection between algebraic statistics and particle physics that was developed in [165]. The context is scattering amplitudes, where the critical equations for (11.2) – (11.3) are the *scattering equations*. We consider the *CEGM model*, due to Cachazo and his collaborators [37, 38]. The role of the data vector **u** is played in physics by the *Mandelstam invariants*. This theory rests on the space X^o of m labeled points in general position in $\mathbb{P}^{k-1}$, up to projective transformations.

Consider the Grassmannian $\mathrm{Gr}(k-1, \mathbb{P}^{m-1})$ in its Plücker embedding into $\mathbb{P}^{\binom{m}{k}-1}$. The torus $(\mathbb{C}^*)^m$ acts on this by scaling the columns of $k \times m$ matrices representing subspaces. Let $\mathrm{Gr}(k-1, \mathbb{P}^{m-1})^o$ be the open Grassmannian where all Plücker coordinates are non-zero.

Definition 11.14 The CEGM model is the $(k-1)(m-k-1)$-dimensional manifold

$$X^o \quad = \quad \mathrm{Gr}(k-1, \mathbb{P}^{m-1})^o / (\mathbb{C}^*)^m. \tag{11.11}$$

Example 11.15 ($k = 2$) For $k = 2$, the very affine variety in (11.11) has dimension $m - 3$, and it is the moduli space of m distinct labeled points on the complex projective line $\mathbb{P}^1$. This space is ubiquitous in algebraic geometry, where it is known as $\mathcal{M}_{0,m}$. The punchline of our discussion here is that we interpret the moduli space $\mathcal{M}_{0,m}$ as a statistical model. And, we then argue that its ML degree is equal to $(m - 3)!$. For instance, if $m = 4$ then $X^o = \mathcal{M}_{0,4}$ is the Riemann sphere $\mathbb{P}^1$ with three points removed. The signed Euler characteristic of this surface is one, and Theorem 11.9 applies.   ◇

Proposition 11.16 *The configuration space X^o in (11.11) is a very affine variety, with coordinates given by the $k \times k$ minors of the $k \times m$ matrix*

$$M_{k,m} \quad = \quad \begin{bmatrix} A & B \end{bmatrix},$$

where the matrices $A \in \mathbb{R}^{k \times k}$ and $B \in \mathbb{R}^{k \times (m-k)}$ are given by

$$A = \begin{bmatrix} 0 & 0 & 0 & \cdots & 0 & (-1)^k \\ 0 & 0 & 0 & \cdots & (-1)^{k-1} & 0 \\ \vdots & \vdots & \vdots & \ddots & \vdots & \vdots \\ 0 & 0 & -1 & \cdots & 0 & 0 \\ 0 & 1 & 0 & \cdots & 0 & 0 \\ -1 & 0 & 0 & \cdots & 0 & 0 \end{bmatrix}$$

and

$$B = \begin{bmatrix} 1 & 1 & 1 & \cdots & 1 \\ 1 & x_{1,1} & x_{1,2} & \cdots & x_{1,m-k-1} \\ \vdots & \vdots & \vdots & \ddots & \vdots \\ 1 & x_{k-3,1} & x_{k-3,2} & \cdots & x_{k-3,m-k-1} \\ 1 & x_{k-2,1} & x_{k-2,2} & \cdots & x_{k-2,m-k-1} \\ 1 & x_{k-1,1} & x_{k-1,2} & \cdots & x_{k-1,m-k-1} \end{bmatrix}.$$

We denote by $p_{i_1 i_2 \cdots i_k}$ the $k \times k$ minor of $M_{k,m}$ with columns $i_1 < i_2 < \cdots < i_k$.

Following [2, Equation (4)], the antidiagonal matrix in the left $k \times k$ block of $M_{k,m}$ is chosen so that each unknown $x_{i,j}$ is precisely equal to the minor $p_{i_1 i_2 \cdots i_k}$ for some $i_1 < i_2 < \cdots < i_k$. No signs are needed.

Definition 11.17 The *scattering potential* for the CEGM model is the function

$$\ell_{\mathbf{u}} \;\; = \;\; \sum_{i_1 i_2 \cdots i_k} u_{i_1 i_2 \cdots i_k} \cdot \log(p_{i_1 i_2 \cdots i_k}). \tag{11.12}$$

This is a multi-valued function on the very affine complex variety X^o. Here, $\mathbf{u} = (u_{i_1 i_2 \cdots i_k})$ is the data vector (its coordinates are called Mandelstam invariants in physics) and the $p_{i_1 i_2 \cdots i_k}$ are the coordinates of the open Grassmannian in the Plücker embedding.

The critical point equations, known as *scattering equations* [2, equation (7)], are given by

$$\frac{\partial \ell_{\mathbf{u}}}{\partial x_{i,j}} = 0 \qquad \text{for } 1 \le i \le k-1 \text{ and } 1 \le j \le m - k - 1. \tag{11.13}$$

These are equations of rational functions. Solving these equations is the agenda in the articles [37, 38, 165].

Corollary 11.18 *The number of complex solutions to (11.13) is the ML degree of the CEGM model X^o. This number equals the signed Euler characteristic $(-1)^{(k-1)(m-k-1)} \cdot \chi(X^o)$.*

Example 11.19 ($k = 2, m = 6$) The very affine threefold $X^o = \mathcal{M}_{0,6}$ sits in $(\mathbb{C}^*)^9$ via

$$p_{24} = x_1, \; p_{25} = x_2, \; p_{26} = x_3, \; p_{34} = x_1 - 1, \; p_{35} = x_2 - 1,$$
$$p_{36} = x_3 - 1, \; p_{45} = x_2 - x_1, \; p_{46} = x_3 - x_1, \; p_{56} = x_3 - x_2.$$

These nine coordinates on $X^o \subset (\mathbb{C}^*)^9$ are the non-constant 2×2 minors of our matrix

$$M_{2,6} \;\; = \;\; \begin{bmatrix} 0 & 1 & 1 & 1 & 1 & 1 \\ -1 & 0 & 1 & x_1 & x_2 & x_3 \end{bmatrix}.$$

The scattering potential is the analogue to the log-likelihood function in statistics:

$$\ell_{\mathbf{u}} \;\; = \;\; u_{24} \log(p_{24}) + u_{25} \log(p_{25}) + \cdots + u_{56} \log(p_{56}).$$

This function has six critical points in X^o. Hence, MLdegree$(X^o) = -\chi(X^o) = 6$. ◇

We now examine the number of critical points of the scattering potential (11.12).

Theorem 11.20 *The known values of the ML degree for the CEGM model (11.11) are as follows. For $k = 2$, the ML degree equals $(m-3)!$ for all $m \ge 4$. For $k = 3$, the ML degree equals $2, 26, 1272, 188112, 74570400$ when the number of points is $m = 5, 6, 7, 8, 9$. For $k = 4$ and $m = 8$, the ML degree equals 5211816.*

Proof We refer to [2, Example 2.2], [2, Theorem 5.1], and [2, Theorem 6.1] for $k = 2, 3, 4$. □

Knowing these ML degrees for the CEGM model helps in solving the scattering equations reliably. It was demonstrated in [2, 165] how this can be done in practice with the software HomotopyContinuation.jl [31]. For instance, [165, Table 1] discusses the computation of the $10! = 3628800$ critical points for $k = 2$ and $m = 13$. See [2, Section 6] for the solution in the challenging case $k = 4$ and $m = 8$.

One purpose of this short section was to demonstrate that ML degrees of very affine varieties X^o appear in many scenarios, notably in physics, well beyond statistical models. By connecting these scenarios to algebraic statistics, both sides benefit. Metric algebraic geometry offers a framework for developing such connections. See also [7] for connecting ML estimation with norm minimization over a group orbit in invariant theory.

11.4 Gaussian Models

We now change the topic by turning to statistical models for Gaussian random variables. Let PD_n denote the set of positive-definite symmetric $n \times n$ matrices, i.e. matrices all of whose eigenvalues are positive. This is an open convex cone in the real vector space $S^2(\mathbb{R}^n)$ of symmetric $n \times n$ matrices, which has dimension $\binom{n+1}{2}$. This cone now plays the role which was played by the simplex Δ_n^o when we discussed discrete models.

Given a mean vector $\mu \in \mathbb{R}^n$ and a covariance matrix $\Sigma \in PD_n$, the associated *Gaussian distribution* is supported on $\mathbb{R}^n$. Its density has the familiar "bell shape"; it is the function

$$ f_{\mu,\Sigma}(\mathbf{x}) \; := \; \frac{1}{\sqrt{(2\pi)^n \, \det(\Sigma)}} \cdot \exp\left(-\tfrac{1}{2}(\mathbf{x} - \mu)^\top \Sigma^{-1}(\mathbf{x} - \mu) \right). $$

We fix a model $Y \subset \mathbb{R}^n \times PD_n$ defined by polynomial equations in (μ, Σ). Suppose we are given N samples $U^{(1)}, \ldots, U^{(N)}$. These samples are vectors in $\mathbb{R}^n$. These data are summarized in the *sample mean* $\bar{U} = \frac{1}{N} \sum_{i=1}^{N} U^{(i)}$ and in the *sample covariance matrix*

$$ S \; = \; \frac{1}{N} \sum_{i=1}^{N} (U^{(i)} - \bar{U})(U^{(i)} - \bar{U})^\top. $$

With this data matrix, the *log-likelihood* is the following function in the unknowns (μ, Σ):

$$ \ell(\mu, \Sigma) \; = \; -\frac{N}{2} \cdot \left[\log \det(\Sigma) \; + \; \mathrm{trace}(S\Sigma^{-1}) \; + \; (\bar{U} - \mu)^\top \Sigma^{-1}(\bar{U} - \mu) \right]. \qquad (11.14) $$

The task of likelihood inference is to maximize this function subject to $(\mu, \Sigma) \in Y$.

There are two extreme cases. First, consider a model where Σ is fixed to be the identity matrix I_n. Then $Y = X \times \{I_n\}$ and we are supposed to minimize the Euclidean distance from the sample mean $\bar{U}$ to the variety X in $\mathbb{R}^n$. This is precisely the earlier ED problem.

We instead focus on the second case, when $Y = \mathbb{R}^n \times X$ for some subvariety X of $S^2(\mathbb{R}^n)$. Then, any maximum $(\mu, \Sigma) \in Y$ of (11.14) satisfies $\mu = \bar{U}$, and so the problem is reduced to estimating the covariance matrix $\Sigma \in X$. Hence, by shifting all samples until $\bar{U}$ becomes $\mathbf{0}$, we may assume that our model has the form $Y = \{\mathbf{0}\} \times X$ and our task is

Minimize the function $\Sigma \mapsto \log \det(\Sigma) + \text{trace}(S\Sigma^{-1})$ subject to $\Sigma \in X$. (11.15)

Using the concentration matrix $K = \Sigma^{-1}$, we can write this equivalently as follows:

Maximize the function $K \mapsto \log \det(K) - \text{trace}(SK)$ subject to $K \in X^{-1}$. (11.16)

Here, the variety X^{-1} is the Zariski closure of the set of inverses of all matrices in X.

Remark 11.21 The optimization problem (11.15)-(11.16) has a metric interpretation as in (11.3). Namely, we can define the KL divergence between two continuous probability distributions on $\mathbb{R}^n$ by replacing the sum in (11.1) with the corresponding integral over $\mathbb{R}^n$. That way, we obtain a certain kind of distance between the unknown matrix Σ and the given sample covariance matrix S.

The critical equations for (11.15)-(11.16) can be written as polynomials since the partial derivatives of the logarithm are rational functions. These equations have finitely many complex solutions. Their number is the ML degree of the statistical model X^{-1}.

In the remainder of this section, we focus on Gaussian statistical models that are described by linear constraints on either the covariance matrix or its inverse, which is the concentration matrix. We consider a linear space of symmetric matrices (LSSM),

$$\mathcal{L} \subset S^2(\mathbb{R}^n),$$

whose general element is assumed to be invertible. We are interested in the models $X^{-1} = \mathcal{L}$ and $X = \mathcal{L}$. It is convenient to use primal-dual coordinates (Σ, K) to write the respective critical equations. For a proof of the next result, see [166, Propositions 3.1 and 3.3].

Proposition 11.22 *Fix an LSSM $\mathcal{L}$ and its orthogonal complement $\mathcal{L}^{\perp}$ under the Euclidean inner product $\langle X, Y \rangle = \text{Trace}(X^{\top}Y)$. The critical equations for the linear concentration model $X^{-1} = \mathcal{L}$ are*

$$K \in \mathcal{L} \text{ and } K\Sigma = I_n \text{ and } \Sigma - S \in \mathcal{L}^{\perp}. \tag{11.17}$$

The critical equations for the linear covariance model $X = \mathcal{L}$ are

$$\Sigma \in \mathcal{L} \text{ and } K\Sigma = I_n \text{ and } KSK - K \in \mathcal{L}^{\perp}. \tag{11.18}$$

The system (11.17) is linear in the unknown matrix K, whereas the last group of equations in (11.18) is quadratic in K. The numbers of complex solutions are the ML degree of $\mathcal{L}$ and the *reciprocal ML degree* of $\mathcal{L}$. The former is smaller than the latter.

Example 11.23 Let $n = 4$ and let $\mathcal{L}$ be a generic LSSM of dimension k. The degrees are:

$k = \dim(\mathcal{L})$:	2	3	4	5	6	7	8	9
ML degree :	3	9	17	21	21	17	9	3
reciprocal ML degree :	5	19	45	71	81	63	29	7

These numbers and many more appear in [166, Table 1]. ◇

ML degrees and reciprocal ML degrees of linear spaces of symmetric matrices have been studied intensively in the recent literature, both for generic and special spaces $\mathcal{L}$. See [6, 22, 67] and the references therein. We now present an important result due to Manivel, Michałek, Monin, Seynnaeve, Vodička, and Wiśniewski. Theorem 11.24 paraphrases highlights from their articles [126, 132].

Theorem 11.24 *The ML degree of a generic linear subspace $\mathcal{L}$ of dimension k in $S^2(\mathbb{R}^n)$ is the number of quadrics in $\mathbb{P}^{n-1}$ that pass through $\binom{n+1}{2} - k$ general points and are tangent to $k - 1$ general hyperplanes. For fixed k, this number is a polynomial in n of degree $k - 1$.*

Proof The first statement is [132, Corollary 2.6 (4)], here interpreted classically in terms of Schubert calculus. For a detailed discussion, see the introduction of the article [126]. The second statement appears in [126, Theorem 1.3 and Corollary 4.13].   □

Example 11.25 Let $n = 4$. Fix $10 - k$ points and $k - 1$ planes in $\mathbb{P}^3$. We are interested in all quadratic surfaces that contain the points and are tangent to the planes. These points and planes impose 9 constraints on $\mathbb{P}(S^2(\mathbb{C}^4)) \simeq \mathbb{P}^9$. Passing through a point imposes a linear equation. Being tangent to a plane is a cubic constraint on $\mathbb{P}^9$. Bézout's Theorem suggests that there could be 3^{k-1} solutions. This is correct for $k \leq 3$, but it overcounts for $k \geq 4$. Indeed, in Example 11.23, we see $17, 21, 21, \ldots$ instead of $27, 81, 243, \ldots$ ◇

The intersection theory approach in [126, 132] leads to formulas for the ML degrees of linear Gaussian models. From this, we obtain provably correct numerical methods for maximum likelihood estimation, based on homotopy continuation. Namely, after computing all critical points as in [166], we can certify them with interval arithmetic as in [30]. Since the ML degree is known beforehand, one can be sure that all solutions have been found.

Chapter 12
Tensors

Tensors are generalizations of matrices. They can be viewed as tables of higher dimension. A 2×2-matrix is a table that contains four numbers aligned in two directions, and each direction has dimension 2. Similarly, a $2 \times 2 \times 2$-tensor is a table with eight numbers aligned in three directions where each direction has dimension 2.

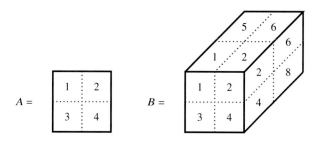

Fig. 12.1: A 2×2 matrix A and a $2 \times 2 \times 2$ tensor B.

In this chapter, we offer an introduction to tensors, with a view toward metric algebraic geometry. We present the notion of tensor rank, along with the Segre variety and the Veronese variety. These arise from tensors of bounded rank. We then turn to the spectral theory of tensors, and we show that eigenvectors and singular vectors arise naturally in the study of Euclidean distance problems. In the last section, we study volumes in complex projective space, and we determine these volumes for Segre and Veronese varieties.

12.1 Tensors and their Rank

For positive integers $n_1, \ldots, n_d$, the vector space of real $n_1 \times \cdots \times n_d$ tensors is denoted by

$$\mathbb{R}^{n_1 \times \cdots \times n_d} := \left\{ A = \left(a_{i_1, \ldots, i_d} \right)_{1 \le i_1 \le n_1, \ldots, 1 \le i_d \le n_d} \mid a_{i_1, \ldots, i_d} \in \mathbb{R} \right\}.$$

P. Breiding et al., *Metric Algebraic Geometry*, Oberwolfach Seminars 53,
https://doi.org/10.1007/978-3-031-51462-3_12

In the literature, elements in the vector space $\mathbb{R}^{n_1 \times \cdots \times n_d}$ are also called *hypermatrices* or *Cartesian tensors*, because our definition of a tensor as a multidimensional array relies on a basis. The dimension of the vector space $\mathbb{R}^{n_1 \times \cdots \times n_d}$ is $n_1 \cdots n_d$. We also consider the space of complex tensors $\mathbb{C}^{n_1 \times \cdots \times n_d}$. The number d is called the *order* of the tensor. Tensors of order one are vectors. Tensors of order two are matrices.

A common notation for order-three tensors is by writing them as a tuple of matrices.

Example 12.1 The tensor $B = (b_{ijk})_{1 \leq i,j,k \leq 2}$ from Figure 12.1 can be written as

$$B = \begin{bmatrix} 1 & 2 & 5 & 6 \\ 3 & 4 & 7 & 8 \end{bmatrix} \in \mathbb{R}^{2 \times 2 \times 2}.$$

The left matrix contains the entries b_{ij1} and the matrix on the right contains the b_{ij2}.   ◇

The Euclidean inner product of two tensors $A = (a_{i_1,\ldots,i_d})$ and $B = (b_{i_1,\ldots,i_d})$ is defined as

$$\langle A, B \rangle := \sum_{i_1=1}^{n_1} \cdots \sum_{i_d=1}^{n_d} a_{i_1,\ldots,i_d} \cdot b_{i_1,\ldots,i_d}.$$

The resulting norm on $\mathbb{R}^{n_1 \times \cdots \times n_d}$ is given by the familiar formula $\|A\| = \sqrt{\langle A, A \rangle}$.

One way to create a tensor is to take the *outer product* of d vectors $\mathbf{v}_i \in \mathbb{R}^{n_i}$, where $1 \leq i \leq d$. This tensor is defined by taking all $n_1 n_2 \cdots n_d$ possible products between the entries of distinct vectors:

$$\mathbf{v}_1 \otimes \cdots \otimes \mathbf{v}_d := \left((\mathbf{v}_1)_{i_1} \cdots (\mathbf{v}_d)_{i_d} \right)_{1 \leq i_1 \leq n_1, \ldots, 1 \leq i_d \leq n_d}. \tag{12.1}$$

The outer products of the standard basis vectors $\mathbf{e}_{i_j} \in \mathbb{R}^{n_j}$ give the standard basis of $\mathbb{R}^{n_1 \times \cdots \times n_d}$. In particular, we have $(a_{i_1,\ldots,i_d}) = \sum_{i_1=1}^{n_1} \cdots \sum_{i_d=1}^{n_d} a_{i_1,\ldots,i_d} \, \mathbf{e}_{i_1} \otimes \cdots \otimes \mathbf{e}_{i_d}$.

If $n := n_1 = \cdots = n_d$ are all equal, then we write the tensor space as $(\mathbb{R}^n)^{\otimes d} := \mathbb{R}^{n \times \cdots \times n}$. Each permutation $\pi \in \mathfrak{S}_d$ acts on outer products in $(\mathbb{R}^n)^{\otimes d}$ by permuting the factors: $\pi(\mathbf{v}_1 \otimes \cdots \otimes \mathbf{v}_d) = \mathbf{v}_{\pi(1)} \otimes \cdots \otimes \mathbf{v}_{\pi(d)}$. This yields an action of the symmetric group $\mathfrak{S}_d$ on $(\mathbb{R}^n)^{\otimes d}$ via linear extension. For $d = 2$, this is transposition of matrices.

Definition 12.2 A tensor $A \in (\mathbb{R}^n)^{\otimes d}$ is called *symmetric* if $\pi(A) = A$ for all $\pi \in \mathfrak{S}_d$. We use the following notation for the vector space of symmetric tensors:

$$S^d(\mathbb{R}^n) := \{A \in (\mathbb{R}^n)^{\otimes d} \mid A \text{ is symmetric}\}.$$

Being a symmetric tensor is a linear condition. Hence, the space $S^d(\mathbb{R}^n)$ of symmetric tensors is a linear subspace of $(\mathbb{R}^n)^{\otimes d}$. For $\mathbf{v} \in \mathbb{R}^n$, we write the d-fold outer product as

$$\mathbf{v}^{\otimes d} := \mathbf{v} \otimes \cdots \otimes \mathbf{v} \in S^d(\mathbb{R}^n).$$

Every symmetric tensor $A \in S^d(\mathbb{R}^n)$ determines a homogeneous polynomial of degree d:

$$F_A(\mathbf{x}) := \sum_{i_1=1}^{n} \sum_{i_2=1}^{n} \cdots \sum_{i_d=1}^{n} a_{i_1,i_2,\ldots,i_d} \, x_{i_1} x_{i_2} \cdots x_{i_d}. \tag{12.2}$$

This is a polynomial in n variables $x_1, \ldots, x_n$. The map $A \mapsto F_A$ is an isomorphism between $S^d(\mathbb{R}^n)$ and the vector space of homogeneous polynomials of degree d in n variables. Their common dimension is $\binom{n+d-1}{d}$. We can write the polynomial in (12.2) as

$$F_A(\mathbf{x}) = \langle A, \mathbf{x}^{\otimes d} \rangle.$$

For $A \in S^d(\mathbb{R}^n)$, we have $\|A\| = \|F_A\|_{\mathrm{BW}}$. This is the Bombieri–Weyl metric (Chapter 9).

Example 12.3 Symmetric tensors of order three are in bijection with homogeneous cubics. For instance, symmetric $2 \times 2 \times 2$ tensors correspond to binary cubics:

$$\text{For} \quad A = \left[\begin{smallmatrix} 1 & -1 \\ -1 & 0 \end{smallmatrix} \,\middle|\, \begin{smallmatrix} -1 & 0 \\ 0 & 2 \end{smallmatrix} \right] \in S^3(\mathbb{R}^2), \quad \text{we have} \quad F_A(\mathbf{x}) = x_1^3 - 3x_1^2 x_2 + 2x_2^3.$$

This generalizes the familiar bijection between symmetric matrices and quadratic forms. ⋄

Given a tuple of matrices $(M_1, \ldots, M_d) \in \mathbb{R}^{k_1 \times n_1} \times \cdots \times \mathbb{R}^{k_d \times n_d}$, we define the *multi-linear multiplication* by letting each matrix act on one side. For outer products, we have

$$(M_1, \ldots, M_d).(\mathbf{v}_1 \otimes \cdots \otimes \mathbf{v}_d) := (M_1 \mathbf{v}_1) \otimes \cdots \otimes (M_d \mathbf{v}_d).$$

This action extends by linearity to all of $\mathbb{R}^{n_1 \times \cdots \times n_d}$. For $k_1 = n_1, \ldots, k_d = n_d$, this induces a representation of $\mathrm{GL}(n_1) \times \cdots \times \mathrm{GL}(n_d)$ into the general linear group of $\mathbb{R}^{n_1 \times \cdots \times n_d}$. If $d = 2$, then $(M_1, M_2).A = M_1 A M_2^\top$. Thus, multilinear multiplication is a generalization of simultaneous left-right multiplication for matrices.

A central topic of this chapter are *rank-one tensors*. A tensor A has rank one if it has the form (12.1). For $d = 2$, this gives us $n_1 \times n_2$ matrices of rank one. The inner product between rank-one tensors is

$$\langle \mathbf{v}_1 \otimes \cdots \otimes \mathbf{v}_d, \mathbf{w}_1 \otimes \cdots \otimes \mathbf{w}_d \rangle = \langle \mathbf{v}_1, \mathbf{w}_1 \rangle \cdots \langle \mathbf{v}_d, \mathbf{w}_d \rangle. \qquad (12.3)$$

Definition 12.4 The (real) *Segre variety* $S_{n_1, \ldots, n_d}$ consists of all $n_1 \times \cdots \times n_d$-tensors of rank one, i.e.

$$S_{n_1, \ldots, n_d} := \{ \mathbf{v}_1 \otimes \cdots \otimes \mathbf{v}_d \mid \mathbf{v}_i \in \mathbb{R}^{n_i}, 1 \le i \le d \}.$$

We simply write S when $n_1, \ldots, n_d$ are clear from the context. The most immediate way to see that S is an algebraic variety goes as follows. We can always *flatten* a tensor $A \in \mathbb{R}^{n_1 \times \cdots \times n_d}$ into d matrices $F_1, \ldots, F_d$, where $F_i \in \mathbb{R}^{n_i \times (\prod_{j \ne i} n_j)}$. Then, we have $A \in S$ if and only if the column span of F_i has dimension at most one for $1 \le i \le d$. This is equivalent to saying that the F_i are all of rank at most one; i.e., their 2×2-minors vanish. The dimension of the Segre variety is $\dim S = n_1 + \cdots + n_d + 1 - d$.

The analogue of the Segre variety for symmetric tensors is the Veronese variety.

Definition 12.5 The *Veronese variety* is the variety of symmetric tensors of rank one:

$$\mathcal{V} := \mathcal{V}_{n,d} := \{ \mathbf{v}^{\otimes d} \mid \mathbf{v} \in \mathbb{R}^n \}.$$

This is the intersection of the Segre variety S with a linear subspace. Hence, $\mathcal{V}$ is an algebraic variety. Note that $\dim \mathcal{V} = n$.

Example 12.6 Let $\mathbf{v} = (x, y) \in \mathbb{R}^2$. Then, $\mathbf{v}^{\otimes 3}$ comprises all monomials of degree three:

$$\mathbf{v}^{\otimes 3} = \begin{bmatrix} x^3 & x^2 y & | & x^2 y & xy^2 \\ x^2 y & xy^2 & | & xy^2 & y^3 \end{bmatrix}.$$

Here, the Veronese variety $\mathcal{V}$ is a surface. Namely, it is the cone over the twisted cubic curve. In general, the distinct entries of $\mathbf{v}^{\otimes d}$ are all monomials of degree d in the entries of the vector $\mathbf{v}$. The Veronese variety $\mathcal{V}$ is parametrized by these monomials.   ◇

The Segre variety induces a notion of rank of a tensor $A \in \mathbb{R}^{n_1 \times \cdots \times n_d}$. Namely, we define

$$\operatorname{rank}(A) := \min\{r \geq 0 \mid \text{there exists } A_1, \ldots, A_r \in \mathcal{S} \text{ with } A = A_1 + \cdots + A_r\}.$$

For matrices ($d = 2$), this is the usual matrix rank. We denote tensors of rank at most r by

$$\mathcal{R}_r := \mathcal{R}_{r, n_1, \ldots, n_d} := \{A \in \mathbb{R}^{n_1 \times \cdots \times n_d} \mid \operatorname{rank}(A) \leq r\}.$$

A tensor of order three and rank r can be visualized as follows:

In the case of matrices, the set $\mathcal{R}_r$ is a variety. It is the common zero set of all $(r+1) \times (r+1)$-minors of the matrix. For $d \geq 3$ and $r \geq 2$, however, $\mathcal{R}_r$ is no longer a variety. This is implied by the following result due to de Silva and Lim [53].

Proposition 12.7 *The set $\mathcal{R}_2$ of tensors of rank at most two is generally not closed in the Euclidean topology.*

Proof Fix $d = 3$ and $n_1 = n_2 = n_3 = 2$. Consider linearly independent pairs $\{\mathbf{x}_1, \mathbf{x}_2\}$, $\{\mathbf{y}_1, \mathbf{y}_2\}$, and $\{\mathbf{z}_1, \mathbf{z}_2\}$ in $\mathbb{R}^2$. For any $\varepsilon > 0$, we define a rank-two tensor $A_\varepsilon \in \mathcal{R}_2$ as follows:

$$A_\varepsilon := \varepsilon^{-1} (\mathbf{x}_1 + \varepsilon\mathbf{x}_2) \otimes (\mathbf{y}_1 + \varepsilon\mathbf{y}_2) \otimes (\mathbf{z}_1 + \varepsilon\mathbf{z}_2) - \varepsilon^{-1} \mathbf{x}_1 \otimes \mathbf{y}_1 \otimes \mathbf{z}_1. \tag{12.4}$$

The limit tensor equals $A := \lim_{\varepsilon \to 0} A_\varepsilon = \mathbf{x}_1 \otimes \mathbf{y}_1 \otimes \mathbf{z}_2 + \mathbf{x}_1 \otimes \mathbf{y}_2 \otimes \mathbf{z}_1 + \mathbf{x}_2 \otimes \mathbf{y}_1 \otimes \mathbf{z}_1$. We claim that A has rank three. For $\mathbf{h} \in \mathbb{R}^2$, we apply multilinear multiplication by $X := (I_2, I_2, \mathbf{h}^\top)$ to A. This gives the 2×2-matrix

$$X.A = \langle \mathbf{h}, \mathbf{z}_2 \rangle \cdot \mathbf{x}_1 \otimes \mathbf{y}_1 + \langle \mathbf{h}, \mathbf{z}_1 \rangle \cdot \mathbf{x}_1 \otimes \mathbf{y}_2 + \langle \mathbf{h}, \mathbf{z}_1 \rangle \cdot \mathbf{x}_2 \otimes \mathbf{y}_1. \tag{12.5}$$

Choosing $\mathbf{h}$ with $\langle \mathbf{h}, \mathbf{z}_1 \rangle \neq 0$ yields a matrix of rank two, and hence A has rank at least two. We suppose now that $A = \mathbf{u}_1 \otimes \mathbf{v}_1 \otimes \mathbf{w}_1 + \mathbf{u}_2 \otimes \mathbf{v}_2 \otimes \mathbf{w}_2$ has rank two. Among the three pairs of vectors, at least two pairs are linearly independent. Suppose the pairs $(\mathbf{u}_1, \mathbf{u}_2)$ and $(\mathbf{v}_1, \mathbf{v}_2)$ are linearly independent. Then

$$X.A = \langle \mathbf{h}, \mathbf{w}_1 \rangle \cdot \mathbf{u}_1 \otimes \mathbf{v}_1 + \langle \mathbf{h}, \mathbf{w}_2 \rangle \cdot \mathbf{u}_2 \otimes \mathbf{v}_2. \tag{12.6}$$

Let us now pick a non-zero $\mathbf{h}$ such that $\langle \mathbf{h}, \mathbf{z}_1 \rangle = 0$. By (12.5), the matrix $X.A$ has rank one and hence $\langle \mathbf{h}, \mathbf{w}_1 \rangle = 0$ or $\langle \mathbf{h}, \mathbf{w}_2 \rangle = 0$ by (12.6). We may assume $\langle \mathbf{h}, \mathbf{w}_1 \rangle = 0$. This implies that $\mathbf{w}_1$ is a multiple of $\mathbf{z}_1$, since both are perpendicular to $\mathbf{h}$. We distinguish two cases. First, if the pair $(\mathbf{w}_1, \mathbf{w}_2)$ is linearly independent, there is another $\mathbf{h}'$ with $\langle \mathbf{h}', \mathbf{w}_1 \rangle \neq 0$ but $\langle \mathbf{h}', \mathbf{w}_2 \rangle = 0$. Then, $\langle \mathbf{h}', \mathbf{z}_1 \rangle \neq 0$ and so $X.A$ has rank two by (12.5) and it has rank one by (12.6); a contradiction. Second, if $\mathbf{w}_1$, $\mathbf{w}_2$ and $\mathbf{z}_1$ are multiples of one another, then the tensor A is of the form $A = (\lambda_1 \mathbf{u}_1 \otimes \mathbf{v}_1 + \lambda_2 \mathbf{u}_2 \otimes \mathbf{v}_2) \otimes \mathbf{z}_1$ (i.e., its "third row space" is spanned by $\mathbf{z}_1$). This is a contradiction to $A = \mathbf{x}_1 \otimes \mathbf{y}_1 \otimes \mathbf{z}_2 + (\mathbf{x}_1 \otimes \mathbf{y}_2 \otimes + \mathbf{x}_2 \otimes \mathbf{y}_1) \otimes \mathbf{z}_1$. In both cases, we have derived a contradiction, so A cannot have rank two, but it must have rank three. Finally, the same proof extends to larger tensor formats. Namely, we can simply embed A into a larger tensor format while retaining the relevant rank properties. $\qquad \square$

A decomposition of a tensor A of the form $A = A_1 + \cdots + A_r$ with $A_i \in S$ is called a *rank-r decomposition*. In the signal processing literature, it is also called *canonical polyadic decomposition*. One appealing property of higher-order tensors is *identifiability*. That is, many tensor decompositions are actually unique (while matrices of rank at least two do not have a unique decomposition). The following is [48, Theorem 1.1] and [149, Lemma 28].

Theorem 12.8 Let $d \geq 3$ and $n_1 \geq \cdots \geq n_d$ with $\prod_{i=1}^d n_i \leq 15000$, and set

$$r_0 = \left\lceil \frac{\dim \mathbb{R}^{n_1 \times \cdots \times n_d}}{\dim S} \right\rceil = \left\lceil \frac{n_1 \cdots n_d}{1 + \sum_{i=1}^d (n_i - 1)} \right\rceil.$$

Suppose that $r < r_0$ and the tuple $(n_1, \ldots, n_d, r)$ is not among the following special cases:

$(n_1, \ldots, n_d)$	r
$(4, 4, 3)$	5
$(4, 4, 4)$	6
$(6, 6, 3)$	8
$(n, n, 2, 2)$	$2n - 1$
$(2, 2, 2, 2, 2)$	5
$n_1 > \prod_{i=2}^d n_i - \sum_{i=2}^d (n_i - 1)$	$r \geq \prod_{i=2}^d n_i - \sum_{i=2}^d (n_i - 1)$

Then, a general tensor in $\mathcal{R}_r$ has a unique rank-r decomposition.

Remark 12.9 There are other notions of tensor rank, and these give alternative decompositions. Let $\mathbf{k} = (k_1, \ldots, k_d)$ be a vector of integers with $1 \leq k_i \leq n_i$. Let $A \in \mathbb{R}^{n_1 \times \cdots \times n_d}$ and $F_1, \ldots, F_d$ be the flattenings of A. Then, we say that A has *multilinear rank* (at most) $\mathbf{k}$ if $\text{rank}(F_j) \leq k_j$ for all $1 \leq j \leq d$. A *block term decomposition* of the tensor A has the form $A = A_1 + \cdots + A_r$, where each A_i has low multilinear rank $\mathbf{k}$. For order-three tensors, a block term decomposition can be visualized as follows:

The rank decomposition is the special case $\mathbf{k} = (1, \ldots, 1)$. In statistics, this corresponds to mixtures of independence models. Block term decompositions represent mixtures of discrete probability distributions that are more complex. Identifiability for block term decompositions is less studied than for rank decompositions. Some results, primarily for the case $d = 3$, can be found in [118] and references therein. ⋄

12.2 Eigenvectors and Singular Vectors

In this section, we study the Euclidean distance (ED) problem for the Segre variety and the Veronese variety. We start with the observation that each of them is the image of a polynomial map. The maps are

$$\psi : \mathbb{R}^{n_1} \times \cdots \times \mathbb{R}^{n_d} \to S, \quad (\mathbf{v}_1, \ldots, \mathbf{v}_d) \mapsto \mathbf{v}_1 \otimes \cdots \otimes \mathbf{v}_d, \quad \text{and}$$
$$v : \mathbb{R}^n \to \mathcal{V}, \qquad\qquad \mathbf{v} \mapsto \mathbf{v}^{\otimes d}. \tag{12.7}$$

We first consider the ED problem for the Segre variety S. Given any tensor $A \in \mathbb{R}^{n_1 \times \cdots \times n_d}$, we seek the rank one tensor B which is closest to A in Euclidean norm. Thus, we must solve the optimization problem

$$\min_{B \in S} \|A - B\| = \min_{\mathbf{v}_i \in \mathbb{R}^{n_i}} \|A - \psi(\mathbf{v}_1, \ldots, \mathbf{v}_d)\|. \tag{12.8}$$

Consider the best rank r approximation for a tensor A of order $d \geq 3$. This means computing $\min_{B \in \mathcal{R}_r} \|A - B\|$. For $r = 1$, this is precisely (12.8), and we shall present the solution in this section. For $r \geq 2$, this problem can be ill-posed (i.e., no minimizer exists). In fact, Proposition 12.7 implies that this problem is ill-posed for all tensors A in a full-dimensional open subset of $\mathbb{R}^{n_1 \times \cdots \times n_d}$. This was first noted by de Silva and Lim [53]. This is different for matrices, where the Eckart–Young Theorem (Theorem 2.9) provides an explicit algorithm for computing the minimizer. Likewise, for $r = 1$, the image of the Segre map ψ is closed, and hence the problem (12.8) has a solution $B \in S$. Our goal is to identify this B.

The critical values of the distance function $S \to \mathbb{R}, B \mapsto \|A - B\|$ are in one-to-one correspondence to the critical values of

$$\mathbb{R}^{n_1} \times \cdots \times \mathbb{R}^{n_d} \to S, \quad (\mathbf{v}_1, \ldots, \mathbf{v}_d) \mapsto \|A - \psi(\mathbf{v}_1, \ldots, \mathbf{v}_d)\|,$$

up to scaling $(\mathbf{v}_1, \ldots, \mathbf{v}_d) \mapsto (t_1\mathbf{v}_1, \ldots, t_d\mathbf{v}_d)$ with $t_1 \cdots t_d = 1$. Our goal is therefore to find the critical points of the function that maps the tuple $(\mathbf{v}_1, \ldots, \mathbf{v}_d)$ to

$$\|A - \psi(\mathbf{v}_1, \ldots, \mathbf{v}_d)\|^2 = \|A\|^2 - 2\langle A, \psi(\mathbf{v}_1, \ldots, \mathbf{v}_d)\rangle + \|\psi(\mathbf{v}_1, \ldots, \mathbf{v}_d)\|^2. \tag{12.9}$$

We compute the gradient of this function and set it equal to zero. This yields the critical points B we are interested in. By (12.3), the norm of ψ can be written as

$$\|\psi(\mathbf{v}_1, \ldots, \mathbf{v}_d)\|^2 = \|\mathbf{v}_1\|^2 \cdots \|\mathbf{v}_d\|^2.$$

The gradient of (12.9) with respect to $\mathbf{v}_k$ therefore equals

$$-2\tfrac{\mathrm{d}}{\mathrm{d}\mathbf{v}_k}\langle A, \psi(\mathbf{v}_1,\ldots,\mathbf{v}_d)\rangle + \tfrac{\mathrm{d}}{\mathrm{d}\mathbf{v}_k}(\|\mathbf{v}_1\|^2\cdots\|\mathbf{v}_d\|^2)$$

$$= -2\begin{bmatrix}\langle A, \mathbf{v}_1\otimes\cdots\otimes\mathbf{v}_{k-1}\otimes\mathbf{e}_1\otimes\mathbf{v}_{k+1}\otimes\cdots\otimes\mathbf{v}_d\rangle\\ \vdots \\ \langle A, \mathbf{v}_1\otimes\cdots\otimes\mathbf{v}_{k-1}\otimes\mathbf{e}_{n_k}\otimes\mathbf{v}_{k+1}\otimes\cdots\otimes\mathbf{v}_d\rangle\end{bmatrix} + \left(2\prod_{j\neq k}\|\mathbf{v}_j\|^2\right)\mathbf{v}_k. \tag{12.10}$$

Here $\mathbf{e}_1,\ldots,\mathbf{e}_{n_k}$ is the standard basis of $\mathbb{R}^{n_k}$. Abbreviating $\sigma_k := \prod_{j\neq k}\|\mathbf{v}_j\|^2$, we write the critical equations $\tfrac{\mathrm{d}}{\mathrm{d}\mathbf{v}_k}\|A - \psi(\mathbf{v}_1,\ldots,\mathbf{v}_d)\|^2 = 0$ from (12.10) in compact form:

$$A\bullet\otimes_{j\neq k}\mathbf{v}_j = \sigma_k\cdot\mathbf{v}_k \qquad \text{for } k = 1,\ldots,d. \tag{12.11}$$

For $d = 3$, we can visualize this as follows:

$$\tag{12.12}$$

Definition 12.10 Let $A \in \mathbb{R}^{n_1\times\cdots\times n_d}$. We say that $(\mathbf{v}_1,\ldots,\mathbf{v}_d) \in \mathbb{C}^{n_1}\times\cdots\times\mathbb{C}^{n_d}$, $\mathbf{v}_i \neq 0$, is a *singular vector tuple* for A if the equation (12.11) is valid for some $\sigma_1,\ldots,\sigma_d \in \mathbb{C}$. If this holds, then $(t_1\mathbf{v}_1,\ldots,t_d\mathbf{v}_d)$ satisfies the same condition, for all $t_1,\ldots,t_d \in \mathbb{C}\setminus\{0\}$. Thus, singular vector tuples are considered only up to scaling.

For $d = 2$, this coincides with the classic definition for rectangular matrices. The singular vector pair consists of the left singular vector and the right singular vector. The singular value decomposition implies that a general matrix $A \in \mathbb{R}^{n_1\times n_2}$ has precisely $\min\{n_1, n_2\}$ real singular vector pairs. For higher-order tensors, not all singular vectors need to be real.

Proposition 12.11 *The critical points (over $\mathbb{C}$) of the ED problem for the Segre variety S are the tuples of singular vectors. Hence, the Euclidean distance degree of S equals the number of singular vector tuples.*

Proof The computation in (12.10) shows that every critical point of the optimization problem (12.8) is a singular vector tuple of the tensor A. Conversely, consider a solution of (12.11). Multiplying (12.11) from the left with $\mathbf{v}_k^\top$ yields that $\sigma_k\cdot\|\mathbf{v}_k\|^2 = A\bullet\otimes_{j=1}^d\mathbf{v}_j$ is independent of k. Thus, $t := \sigma_k\cdot\|\mathbf{v}_k\|^2/\prod_{j=1}^d\|\mathbf{v}_j\|^2$ is independent of k. Now, rescaling $(\mathbf{v}_1,\ldots,\mathbf{v}_d)$ by $(t_1,\ldots,t_d)$ such that $t_1\cdot t_2\cdots t_d = t$ yields a new solution of (12.11) that satisfies $\sigma_k = \prod_{j\neq k}\|\mathbf{v}_j\|^2$ for all k. Hence, the critical points are in bijection with the singular vector tuples, assuming that these are viewed as points in $\mathbb{P}^{n_1-1}\times\cdots\times\mathbb{P}^{n_d-1}$. ∎

Example 12.12 ($d = 3$) Fix an $n_1\times n_2\times n_3$ tensor $A = (a_{ijk})$. Our unknowns are the vectors $\mathbf{v}_1 = (x_1,\ldots,x_{n_1}) \in \mathbb{P}^{n_1-1}$, $\mathbf{v}_2 = (y_1,\ldots,y_{n_2}) \in \mathbb{P}^{n_2-1}$, and $\mathbf{v}_3 = (z_1,\ldots,z_{n_3}) \in \mathbb{P}^{n_3-1}$. The singular vector equations (12.11) are

$$\sum_{j=1}^{n_2}\sum_{k=1}^{n_3} a_{ijk}\,y_j z_k = \sigma_1 x_i \qquad \text{for} \quad i = 1, 2, \ldots, n_1,$$
$$\sum_{i=1}^{n_1}\sum_{k=1}^{n_3} a_{ijk}\,x_i z_k = \sigma_2 y_j \qquad \text{for} \quad j = 1, 2, \ldots, n_2,$$
$$\sum_{i=1}^{n_1}\sum_{j=1}^{n_2} a_{ijk}\,x_i y_j = \sigma_3 z_k \qquad \text{for} \quad k = 1, 2, \ldots, n_3.$$

To work modulo scaling, we set $x_1 = y_1 = z_1 = 1$. Then this is a square system of polynomial equations, and it has finitely many solutions for generic tensors $A = (a_{ijk})$. For instance, the number of solutions is 6 when $n_1 = n_2 = n_3 = 2$ and it is 37 when $n_1 = n_2 = n_3 = 3$. This will be explained by Theorem 12.13.

To obtain the critical points of our ED problem (12.8), we retain x_1, y_1, z_1 as unknowns, and we add the constraints $\sigma_k = \prod_{j \neq k} \|\mathbf{v}_j\|^2$ for $k = 1, 2, 3$. Explicitly, for $n_1 = n_2 = n_3 = 2$, these constraints are the polynomial equations

$$(x_1^2 + x_2^2)(y_1^2 + y_2^2) = \sigma_3, \quad (x_1^2 + x_2^2)(z_1^2 + z_2^2) = \sigma_2, \quad \text{and} \quad (y_1^2 + y_2^2)(z_1^2 + z_2^2) = \sigma_1.$$

For each singular vector triple, we obtain a curve of solutions that maps to the same critical tensor B of rank one, up to scaling. For instance, for $n_1 = n_2 = n_3 = 2$, there are six such curves. Each of them has degree 12. In other words, the formulation on the right of (12.8) requires some care because the rank one factorization (12.1) is not unique. By using singular vector tuples up to scaling, we are on the safe side.   ◇

The following theorem of Friedland and Ottaviani [68] gives a formula, in terms of $n_1, n_2, \ldots, n_d$, for the number of singular vector tuples, and so for the ED degree of S.

Theorem 12.13 *Let $A \in \mathbb{R}^{n_1 \times \cdots \times n_d}$ be a general tensor. The number of singular vector tuples of A, counted up to scaling $(\mathbf{v}_1, \ldots, \mathbf{v}_d) \mapsto (t_1 \mathbf{v}_1, \ldots, t_d \mathbf{v}_d)$, is the coefficient of $x_1^{n_1 - 1} \cdots x_d^{n_d - 1}$ in the polynomial*

$$\prod_{i=1}^{d} \frac{f_i^{n_i} - x_i^{n_i}}{f_i - x_i}, \quad \text{where} \quad f_i := \sum_{j \neq i} x_j.$$

Example 12.14 The formula in Theorem 12.13 for binary tensors $A \in \mathbb{R}^{2 \times \cdots \times 2}$ gives

$$\prod_{i=1}^{d} \frac{f_i^2 - x_i^2}{f_i - x_i} = \prod_{i=1}^{d} (f_i + x_i) = (x_1 + x_2 + \cdots + x_d)^d.$$

The coefficient of $x_1 x_2 \cdots x_d$ in this polynomial equals $d!$. In particular, the ED degree of the Segre variety S in $\mathbb{R}^{2 \times 2 \times 2}$ is found to be $3! = 6$.   ◇

We turn from general tensors to symmetric tensors, and we consider the ED problem for the Veronese variety $\mathcal{V}$ in Definition 12.5. Now A is a symmetric tensor, all $\mathbf{v}_i$ are equal to the same vector $\mathbf{v}$ in $\mathbb{R}^n$, and the gradient in (12.10) does not depend on k. Setting $\lambda := \sigma_k$, the critical equations for the ED problem on $\mathcal{V}$ are

$$A \bullet \mathbf{v}^{\otimes(d-1)} = \lambda \mathbf{v}. \tag{12.13}$$

Similarly to (12.12), we visualize this equation for symmetric tensors of order $d = 3$:

Definition 12.15 Let $A \in S^d(\mathbb{R}^n)$. We call $\mathbf{v} \in \mathbb{C}^n \backslash \{\mathbf{0}\}$ an *eigenvector* of A if (12.13) holds for some $\lambda \in \mathbb{C}$. The pair $(\mathbf{v}, \lambda)$ is called an *eigenpair* of the symmetric tensor A.

Eigenpairs have another interesting interpretation, next to being critical points for the Euclidean distance problem on the Veronese variety $\mathcal{V}$. Recall that every symmetric tensor $A \in S^d(\mathbb{R}^n)$ corresponds uniquely to a homogeneous polynomial $F_A(\mathbf{x}) = \langle A, \mathbf{x}^{\otimes d} \rangle$ of degree d in n variables. Under this interpretation, the vector $A \bullet \mathbf{x}^{\otimes(d-1)}$ on the left in (12.13) is the gradient vector $\nabla F_A(\mathbf{x}) = (\partial F_A / \partial x_1, \ldots, \partial F_A / \partial x_n)^\top$. Therefore, eigenpairs correspond to fixed points of the gradient map $\mathbb{P}^{n-1} \dashrightarrow \mathbb{P}^{n-1}$, $\mathbf{x} \mapsto \nabla F_A(\mathbf{x})$.

Example 12.16 ($n = 3$) Symmetric tensors of order d and format $3 \times 3 \times \cdots \times 3$ correspond to ternary forms $F(x_1, x_2, x_3)$ of degree d. The eigenvectors of such a tensor are solutions of

$$\partial F / \partial x_1 = \lambda x_1, \quad \partial F / \partial x_2 = \lambda x_2, \quad \partial F / \partial x_3 = \lambda x_3.$$

Eliminating λ is equivalent to the statement that the following matrix has rank one:

$$\begin{bmatrix} x_1 & x_2 & x_3 \\ \partial F / \partial x_1 & \partial F / \partial x_2 & \partial F / \partial x_3 \end{bmatrix}.$$

This is a dependent system of three homogeneous equations of degree d in three variables. The number of solutions $(x_1 : x_2 : x_3)$ in $\mathbb{P}^2$ equals $d^2 - d + 1$. For $d = 2$, these are just the three eigenvectors of a symmetric 3×3 matrix. For $d = 3$, one interesting example is $F = x_1^3 + x_2^3 + x_3^3$. The corresponding $3 \times 3 \times 3$ tensor has 7 eigenvectors, namely the elements of $\{0, 1\}^3 \backslash \{\mathbf{0}\}^3$. How about $F = x_1^d + x_2^d + x_3^d$ for $d \geq 4$? ◇

We now remove the assumption that A is symmetric. Definition 12.15 makes sense for any $A \in (\mathbb{R}^n)^{\otimes d}$. We lose the interpretations above, but we gain a general definition. The number of eigenpairs of a general tensor was found by Cartwright and Sturmfels in [39].

Theorem 12.17 *Let $A \in (\mathbb{R}^n)^{\otimes d}$ be a general tensor. The number of complex eigenvectors of A, up to scaling $\mathbf{v} \mapsto t\mathbf{v}$, is the sum*

$$\sum_{i=0}^{n-1} (d-1)^i.$$

Proof For $d = 2$, the sum gives n, which is the number of eigenvectors of a general matrix $A \in \mathbb{R}^{n \times n}$. Hence, we assume $d > 2$ from now on. We use the same proof strategy as for Corollary 3.15. Consider the symmetric tensor A with

$$F_A(\mathbf{x}) = x_1^d + x_2^d + \cdots + x_n^d.$$

The equations for its eigenpairs are

$$x_1^{d-1} - \xi^{d-2} x_1 = x_2^{d-1} - \xi^{d-2} x_2 = \cdots = x_n^{d-1} - \xi^{d-2} x_n = 0,$$

where $\lambda = \xi^{d-2}$ is the eigenvalue. We count that this system of homogeneous polynomial equations has precisely $(d-1)^n$ regular solutions in $\mathbb{P}^n$. By Bézout's theorem, a general system of n polynomials with degrees $(d-1, \ldots, d-1)$ also has $(d-1)^n$ regular zeros. The Parameter Continuation Theorem 3.18 then implies that the analogous equations for a general tensor $A \in (\mathbb{R}^n)^{\otimes d}$ also have $(d-1)^n$ regular zeros. We remove the extraneous solution in $\mathbb{P}^n$ that is given by $\xi = 1$ and $\mathbf{x} = 0$. The remaining $(d-1)^n - 1$ solutions come in clusters of $d-2$, where ξ runs over the $(d-2)$-th roots of unity. After taking the image of the projection $(\xi, \mathbf{x}) \mapsto \mathbf{x}$, the number of distinct points $\mathbf{x}$ in $\mathbb{P}^{n-1}$ is the quotient

$$\frac{(d-1)^n - 1}{d - 2} = \sum_{i=0}^{n-1} (d-1)^i.$$

We conclude that a general tensor of format $n \times \cdots \times n$ has $\sum_{i=0}^{n-1} (d-1)^i$ eigenvectors. $\square$

For matrices, every n-tuple of linearly independent vectors can be eigenvectors of a matrix. For tensors, this is not so. Abo, Sturmfels, and Seigal proved in [1] that a set of d points $\mathbf{v}_1, \ldots, \mathbf{v}_d$ in $\mathbb{C}^2$ is the eigenconfiguration of a symmetric tensor in $S^d(\mathbb{C}^2)$ if and only if d is odd, or $d = 2k$ is even and the differential operator $(\frac{\partial^2}{\partial x^2} + \frac{\partial^2}{\partial y^2})^k$ annihilates the binary form $f_{\mathbf{v}_i}(x, y) := \prod_{i=1}^d (b_i x - a_i y)$, where $\mathbf{v}_i = (a_i, b_i)$, for all $i = 1, \ldots, d$. The extension of this result to any number of variables was studied by Beorchia, Galuppi, and Venturello in [17]. This study of eigenschemes continues to be an area of active research.

In this section, we showed that the ED problem for Segre varieties and Veronese varieties leads directly to singular vectors and eigenvectors of tensors. By the ED duality in Theorem 2.23, it is equivalent to study the ED problem for the dual hypersurfaces, which are the hyperdeterminant and the discriminant, respectively. This, in turn, takes us to the geometry of condition numbers, as discussed in Chapter 9. A natural generalization arises from the Segre–Veronese varieties, whose polar degrees we saw in Theorem 5.18.

12.3 Volumes of Rank-One Varieties

We now shift gears, and we examine the metric geometry of the Segre variety and the Veronese variety. Both are now taken to be complex projective varieties. To this end, we pass to complex projective space $\mathbb{P}^N$, where $N = n_1 \cdots n_d - 1$, or $N = n^d - 1$ if the n_i are all equal. Our two complex projective varieties are

$$\begin{aligned}
S_{\mathbb{P}} &:= \{ \mathbf{v}_1 \otimes \cdots \otimes \mathbf{v}_d \in \mathbb{P}^N \mid \mathbf{v}_i \in \mathbb{P}^{n_i - 1}, 1 \leq i \leq d \} \quad \text{and} \\
V_{\mathbb{P}} &:= \{ \mathbf{v}^{\otimes d} \in \mathbb{P}^N \mid \mathbf{v}_i \in \mathbb{P}^{n-1} \}.
\end{aligned} \tag{12.14}$$

Thus, $S_{\mathbb{P}}$ consists of rank-one tensors $\mathbf{v}_1 \otimes \cdots \otimes \mathbf{v}_d$, where $\mathbf{v}_i \in \mathbb{C}^{n_i}$, taken up to scaling by $\mathbb{C}^*$. Similarly, $V_{\mathbb{C}}$ consists of all vectors $\mathbf{v}^{\otimes d}$ for $\mathbf{v} \in \mathbb{C}^n$, again taken up to scaling by $\mathbb{C}^*$.

Our aim is to compute the volumes of $S_{\mathbb{P}}$ and $V_{\mathbb{P}}$. The first crucial step is to explain what this means. In Section 6.3 and Chapter 14, we discuss volumes of real semialgebraic sets. But here, the situation is different. In this section, our objects are lower-dimensional submanifolds in the complex projective space $\mathbb{P}^N$. How does one measure volumes in $\mathbb{P}^N$? How is the volume of a submanifold defined and computed?

We begin with the Hermitian inner product on the tensor space $\mathbb{C}^{n_1 \times \cdots \times n_d}$. This is

$$\langle A, B \rangle_{\mathbb{C}} := \sum_{i_1=1}^{n_1} \cdots \sum_{i_d=1}^{n_d} \overline{a_{i_1,\ldots,i_d}} \cdot b_{i_1,\ldots,i_d}.$$

We view $\mathbb{C}^{n_1 \times \cdots \times n_d}$ as a real Euclidean space of dimension $2n_1 \cdots n_d$, where the Euclidean inner product is $(A, B) \mapsto \mathrm{Re}(\langle A, B \rangle_{\mathbb{C}})$. The same definitions make sense in $\mathbb{C}^n$, now identified with $\mathbb{R}^{2n}$, and we define the unit sphere in the Euclidean space $\mathbb{C}^n$ as

$$\mathbb{S}^{2n-1} := \{\mathbf{a} \in \mathbb{C}^n \mid \langle \mathbf{a}, \mathbf{a} \rangle_{\mathbb{C}} = 1\}.$$

This sphere is a real manifold of real dimension $\dim_{\mathbb{R}} \mathbb{S}^{2n-1} = 2n - 1$. The projection $\pi : \mathbb{S}^{2n-1} \to \mathbb{P}^{n-1}$ sends a vector $\mathbf{a} \in \mathbb{S}^{2n-1}$ to its class in complex projective space. The fiber of the quotient map π has real dimension one: it is a circle. The tangent space of the sphere $\mathbb{S}^{2n-1}$ at a point $\mathbf{a}$ is the real vector space

$$T_{\mathbf{a}}\mathbb{S}^{2n-1} = \{\mathbf{t} \in \mathbb{C}^n \mid \mathrm{Re}(\langle \mathbf{a}, \mathbf{t} \rangle_{\mathbb{C}}) = 0\} \simeq \mathbb{R}^{2n-1}.$$

Lemma 12.18 *Let $\mathbf{t} \in \mathbb{C}^n$. We have $\mathrm{Re}(\langle \mathbf{a}, \mathbf{t} \rangle_{\mathbb{C}}) = 0$ if and only if $\mathbf{t}$ is a point in the real hyperplane spanned by the real line through $\sqrt{-1} \cdot \mathbf{a}$ and the codimension-two linear space defined by $\langle \mathbf{a}, \mathbf{t} \rangle_{\mathbb{C}} = 0$.*

Proof The space of all $\mathbf{t} \in \mathbb{C}^n$ such that $\mathrm{Re}(\langle \mathbf{a}, \mathbf{t} \rangle_{\mathbb{C}}) = 0$ is a linear space of real codimension one in $\mathbb{C}^n \cong \mathbb{R}^{2n}$. If $\mathbf{t}$ is a real multiple of $\sqrt{-1} \cdot \mathbf{a}$ or $\langle \mathbf{a}, \mathbf{t} \rangle_{\mathbb{C}} = 0$, then we have $\mathrm{Re}(\langle \mathbf{a}, \mathbf{t} \rangle_{\mathbb{C}}) = 0$. The other inclusion follows from comparing dimensions. $\qquad\square$

For $\mathbf{x} = \pi(\mathbf{a}) \in \mathbb{P}^{n-1}$, the fiber under π is $\pi^{-1}(\mathbf{x}) = \{\exp(\sqrt{-1}\varphi) \cdot \mathbf{a} \mid \varphi \in \mathbb{R}\}$. Its tangent space at $\mathbf{a}$ is spanned by $(\frac{d}{d\varphi} \exp(\sqrt{-1}\varphi) \cdot \mathbf{a})|_{\varphi=0} = \sqrt{-1} \cdot \mathbf{a}$. Consequently, Lemma 12.18 implies that the tangent space of the sphere $T_{\mathbf{a}}\mathbb{S}^{2n-1}$ has an orthogonal decomposition into the tangent space of the fiber $T_{\mathbf{a}}\pi^{-1}(\mathbf{x}) = \mathbb{R} \cdot \sqrt{-1}\,\mathbf{a}$ and the linear space defined by $\langle \mathbf{a}, \mathbf{t} \rangle_{\mathbb{C}} = 0$. Therefore, we can view the following as the tangent space of projective space at the point $\mathbf{x} = \pi(\mathbf{a})$, where $\mathbf{a} \in \mathbb{S}^{2n-1}$:

$$T_{\mathbf{x}}\mathbb{P}^{n-1} = \{\mathbf{t} \in \mathbb{C}^n \mid \langle \mathbf{a}, \mathbf{t} \rangle_{\mathbb{C}} = 0\} \simeq \mathbb{R}^{2n-2}. \tag{12.15}$$

In conclusion, the complex projective space $\mathbb{P}^{n-1}$ is a Riemannian manifold with respect to the Euclidean structure $\mathrm{Re}(\langle \mathbf{a}, \mathbf{b} \rangle)_{\mathbb{C}}$. This induces a notion of volume for subsets of $\mathbb{P}^{n-1}$.

Our metric structures on $\mathbb{S}^{2n-1}$ and $\mathbb{P}^{n-1}$ have the property that the projection π is a Riemannian submersion. This implies that the m-dimensional real volume of a measurable subset $U \subset \mathbb{P}^{n-1}$ is

$$\mathrm{vol}_m(U) = \frac{1}{2\pi} \mathrm{vol}_{m+1}(\pi^{-1}(U)),$$

since the preimage $\pi^{-1}(\mathbf{x})$ for $\mathbf{x} \in \mathbb{P}^{n-1}$ is a circle.

The volume of projective space itself is

$$\mathrm{vol}_{2(n-1)}(\mathbb{P}^{n-1}) \;=\; \frac{1}{2\pi}\mathrm{vol}_{2n-1}(\mathbb{S}^{2n-1}) = \frac{\pi^{n-1}}{(n-1)!}. \tag{12.16}$$

In the following, we sometimes omit the subscript from vol when the dimension is clear from the context. For the next proposition, we abbreviate $m := \dim_{\mathbb{R}} S_{\mathbb{P}}$.

Proposition 12.19 *The m-dimensional volume of the Segre variety $S_{\mathbb{P}}$ in (12.14) is*

$$\mathrm{vol}(S_{\mathbb{P}}) \;=\; \mathrm{vol}(\mathbb{P}^{n_1-1}) \cdots \mathrm{vol}(\mathbb{P}^{n_d-1}).$$

Proof The Segre map from (12.7) for complex projective space is the smooth embedding

$$\psi_{\mathbb{P}} : \mathbb{P}^{n_1-1} \times \cdots \times \mathbb{P}^{n_d-1} \to S_{\mathbb{P}}.$$

Let $(\mathbf{x}_1,\ldots,\mathbf{x}_d) \in \mathbb{P}^{n_1-1} \times \cdots \times \mathbb{P}^{n_d-1}$ and $\mathbf{a}_i \in \mathbb{S}^{2n_i-1}$ with $\pi(\mathbf{a}_i) = \mathbf{x}_i$ for $i = 1,\ldots,d$. We also fix elements $\mathbf{t}_i \in \mathbb{C}^{n_i}$ with $\langle \mathbf{a}_i, \mathbf{t}_i \rangle_{\mathbb{C}} = 0$ for all i. These specify vectors in the various tangent spaces (12.15). The derivative of the Segre map $\psi_{\mathbb{P}}$ at our point takes $(\mathbf{t}_1,\ldots,\mathbf{t}_d) \in T_{(\mathbf{x}_1,\ldots,\mathbf{x}_d)}(\mathbb{P}^{n_1-1} \times \cdots \times \mathbb{P}^{n_d-1})$ to

$$\theta \;:=\; \mathbf{t}_1 \otimes \mathbf{a}_2 \otimes \cdots \otimes \mathbf{a}_d \,+\, \mathbf{a}_1 \otimes \mathbf{t}_2 \otimes \cdots \otimes \mathbf{a}_d \,+\, \cdots \,+\, \mathbf{a}_1 \otimes \mathbf{a}_2 \otimes \cdots \otimes \mathbf{t}_d.$$

It follows from (12.3) that the terms in this sum are pairwise orthogonal. Therefore,

$$\|\theta\|^2 \;=\; \|\mathbf{t}_1\|^2 + \|\mathbf{t}_2\|^2 + \cdots + \|\mathbf{t}_d\|^2.$$

Hence, the derivative of $\psi_{\mathbb{P}}$ preserves norms, and $\psi_{\mathbb{P}}$ is a volume-preserving embedding.□

Proposition 12.20 *The $2(n-1)$-dimensional volume of the Veronese variety in (12.14) is*

$$\mathrm{vol}(\mathcal{V}_{\mathbb{P}}) \;=\; d^{n-1} \cdot \mathrm{vol}(\mathbb{P}^{n-1}).$$

Proof The proof is similar to that of Proposition 12.19. We denote the projective Veronese map by $v_{\mathbb{P}} : \mathbb{P}^{n-1} \to \mathcal{V}_{\mathbb{P}}$. Just like the Segre map, $v_{\mathbb{P}}$ is a smooth embedding. Let $\mathbf{x} \in \mathbb{P}^{n-1}$ be a point in projective space and fix a representative $\mathbf{a} \in \mathbb{S}^{2n-1}$ for $\mathbf{x}$; i.e., $\pi(\mathbf{a}) = \mathbf{x}$. Let $\mathbf{t} \in \mathbb{C}^n$ with $\langle \mathbf{a}, \mathbf{t} \rangle_{\mathbb{C}} = 0$. The derivative of $v_{\mathbb{P}}$ at $\mathbf{a}$ maps $\mathbf{t} \in T_{\mathbf{x}}\mathbb{P}^{n-1}$ to

$$\theta \;:=\; \mathbf{t} \otimes \mathbf{a} \otimes \cdots \otimes \mathbf{a} + \mathbf{a} \otimes \mathbf{t} \otimes \cdots \otimes \mathbf{a} + \cdots + \mathbf{a} \otimes \mathbf{a} \otimes \cdots \otimes \mathbf{t}.$$

It follows from (12.3) that the terms in this sum are pairwise orthogonal, so $\|\theta\|^2 = d\|\mathbf{t}\|^2$. This shows that the derivative of $v_{\mathbb{P}}$ scales norms by a factor of $\sqrt{d}$. We get the formula

$$\mathrm{vol}(\mathcal{V}_{\mathbb{P}}) \;=\; \left(\sqrt{d}\,\right)^{\dim_{\mathbb{R}} \mathbb{P}^{n-1}} \cdot \mathrm{vol}(\mathbb{P}^{n-1}) \;=\; d^{n-1} \cdot \mathrm{vol}(\mathbb{P}^{n-1}),$$

which relates the volume of the Veronese variety to that of the underlying projective space.□

The volume of a complex projective variety is closely related to its degree. This is the content of Howard's *Kinematic Formula* [94], which concerns the average volume of intersections of submanifolds in homogeneous spaces. For complex projective space, we apply [94, Theorem 3.8 & Corollary 3.9] as follows. Let $M \subset \mathbb{P}^N$ be a smooth manifold of complex dimension m. Here, the Kinematic Formula states

$$\mathbb{E}_U \, \#(M \cap U \cdot (\mathbb{P}^{N-m} \times \{0\}^m)) = \frac{\mathrm{vol}_{2m}(M)}{\mathrm{vol}_{2m}(\mathbb{P}^m)},$$

where the expectation refers to the probability Haar measure on the unitary group $U(N+1)$.

If $X \subset \mathbb{P}^N$ is a smooth variety of complex dimension m, then the number of intersection points $\#(X \cap U \cdot (\mathbb{P}^{N-m} \times \{0\}^m))$ equals the degree of X for almost all $U \in U(N+1)$. Thus, we take the expected value of a constant function:

$$\deg(X) = \frac{\mathrm{vol}_{2m}(X)}{\mathrm{vol}_{2m}(\mathbb{P}^m)}, \qquad \text{where } m = \dim_{\mathbb{C}}(X). \tag{12.17}$$

Using Propositions 12.19 and 12.20, our volume computation yields the following result:

Corollary 12.21 *The degrees of the Segre variety and the Veronese variety are equal to*

(a) $\deg(\mathcal{S}_\mathbb{P}) = \dfrac{(n_1 + \cdots + n_d - d)!}{(n_1 - 1)! \cdots (n_d - 1)!}$,

(b) $\deg(\mathcal{V}_\mathbb{P}) = d^{n-1}$.

Proof The second formula follows directly from Proposition 12.20 and (12.17). For the first formula, we recall from (12.16) the volume of projective space: $\mathrm{vol}_{2(n-1)}(\mathbb{P}^{n-1}) = \frac{\pi^{n-1}}{(n-1)!}$. We set $m := \dim_{\mathbb{C}} \mathcal{S}_\mathbb{P}$. Using Proposition 12.19 and (12.17) we then have

$$\deg(\mathcal{S}_\mathbb{P}) = \frac{\mathrm{vol}(\mathbb{P}^{n_1-1}) \cdots \mathrm{vol}(\mathbb{P}^{n_d-1})}{\mathrm{vol}(\mathbb{P}^m)} = \frac{\pi^{\sum_{i=1}^d (n_i-1)}}{\pi^m} \frac{m!}{(n_1 - 1)! \cdots (n_d - 1)!}$$

$$= \frac{(n_1 + \cdots + n_d - d)!}{(n_1 - 1)! \cdots (n_d - 1)!},$$

where we have used that $m = n_1 + \cdots + n_d - d$. □

Remark 12.22 By (12.2), a linear equation on the Veronese variety $\mathcal{V}_\mathbb{P}$ is a homogeneous polynomial equation of degree d on projective space $\mathbb{P}^{n-1}$. Thus, $\deg(\mathcal{V}_\mathbb{P}) = d^{n-1}$ means that $n-1$ general homogeneous polynomials of degree d have d^{n-1} zeros.

Remark 12.23 Howard's Kinematic Formula [94] also provides the following result for real projective space $\mathbb{P}^N_{\mathbb{R}}$. Let $M \subset \mathbb{P}^N_{\mathbb{R}}$ be a real submanifold of real dimension m. Then, the Kinematic Formula implies that $\mathbb{E}_U \, \#(M \cap U \cdot (\mathbb{P}^{N-m}_{\mathbb{R}} \times \{0\}^m)) = \mathrm{vol}_m(M)/\mathrm{vol}_m(\mathbb{P}^m_{\mathbb{R}})$, where the expectation refers to the probability measure on the real orthogonal group $O(n+1)$. Thus, the volume of real projective varieties can be interpreted as an "average degree". We can use the same proof strategies as above to show that the projective volume of the real Segre variety S is equal to $\mathrm{vol}(\mathbb{P}^{n_1-1}_{\mathbb{R}}) \cdots \mathrm{vol}(\mathbb{P}^{n_d-1}_{\mathbb{R}})$ and the projective volume of the real Veronese V is $\sqrt{d^{n-1}} \cdot \mathrm{vol}(\mathbb{P}^{n-1}_{\mathbb{R}})$. The latter result was first derived by Edelman and Kostlan in their seminal paper [66]. They use this to find the average number of real zeros of a system of homogeneous polynomials of degree d in n variables.

Chapter 13
Computer Vision

The field of computer vision studies how computers can gain understanding from images and videos, similar to human cognitive abilities. One of the classical challenges is to reconstruct a 3D object from images taken by several unknown cameras. While the resulting questions in multiview geometry [80] have a long history in computer vision, recent years have seen a confluence with ideas and algorithms from algebraic geometry. This led to the development of a new subject area called *Algebraic Vision* [104]. In this chapter, we present an introduction to this area from the perspective of metric algebraic geometry.

13.1 Multiview Varieties

A pinhole camera consists of a point $\mathbf{c}$, called *camera center*, and an image plane H in 3-space. Taking an image of a 3D point $\mathbf{x}$ is modeled by the following process. One draws a line between $\mathbf{x}$ and the camera center $\mathbf{c}$. The intersection point of that line with H is the image point. This is shown in Figure 13.1. The common algebraic model for such a camera is a surjective linear projection $C : \mathbb{P}^3 \dashrightarrow \mathbb{P}^2$, given by a full-rank 3×4 matrix A, i.e., $C(\mathbf{x}) = A\mathbf{x}$. Thus, we work in projective geometry, with homogeneous coordinates. We take 3D points in projective space $\mathbb{P}^3$, and we identify the image plane H with the projective plane $\mathbb{P}^2$. For arbitrary A, such cameras are called *projective* or *uncalibrated*.

In many applications, one does not allow arbitrary matrices A because one assumes several internal camera parameters, such as focal length, to be known. *Calibrated cameras* are those that can be obtained by rotating and translating a camera in standard position; see Figure 13.1. The standard camera is the linear projection $\mathbb{P}^3 \dashrightarrow \mathbb{P}^2$, $[x_1 : x_2 : x_3 : x_4] \mapsto [x_1 : x_2 : x_3]$, given by the 3×4 matrix $A = \begin{bmatrix} I_3 \mid 0 \end{bmatrix}$. Thus, all calibrated cameras are given by matrices $A = \begin{bmatrix} R \mid t \end{bmatrix}$, where $R \in SO(3)$ and $t \in \mathbb{R}^3$. Between uncalibrated and calibrated cameras, there are also partially calibrated camera models; see [80, Chapter 6].

For 3D reconstruction, one needs at least two cameras. The *joint camera map* is

$$\Phi : C_m \times X \dashrightarrow Y. \tag{13.1}$$

© The Author(s) 2024
P. Breiding et al., *Metric Algebraic Geometry*, Oberwolfach Seminars 53,
https://doi.org/10.1007/978-3-031-51462-3_13

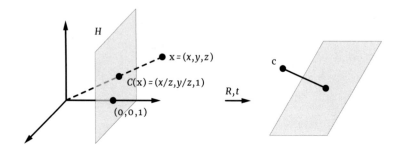

Fig. 13.1: Calibrated cameras: in standard position (on the left) and arbitrary (on the right).

This map takes an m-tuple of cameras $(C_1, \ldots, C_m) \in \mathcal{C}_m$ and an algebraic 3D object $\mathbf{x} \in X$ to the m images $(C_1(\mathbf{x}), \ldots, C_m(\mathbf{x})) \in Y$. Here, X and Y are appropriate spaces of objects and their images, respectively. For instance, the 3D object $\mathbf{x}$ could be a single point (i.e., $X = \mathbb{P}^3$), a line (i.e., $X = \mathrm{Gr}(1, \mathbb{P}^3)$), an arrangement of points and lines with prescribed incidences, a curve, or a surface. The image space Y parametrizes m-tuples of points, lines, or other arrangments in $\mathbb{P}^2$. For instance, if $X = \mathbb{P}^3$ then $Y = (\mathbb{P}^2)^m$.

Formally, the task of 3D reconstruction from unknown cameras means to compute fibers under the joint camera map Φ. Thus, there are unknown objects in 3-space and m unknown cameras. We get to see the planar projections of the objects. The task is to find the original objects and the cameras with which the images were taken.

Example 13.1 For $m = 2$ projective cameras observing k points, the joint camera map is

$$\Phi : \ (\mathbb{P}\mathbb{R}^{3\times4})^2 \times (\mathbb{P}^3_{\mathbb{R}})^k \ \dashrightarrow \ (\mathbb{P}^2_{\mathbb{R}})^k \times (\mathbb{P}^2_{\mathbb{R}})^k,$$
$$(A_1, A_2, \mathbf{x}_1, \ldots, \mathbf{x}_k) \ \mapsto \ (A_1\mathbf{x}_1, \ldots, A_1\mathbf{x}_k, A_2\mathbf{x}_1, \ldots, A_2\mathbf{x}_k). \tag{13.2}$$

It is defined whenever none of the 3D points $\mathbf{x}_i$ is the kernel of A_1 or A_2. The kernel of the matrix A_i is the *camera center* of C_i. We note that Φ maps a space of dimension $22 + 3k$ to a space of dimension $4k$. The general non-empty fiber has dimension at least 15. This is given by transforming $A_i \mapsto A_i g$ and $\mathbf{x}_j \mapsto g^{-1}\mathbf{x}_j$ for some $g \in \mathrm{PGL}(4)$. For large k, the image of Φ is a proper subvariety of its codomain. For small k, the map Φ is dominant. The transition occurs for $k = 7$. Indeed, 7 is the largest k such that the map Φ is dominant. We will study this issue of dominance transition in Section 13.3.

For calibrated cameras, given by matrices $A_i = [\, R_i \,|\, t_i \,]$, the joint camera map becomes

$$\Phi : (\mathrm{SO}(3) \times \mathbb{R}^3)^2 \times (\mathbb{P}^3_{\mathbb{R}})^k \ \dashrightarrow \ (\mathbb{P}^2_{\mathbb{R}})^k \times (\mathbb{P}^2_{\mathbb{R}})^k. \tag{13.3}$$

Here, the domain has dimension $12 + 3k$. The dominance transition occurs at $k = 5$. $\diamond$

We now change the problem, and we assume that the cameras are fixed and known. We are thus fixing a camera configuration $C = (C_1, \ldots, C_m) \in C_m$, and we consider the specialization $\Phi_C : X \dashrightarrow Y$ of the joint camera map. The Zariski closure of the image of that map is the *multiview variety* $\mathcal{M}_C$ of the cameras C.

In what follows, we focus on the multiview variety of a single point in 3-space. This means that $X = \mathbb{P}^3$ and $Y = (\mathbb{P}^2)^m$. In this case, the multiview variety $\mathcal{M}_C$ is the closure of the image of the map given by multiplying the same vector with m distinct matrices:

$$\Phi_C : \mathbb{P}^3 \dashrightarrow (\mathbb{P}^2)^m, \quad \mathbf{x} \mapsto (A_1\mathbf{x}, A_2\mathbf{x}, \ldots, A_m\mathbf{x}). \tag{13.4}$$

This map is well-defined at all points of $\mathbb{P}^3$ except for the camera centers $\ker(A_i)$. If $m \geq 2$ and not all camera centers coincide, then $\mathcal{M}_C$ is a threefold. For an algebraic geometer, $\mathcal{M}_C$ is the threefold obtained from $\mathbb{P}^3$ by blowing up the camera centers.

From a computational point of view, the first question is to find the implicit description of $\mathcal{M}_C$ as a subvariety of $(\mathbb{P}^2)^m$. Writing $(\mathbf{y}_1, \ldots, \mathbf{y}_m)$ for the $3m$ coordinates of $(\mathbb{P}^2)^m$, we seek multihomogeneous polynomials in the $\mathbf{y}_i$ whose common zero set is equal to $\mathcal{M}_C$.

The desired implicit description is given by the $3m \times (m+4)$ matrix

$$M_A := \begin{bmatrix} A_1 & \mathbf{y}_1 & 0 & \cdots & 0 \\ A_2 & 0 & \mathbf{y}_2 & \cdots & 0 \\ \vdots & & & \ddots & \\ A_m & 0 & 0 & \cdots & \mathbf{y}_m \end{bmatrix}. \tag{13.5}$$

Here, A_i is the 3×4 matrix that defines the camera C_i, and $\mathbf{y}_i$ is an unknown column vector of length 3, serving as homogeneous coordinates for the ith image plane $\mathbb{P}^2$. The following result on the multiview variety for a single point goes back to [88]. For recent ideal-theoretic versions of Proposition 13.2, see [104, Theorem 2] and the references therein.

Proposition 13.2 *The multiview variety for m cameras with at least two distinct centers is*

$$\mathcal{M}_C = \{(\mathbf{y}_1, \ldots, \mathbf{y}_m) \in (\mathbb{P}^2)^m \mid \text{rank } M_A(\mathbf{y}_1, \ldots, \mathbf{y}_m) < m + 4\}.$$

Proof (Idea) A tuple $(\mathbf{y}_1, \ldots, \mathbf{y}_m) \in (\mathbb{P}^2)^m$ is in the image of (13.4) if and only if there exists $\mathbf{x} \in \mathbb{P}^3$ and non-zero scalars λ_i such that $A_i\mathbf{x} = \lambda_i\mathbf{y}_i$ for $i = 1, \ldots, m$. The latter condition is equivalent to the vector $(\mathbf{x}^\top, -\lambda_1, \ldots, -\lambda_m)^\top$ being in the kernel of the matrix M_A in (13.5). Taking the closure, we obtain that the multiview variety $\mathcal{M}_C$ consists of those tuples $(\mathbf{y}_1, \ldots, \mathbf{y}_m)$ such that M_A has a non-zero kernel. $\square$

Example 13.3 For $m = 2$ cameras, the matrix in (13.5) has format 6×6, and it equals

$$M_A = \begin{bmatrix} A_1 & \mathbf{y}_1 & 0 \\ A_2 & 0 & \mathbf{y}_2 \end{bmatrix}.$$

The multiview variety $\mathcal{M}_C$ is the hypersurface in $\mathbb{P}^2 \times \mathbb{P}^2$ defined by $\det(M_A) = 0$. This determinant is a bilinear form. Hence, there exists a 3×3 matrix F such that

$$\det(M_A) = \mathbf{y}_2^\top F \mathbf{y}_1. \tag{13.6}$$

Note that the entries of F depend on the matrices A_i. The matrix F is called the *fundamental matrix*. A computation verifies the following explicit formula:

$$F = \begin{bmatrix} [2356] & -[1356] & [1256] \\ -[2346] & [1346] & -[1246] \\ [2345] & -[1345] & [1245] \end{bmatrix}. \tag{13.7}$$

Here, $[ijkl]$ is the determinant of the 4×4 submatrix with row indices i, j, k, l of the 6×4 matrix $\begin{bmatrix} A_1^\top & A_2^\top \end{bmatrix}^\top$. The matrix F has rank two. In fact, its right kernel is the image of the camera center of C_2 under the camera C_1. Similarly, the left kernel of F is the image of the camera center of C_1 under the camera C_2. A geometric interpretation is shown in Figure 13.2. The two cameras are projections onto the two planes.   ◇

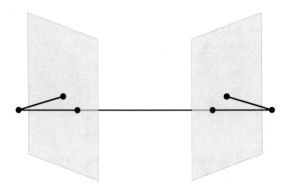

Fig. 13.2: The image of the center of one camera under the other camera (red points) is the kernel of the fundamental matrix.

In computer vision, the term *triangulation* is used quite differently from the usage of that term in geometry and topology. Here, triangulation refers to the task of 3D reconstruction, assuming the configuration C of cameras is known. This task amounts to computing fibers under the specialized joint camera map Φ_C.

Recall that, for $m \geq 2$ cameras, the generic fiber under Φ_C in (13.4) is empty. The measurements $\mathbf{y} := (\mathbf{y}_1, \ldots, \mathbf{y}_m)$ on the images are typically noisy. They do not lie on the multiview variety $\mathcal{M}_C$, but hopefully close it. To triangulate a corresponding space point $\mathbf{x} \in \mathbb{P}^3$, one seeks to compute the point $\tilde{\mathbf{y}}$ on $\mathcal{M}_C$ that is closest to $\mathbf{y}$, and then identify its fiber $\Phi_C^{-1}(\tilde{\mathbf{y}})$. Hence, the algebraic complexity of triangulating a single point from m images is determined by the ED degree of the multiview variety $\mathcal{M}_C$.

In real-life applications, the image measurements $(\mathbf{y}_1, \ldots, \mathbf{y}_m)$ come from an affine chart $(\mathbb{R}^2)^m$ inside $(\mathbb{P}^2)^m$. In that chart, we are working with the *affine multiview variety*

$$\mathcal{M}_C^\circ := \mathcal{M}_C \cap (\mathbb{R}^2)^m.$$

Triangulation thus means solving the ED problem (2.1) for the affine variety $\mathcal{M}_C^\circ$ in $\mathbb{R}^{2m}$.

The algebraic complexity of that problem is the ED degree of the affine multiview variety M_C°. The interest in that ED degree originated in the computer vision community, namely in the articles [79] and [162] titled *How hard is 3-view triangulation really?*.

The computer vision experts Stewénius and Nistér computed the ED degree of M_C° for $m \leq 7$ cameras [77]. Using Gröbner bases, one finds

$$\text{EDdegree}(M_C^\circ) = 6, 47, 148, 336, 638, 1081 \quad \text{for } m = 2, 3, 4, 5, 6, 7. \tag{13.8}$$

That computation was the original motivation for the development of the general notion of ED degrees of algebraic varieties. It started the five-author collaboration on the article [60]. Based on the numbers (13.8), the following cubic polynomial for the ED degree of the multiview variety M_C° for m cameras was conjectured in [60, Example 3.3]. That conjecture was proven by Maxim, Rodriguez, and Wang in [128]. Their result is as follows:

Theorem 13.4 *The ED degree of the affine multiview variety M_C° for $m \geq 2$ cameras in general position is*

$$\frac{9}{2}m^3 - \frac{21}{2}m^2 + 8m - 4.$$

We close this section by explaining the Gröbner basis computation that solves the triangulation problem for small m. The m cameras are given by m matrices $A_i = (a_{ijk})$ of format 3×4. In each image plane $\mathbb{P}^2$, we choose the affine chart by setting the last coordinate to 1. Our data are points $(y_{i1}, y_{i2}, 1)$ in these planes for $i = 1, 2, \ldots, m$. We similarly fix affine coordinates for the unknown point $\mathbf{x} = (x_1, x_2, x_3, 1)$ in the space $\mathbb{P}^3$. Consider the squared Euclidean distance to a point $\Phi_C(\mathbf{x})$ on M_C°:

$$\Delta(\mathbf{x}) = \sum_{i=1}^{m} \sum_{j=1}^{2} \left(\frac{a_{ij1}x_1 + a_{ij2}x_2 + a_{ij3}x_3 + a_{ij4}}{a_{i31}x_1 + a_{i32}x_2 + a_{i33}x_3 + a_{i34}} - y_{ij} \right)^2.$$

The optimization problem (2.1) asks us to compute the point $\mathbf{x}$ in $\mathbb{R}^3$ that minimizes the distance $\Delta(\mathbf{x})$. To model this problem algebraically, we consider the common denominator of our objective function:

$$\text{denom}(\mathbf{x}) := \prod_{i=1}^{m} (a_{i31}x_1 + a_{i32}x_2 + a_{i33}x_3 + a_{i34}).$$

We introduce a new variable z with $z \cdot \text{denom}(\mathbf{x}) = 1$. This ensures that all denominators are non-zero. Let I denote the ideal in the polynomial ring $\mathbb{R}[x_1, x_2, x_3, z]$ that is generated by $z \cdot \text{denom}(\mathbf{x}) - 1$ and the numerators of the partial derivatives $\partial\Delta(\mathbf{x})/\partial x_k$ for $k = 1, 2, 3$. Thus, I represents a system of four polynomial equations in four unknowns. The complex variety $V(I)$ is finite, and its points are precisely the complex critical points of the objective function $\Delta(\mathbf{x})$ in the open set given by $\text{denom}(\mathbf{x}) \neq 0$. We compute a Gröbner bases of I and we consider the set $\mathcal{B}$ of standard monomials. By Proposition 3.7, we have $\#\mathcal{B} = \#V(I)$. That number equals $6, 47, 148, \ldots$ for $m = 2, 3, 4 \ldots$, as promised in Theorem 13.4.

Example 13.5 We consider the dinosaur data set from the Visual Geometry Group in the Department of Engineering Science at the University of Oxford. This data set contains 36

Fig. 13.3: The 36 views of the 4983 3D points from the dinosaur data set in Example 13.5.

Fig. 13.4: The 4983 3D points that have been triangulated from the dinosaur data set in Example 13.5.

cameras as 3×4 matrices and a list of pictures taken from 4983 3D points by these 36 cameras. The data is shown in Figure 13.3. We solve triangulation for $m = 2$ randomly chosen cameras as described above, solving the resulting polynomial system with HomotopyContinuation.jl [31]. The result can be seen in Figure 13.4.   ⋄

13.2 Grassmann Tensors

The fundamental matrix F from Example 13.3 is a projectively invariant representation of two cameras; see Proposition 13.12. Note that F has seven degrees of freedom since F has rank two and is defined up to scale. This number makes sense because each matrix A_i has 11 degrees of freedom, up to scale, and we are taking the quotient modulo the 15-dimensional group $\mathrm{PGL}(4)$. Thus, the dimension of the quotient space is $2 \cdot 11 - 15 = 7$.

In this section, we introduce a class of tensors that generalize the fundamental matrix. These give projectively invariant representations for $m \geq 2$ cameras. We now allow higher-dimensional projections

$$C_i : \mathbb{P}^N \dashrightarrow \mathbb{P}^{n_i}. \tag{13.9}$$

Such projections can be used to model basic dynamics [96, 178]. The *camera center* of the camera C_i is its base locus in $\mathbb{P}^N$. This is represented by the kernel of the corresponding matrix A_i with $N+1$ columns. The approach we shall describe was introduced in computer vision by Hartley and Schaffalitzky [78].

As before, the multiview variety is the closure in $\mathbb{P}^{n_1} \times \cdots \times \mathbb{P}^{n_m}$ of the image of $\mathbb{P}^N$ under the C_i. We are interested in tuples of linear spaces $L_i \subset \mathbb{P}^{n_i}$ whose product meets the multiview variety. To this end, we fix the cameras C_i in (13.9), and we fix $c_i = \mathrm{codim}(L_i)$ with $1 \leq c_i \leq n_i$. Let $\Gamma_{C,c}$ be the closure of

$$\left\{ (L_1, \ldots, L_m) \in \mathrm{Gr}(n_1 - c_1, \mathbb{P}^{n_1}) \times \ldots \times \mathrm{Gr}(n_m - c_m, \mathbb{P}^{n_m}) \mid \right.$$
$$\left. \exists\, \mathbf{x} \in \mathbb{P}^N : C_1(\mathbf{x}) \in L_1, \ldots, C_m(\mathbf{x}) \in L_m \right\}. \tag{13.10}$$

Note that $\Gamma_{C,c}$ is a subvariety in a product of Grassmannians. This projective variety fills its ambient space when the c_i are small. It has high codimension when the c_i are large. The following theorem concerns the sweet spot in the middle.

Theorem 13.6 *For generic cameras C_i, the variety $\Gamma_{C,c}$ is a hypersurface if and only if $c_1 + \cdots + c_m = N + 1$. In that case, its defining equation is multilinear in the Plücker coordinates of the Grassmannians $\mathrm{Gr}(n_i - c_i, \mathbb{P}^{n_i})$.*

Multilinear forms are represented by tensors, just like bilinear forms are represented by matrices. The equation described in Theorem 13.6 is a multilinear form in m sets of unknowns, namely the Plücker coordinates of the Grassmannians $\mathrm{Gr}(n_i - c_i, \mathbb{P}^{n_i})$ for $i = 1, \ldots, m$. The *Grassmann tensor* of the camera configuration C is the order m tensor which represents the multilinear equation of the hypersurface $\Gamma_{C,c}$.

Before we come to the proof of Theorem 13.6, we present a census of all Grassmann tensors in the case of primary interest in computer vision, namely $N = 3$ and $n_i = 2$ for all $i = 1, \ldots, m$. According to Theorem 13.6, a Grassmann tensor exists whenever $c_1 + \cdots + c_m = 4$ and $1 \leq c_i \leq 2$ for $i = 1, \ldots, m$.

Example 13.7 ($m = 2, c_1 = c_2 = 2$) Here, the Grassmann tensor is the fundamental matrix F. The ambient space in (13.10) is $(\mathrm{Gr}(2 - 2, \mathbb{P}^2))^2 = (\mathbb{P}^2)^2$ with coordinates $(\mathbf{y}_1, \mathbf{y}_2)$, one for each of the two image points. The hypersurface $\Gamma_{(C_1,C_2),(2,2)}$ is the multiview variety $\mathcal{M}_{C_1,C_2}$ in Example 13.3. The bilinear form in (13.6) is represented by the 3×3 matrix F. The formula (13.7) can be obtained by computing the determinant of the 6×6 matrix M_A using Laplace expansion with respect to the last two columns. ◇

Example 13.8 ($m = 3, c_1 = 2, c_2 = c_3 = 1$) This Grassmann tensor has format $3 \times 3 \times 3$. It is known as the *trifocal tensor* of three cameras. The ambient space in (13.10) is $\mathrm{Gr}(2-2, \mathbb{P}^2) \times (\mathrm{Gr}(2-1, \mathbb{P}^2))^2 \cong (\mathbb{P}^2)^3$. The first plane $\mathbb{P}^2$ has coordinates $\mathbf{y}_1$. The other factors represent lines in $\mathbb{P}^2$, each spanned by two points $\mathbf{y}_{i1}, \mathbf{y}_{i2}$ for $i = 2, 3$. The Plücker coordinates of the line $\mathbf{y}_{i1} \vee \mathbf{y}_{i2}$ are the 2×2 minors of the 3×2 matrix $[\mathbf{y}_{i1}\ \mathbf{y}_{i2}]$. The determinant of the following 9×9 matrix is a trilinear form in $\mathbf{y}_1$, $\mathbf{y}_{21} \vee \mathbf{y}_{22}$, and $\mathbf{y}_{31} \vee \mathbf{y}_{32}$:

$$\det \begin{bmatrix} A_1 & \mathbf{y}_1 & 0 & 0 & 0 & 0 \\ A_2 & 0 & \mathbf{y}_{21} & \mathbf{y}_{22} & 0 & 0 \\ A_3 & 0 & 0 & 0 & \mathbf{y}_{31} & \mathbf{y}_{32} \end{bmatrix}.$$

There are actually three trifocal tensors, one for each choice for the triple of codimensions $(c_1, c_2, c_3) \in \{(2,1,1), (1,2,1), (1,1,2)\}$. The variety of all trifocal tensors, for one such fixed choice, was studied by Aholt and Oeding in [3]. ◇

Example 13.9 ($m = 4, c_1 = c_2 = c_3 = c_4 = 1$) Here, we get the *quadrifocal tensor* of format $3 \times 3 \times 3 \times 3$. It represents a multilinear form on four sets of lines $\mathbf{y}_{i1} \vee \mathbf{y}_{i2}$, namely the determinant of the 12×12 matrix

$$\det \begin{bmatrix} A_1 & \mathbf{y}_{11} & \mathbf{y}_{12} & 0 & 0 & 0 & 0 & 0 & 0 \\ A_2 & 0 & 0 & \mathbf{y}_{21} & \mathbf{y}_{22} & 0 & 0 & 0 & 0 \\ A_3 & 0 & 0 & 0 & 0 & \mathbf{y}_{31} & \mathbf{y}_{32} & 0 & 0 \\ A_4 & 0 & 0 & 0 & 0 & 0 & 0 & \mathbf{y}_{41} & \mathbf{y}_{42} \end{bmatrix}.$$

The variety of all quadrifocal tensors was studied by Oeding in [139]. ◇

Proof (of Theorem 13.6) This result appears in [78]. We give an independent proof. For general $L_i \in \mathrm{Gr}(n_i - c_i, \mathbb{P}^{n_i})$, the *back-projected space* $\tilde{L}_i := C_i^{-1}(L_i)$ also has codimension c_i, i.e., $\tilde{L}_i \in \mathrm{Gr}(N - c_i, \mathbb{P}^N)$. If $\sum c_i \leq N$, the subspaces $\tilde{L}_1, \ldots, \tilde{L}_m$ always intersect, meaning that $\Gamma_{C,c}$ is equal to its ambient space.

Now suppose $\sum c_i = N + 1$. Due to the genericity of the cameras, the multiview variety $\mathcal{M}_C$ of a single point (i.e., the Zariski closure of the image of $\Phi_C : \mathbb{P}^N \dashrightarrow \mathbb{P}^{n_1} \times \cdots \times \mathbb{P}^{n_m}$) has dimension N. Hence, the product $L_1 \times \cdots \times L_m$ of general subspaces $L_i \subseteq \mathbb{P}^{n_i}$ of codimension c_i does not intersect the multiview variety $\mathcal{M}_C$. Therefore, $\Gamma_{C,c}$ is a proper subvariety. We show that it is a hypersurface and we simultaneously compute its multidegree by intersecting it with generic pencils in $\mathrm{Gr}(n_1 - c_1, \mathbb{P}^{n_1}) \times \cdots \times \mathrm{Gr}(n_m - c_m, \mathbb{P}^{n_m})$.

It suffices to show that every general pencil meets $\Gamma_{C,c}$ in exactly one point. Such a pencil is of the form $L_1 \times \cdots \times \mathcal{L}_k \times \cdots \times L_m$, where L_i (for $i \neq k$) is a general point in the Grassmannian $\mathrm{Gr}(n_i - c_i, \mathbb{P}^{n_i})$, and

$$\mathcal{L}_k = \{L \in \mathrm{Gr}(n_k - c_k, \mathbb{P}^{n_k}) \mid V_k \subset L \subset W_k\},$$

where $V_k, W_k \subseteq \mathbb{P}^{n_k}$ are general subspaces of codimension $c_k + 1$ and $c_k - 1$, respectively. The back-projected spaces $\tilde{L}_1, \ldots, \tilde{W}_k, \ldots, \tilde{L}_m$ intersect in a single point $\mathbf{x} \in \mathbb{P}^N$. That point $\mathbf{x}$ does not lie in the back-projected space $\tilde{V}_k$. Otherwise, for general $L \in \mathcal{L}_k$, the point $L_1 \times \cdots \times L \times \cdots \times L_m$ would be in $\Gamma_{C,c}$. This a contradiction since the codimension of $\Gamma_{C,c}$ is at least one. Thus, there is exactly one $L \in \mathcal{L}_k$ (namely, the span of V_k with $C_k(\mathbf{x})$) which satisfies $L_1 \times \cdots \times L \times \cdots \times L_m \in \Gamma_{C,c}$. This shows that $\Gamma_{C,c}$ is a hypersurface of multidegree $(1, \ldots, 1)$. The multidegree of a hypersurface in a product of Grassmannians equals the multidegree of its equation in Plücker coordinates [71, Chapter 3, Proposition 2.1]. Similar arguments show that $\Gamma_{C,c}$ has codimension ≥ 2 whenever $\sum c_i > N + 1$.   □

By varying the Grassmann tensors over all camera configurations, we obtain a variety that parametrizes camera configurations modulo projective transformations; see Proposition 13.12. In the classical case of three cameras ($N = 3$), this quotient space has dimension $11m - 15$. This approach yields the variety of fundamental matrices for $m = 2$, the trifocal variety [3] for $m = 3$, and the quadrifocal variety [139] for $m = 4$. For $m \geq 5$, we can form quotient varieties by taking appropriate collections of fundamental matrices, trifocal tensors, and quadrifocal tensors. A more systematic representation is furnished by the *Hilbert schemes*, which were studied by Aholt, Sturmfels, and Thomas in [4].

13.3 3D Reconstruction from Unknown Cameras

We now turn to the problem of 3D reconstruction from images taken by *unknown* cameras. This amounts to computing fibers under the joint camera map Φ in (13.1). Typically, a nontrivial group G acts on the fibers of Φ since global 3D transformations that act simultaneously on the cameras and the 3D scene do not change the resulting images. Most real-life applications assume that the cameras are calibrated, so we are naturally led to the rotation group $SO(3)$, which is an important player in metric algebraic geometry.

Example 13.10 For a projective camera $A \in \mathbb{PR}^{3\times 4}$ that observes a point $\mathbf{x} \in \mathbb{P}^3$, the projective linear group $PGL(4)$ acts via $g \mapsto (Ag^{-1}, g\mathbf{x})$ on cameras and points without changing the image $Ag^{-1} \cdot g\mathbf{x} = A\mathbf{x}$. Hence, $PGL(4)$ acts on the fibers of the joint camera map in (13.2), where projective cameras observe k points in $\mathbb{P}^3$. This means that 3D reconstruction is only possible up to a projective transformation.

The action of the group $PGL(4)$ does not map calibrated cameras to calibrated cameras. The largest subgroup G of $GL(4)$ that preserves the structure of calibrated camera matrices $[R \,|\, t] \in SO(3) \times \mathbb{R}^3$ equals

$$G = \left\{ g \in GL(4) \mid g = \begin{bmatrix} R & t \\ 0 & \lambda \end{bmatrix} \text{ for some } R \in SO(3), t \in \mathbb{R}^3, \lambda \in \mathbb{R} \setminus \{0\} \right\}. \quad (13.11)$$

This is the scaled special Euclidean group of $\mathbb{R}^3$. Paraphrasing for computer vision, we conclude that 3D reconstruction with calibrated cameras is only possible up to a proper rigid motion and a non-zero scale.   ◇

The state-of-the-art 3D reconstruction algorithms work with *minimal problems*, where the joint camera map Φ is dominant and its fibers are generically finite modulo the group G. For minimal problems, computing the fibers of Φ means solving a *square* parametrized system of polynomial equations. Such 3D reconstruction problems use the minimal amount of data on the images while having finitely many solutions. The *algebraic degree* of a minimal problem is the number of complex solutions for generic data.

Proposition 13.11 *Consider 3D reconstruction for $m \geq 2$ unknown cameras observing k points. For projective cameras, there are two minimal problems, namely $(m, k) = (2, 7)$ and $(m, k) = (3, 6)$, both of algebraic degree 3. For calibrated cameras, the only minimal problem is $(m, k) = (2, 5)$, with an algebraic degree of 20.*

Proof For $m \geq 2$ projective cameras and k points, the joint camera map Φ takes the quotient space $\left((\mathbb{P}\mathbb{R}^{3 \times 4})^m \times (\mathbb{P}^3_{\mathbb{R}})^k\right)/\mathrm{PGL}(4)$ to the space of images $\left((\mathbb{P}^2)^k\right)^m$. (A comment for experts: the quotient is meant in the sense of *geometric invariant theory*, but we here only need a birational model.) If the 3D reconstruction problem is minimal, then both spaces have the same dimension, which gives $11m + 3k - 15 = 2km$. This equation has precisely two integer solutions $m \geq 2$ and $k \geq 1$, namely $(m, k) = (2, 7)$ and $(m, k) = (3, 6)$. Both yield minimal problems of algebraic degree three [76]. These degrees were found in the 19th century by Hesse [87] and Sturm [163].

For calibrated cameras instead of projective cameras, the domain of the joint camera map Φ changes to $\left((\mathrm{SO}(3) \times \mathbb{R}^3)^m \times (\mathbb{P}^3)^k\right)/G$, where G is the group in (13.11). Now, the two spaces have the same dimension if and only if $6m + 3k - 7 = 2km$. The only relevant integer solution is $(m, k) = (2, 5)$. This yields a minimal problem of algebraic degree 20. This problem has the label 5000_2 in the census of minimal problems found in [64, Table 1]. It is the minimal problem most widely used in practical 3D reconstruction algorithms. □

Later, we shall take a closer look at the systems of polynomial equations arising from Proposition 13.11. First, however, we explain the practical usage of minimal problems for 3D reconstruction in computer vision. Imagine two calibrated cameras that observe 100 points. Then, the joint camera map Φ in (13.3) for $k = 100$ is not dominant. Hence, the fiber under Φ of two noisy images is empty. We would need to find a closest point on the image of Φ, before we can compute any fiber. In other words, we must first solve an ED problem like that in Theorem 13.4. But this is now even harder because the cameras are unknown. In addition, in practical scenarios, it often happens that some of the given point pairs are outliers. Then, solving the ED problem yields incorrect solutions.

In practice, one simply avoids any such ED problem. Instead, one chooses five of the 100 given point pairs and solves the associated minimal problem in Proposition 13.11. This process gets repeated many times. The solutions to the many minimal problems get patched together via *random sample consensus* (RANSAC), until all 100 points and both cameras are reconstructed. For more details, see [104].

Since minimal problems must be solved many times for RANSAC, it is crucial that their formulation as a square polynomial system is efficient. A common strategy for simplification is to first reconstruct the cameras only and afterward recover the 3D scene via triangulation. Hence, instead of solving the full minimal problem at once, we start with a polynomial system whose only unknowns are the camera parameters.

To parametrize the m cameras modulo the group G, we use the Grassmann tensors. We work in arbitrary dimensions, as in (13.10), starting with $G = \mathrm{PGL}(N + 1)$. We consider m surjective projections $C_i : \mathbb{P}^N \dashrightarrow \mathbb{P}^{n_i}$ and integers $c_1, \ldots, c_m$ that satisfy $1 \leq c_i \leq n_i$

and $c_1 + \cdots + c_m = N+1$. The Grassmann tensor $T_{C,c}$ exists, by Theorem 13.6. Namely, $T_{C,c}$ is parametrized by the rational map

$$\gamma_c : \mathbb{P}\mathbb{R}^{(n_1+1)\times(N+1)} \times \cdots \times \mathbb{P}\mathbb{R}^{(n_m+1)\times(N+1)} \dashrightarrow \mathbb{P}\mathbb{R}^{\binom{n_1+1}{c_1}\times\cdots\times\binom{n_m+1}{c_m}}.$$

This map sends the m cameras C_i to their Grassmann tensor, written in Plücker coordinates.

Proposition 13.12 ([78]) *The group* $\mathrm{PGL}(N+1)$ *acts on the fibers of* γ_c *by componentwise right-multiplication. The map* γ_c *becomes birational on the quotient of its domain modulo that group action.*

Example 13.13 Consider the basic scenario in Example 13.7, where we have $N = 3$ and $m = n_1 = n_2 = c_1 = c_2 = 2$. The map γ_c sends two 3×4 matrices A_1, A_2 to the 3×3 matrix F in (13.7). The Zariski closure of the image of γ_c is the hypersurface $\mathcal{F} = V(\det)$ defined by the 3×3 determinant in $\mathbb{P}\mathbb{R}^{3\times4} = \mathbb{P}^8$. It is birational to pairs of projective cameras modulo $\mathrm{PGL}(4)$. This quotient construction reduces the number of camera parameters from 22 to 7. Any pair of image points $(\mathbf{y}_1, \mathbf{y}_2) \in \mathbb{P}^2 \times \mathbb{P}^2$ imposes a linear condition on $\mathcal{F}$, namely $\mathbf{y}_2^\top F \mathbf{y}_1 = 0$ as in (13.6). Reconstructing a fundamental matrix means intersecting the seven-dimensional variety $\mathcal{F}$ with seven hyperplanes. We conclude that observing seven points with two projective cameras is a minimal problem of degree $\deg(\mathcal{F}) = 3$. ◇

We now turn to pairs of calibrated cameras and the scaled special Euclidean group G in (13.11). A real 3×3 matrix E is called *essential* if it has rank two and the two non-zero singular values are equal. The Zariski closure of the set of essential matrices is the *essential variety* $\mathcal{E}$ in $\mathbb{P}\mathbb{R}^{3\times3} \cong \mathbb{P}^8$.

Theorem 13.14 *The map* γ_c *in Example 13.13 is two-to-one when restricted to pairs of calibrated cameras modulo the group G. Its image is the essential variety* $\mathcal{E}$. *This variety has dimension five, degree ten, and is defined by the following ten cubic equations in the entries of an unknown 3×3 matrix:*

$$\det E = 0 \quad and \quad EE^\top E - \frac{1}{2}\mathrm{tr}(EE^\top)E = 0. \tag{13.12}$$

Proof This can be proved by direct computations. The set of essential matrices is a semial-gebraic set of codimension three. Demazure [54] showed that the equations (13.12) hold on that set. With Macaulay2 [73], we can check that these ten homogeneous cubics generate a prime ideal of codimension three. The fact that the map γ_c is two-to-one can be verified by a computation in the coordinates that are used in [64]. □

The reconstruction of calibrated cameras is now analogous to Example 13.13. Any pair of image points $(\mathbf{y}_1, \mathbf{y}_2) \in \mathbb{P}^2 \times \mathbb{P}^2$ imposes a linear condition $\mathbf{y}_2^\top E \mathbf{y}_1 = 0$ on $\mathcal{E}$. Hence, we intersect the five-dimensional variety $\mathcal{E}$ with five hyperplanes. That intersection consists of ten complex matrices E. Each E comes from two pairs of calibrated cameras modulo G. We conclude that observing five points with two calibrated cameras is a minimal problem of algebraic degree $20 = 2 \cdot 10$, confirming Proposition 13.11.

To reiterate: what we described is the standard approach in state-of-the-art reconstruction algorithms. Given two images, one first reconstructs an essential matrix using RANSAC by intersecting the variety $\mathcal{E}$ with five hyperplanes, then one recovers the pair of calibrated cameras, and finally, one reconstructs the 3D points using triangulation. It would be interesting to study this process through the lens of metric algebraic geometry. In particular, following Chapter 9, what is the condition number of intersecting $\mathcal{E}$ with five hyperplanes $\mathbf{y}_2^\top E \mathbf{y}_1 = 0$ given by five image point pairs $(\mathbf{y}_1, \mathbf{y}_2)$? One may hope to utilize those condition numbers in homotopy continuation solvers.

More generally, we can ask the same question for intersecting the variety $\mathcal{G}_c := \overline{\mathrm{im}(\gamma_c)}$ of Grassmann tensors with $\dim \mathcal{G}_c$ hyperplanes given by $(L_1, \ldots, L_m)$ as in (13.10)? The condition number of intersecting a fixed projective variety with varying linear subspaces of complementary dimension was studied in [34]. However, that theory does not immediately apply to our problem since $\mathcal{E}$ is only intersected by *special* linear spaces, namely those that are intersections of five hyperplanes of the form $\{E : \mathbf{y}_2^\top E \mathbf{y}_1 = 0\}$.

Chapter 14
Volumes of Semialgebraic Sets

In this chapter, we study the problem of computing the volume of a semialgebraic subset S of $\mathbb{R}^n$. Being *semialgebraic* means that S is described by a finite Boolean combination of polynomial inequalities. Most of our discussion is limited to semialgebraic sets of the form

$$S = \left\{ \mathbf{x} \in \mathbb{R}^n \mid f(\mathbf{x}) \geq 0 \right\} \quad \text{for some } f \in \mathbb{R}[x_1, \ldots, x_n]. \tag{14.1}$$

This is the special case of only one polynomial inequality. The volume of S is the numerical value of an integral that is described by polynomials. Students encounter such integrals in Multivariable Calculus, and we shall review this elementary perspective in Section 14.1.

The rest of the chapter is much less elementary. Algebraic geometers use the term *period integrals* for real numbers such as vol(S). Our objective is the highly accurate numerical evaluation of period integrals. This is relevant for many fields, including physics (Feynman integrals) and statistics (Bayesian integrals). We present two algebraic methods, developed recently by Lairez [116], Lasserre [86], and their collaborators. Section 14.2 uses linear partial differential equations (D-modules) to represent and evaluate integrals. Section 14.3 explains the evaluation of integrals by means of semidefinite programming (SDP).

14.1 Calculus and Beyond

Suppose we are given a *basic semialgebraic set* in $\mathbb{R}^n$. This means that our set is described by a conjunction of polynomial inequalities; i.e., it has a representation of the form

$$S = \left\{ \mathbf{x} \in \mathbb{R}^n \mid f_1(\mathbf{x}) \geq 0 \text{ and } f_2(\mathbf{x}) \geq 0 \text{ and } \cdots \text{ and } f_k(\mathbf{x}) \geq 0 \right\}. \tag{14.2}$$

Here $f_1, f_2, \ldots, f_k$ are polynomials in n unknowns with real coefficients. Our task is to compute the volume of S, as reliably and accurately as possible, from the input $f_1, f_2, \ldots, f_k$.

The simplest scenario arises when $k = 1$, so our semialgebraic set S is the domain of nonnegativity of one polynomial $f(\mathbf{x}) = f(x_1, \ldots, x_n)$ with real coefficients, as in (14.1).

© The Author(s) 2024
P. Breiding et al., *Metric Algebraic Geometry*, Oberwolfach Seminars 53,
https://doi.org/10.1007/978-3-031-51462-3_14

Writing $d\mathbf{x}$ for the Lebesgue measure on $\mathbb{R}^n$, we wish to evaluate the integral

$$\text{vol}(S) \; = \; \int_S d\mathbf{x}. \tag{14.3}$$

More generally, we can consider integrals $\int_S g(\mathbf{x})d\mathbf{x}$, where $g(\mathbf{x})$ is a rational function, or even an algebraic function. The value of such an integral (if it converges) is a real number, which is called a *period* [115]. Our integrals are special cases of *period integrals*.

We begin with an instance where the volume of S can be found explicitly using calculus.

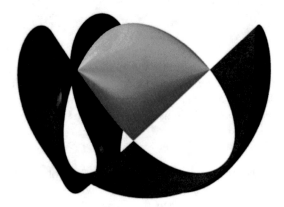

Fig. 14.1: The yellow convex body is the elliptope. It is bounded by Cayley's cubic surface.

Example 14.1 (Elliptope) We consider the *elliptope*. This is the semialgebraic set

$$S \; = \; \{(x, y, z) \in \mathbb{R}^3 \mid M(x, y, z) \text{ is positive semi-definite}\},$$

where $M(x, y, z)$ is the following 3×3-matrix:

$$M(x, y, z) \; = \; \begin{bmatrix} 1 & x & y \\ x & 1 & z \\ y & z & 1 \end{bmatrix}.$$

An equivalent formulation is that S consists of all points (x, y, z) in the cube $[-1, 1]^3$ such that the determinant $\det(M) \; = \; 2xyz - x^2 - y^2 - z^2 + 1$ is nonnegative. It is the region bounded by the yellow surface in Figure 14.1. This picture serves as the logo of the Nonlinear Algebra group at the Max-Planck Institute for Mathematics in the Sciences in Leipzig. The semialgebraic set S happens to be convex. It illustrates several applications of algebraic geometry. In statistics, the convex set S comprises all correlation matrices. In optimization, it is the feasible region of a semidefinite programming problem; see [133, Chapter 12]. Semidefinite programming is our workhorse for approximating volumes in Section 14.3.

We now compute the volume of the elliptope. We begin parameterizing its boundary surface (the yellow surface in Figure 14.1). Solving the equation $\det(M) = 0$ for z with the quadratic formula, we obtain

$$z = xy \pm \sqrt{x^2y^2 - x^2 - y^2 + 1} = xy \pm \sqrt{(1 - x^2)(1 - y^2)} \qquad \text{for } (x, y) \in [-1, 1]^2.$$

The plus sign gives the upper yellow surface and the minus sign gives the lower yellow surface. The volume of the elliptope S is obtained by integrating the difference between the upper function and the lower function over the square. Hence, the desired volume equals

$$\text{vol}(S) = \int_{-1}^{1} \int_{-1}^{1} 2\sqrt{(1 - x^2)(1 - y^2)}\, dx dy = 2 \left[\int_{-1}^{1} \sqrt{1 - t^2}\, dt \right]^2.$$

The univariate integral on the right gives the area below the semicircle with radius 1. We know from trigonometry that this area equals $\pi/2$, where $\pi = 3.14159265....$ We conclude

$$\text{vol}(S) = \frac{\pi^2}{2} = 4.934802202...$$

Thus, our elliptope covers about $(\pi^2/2)/8 = 61.7\%$ of the volume of the cube $[-1, 1]^3$. ◇

The number $\pi^2/2$ we found is an example of a period. It is generally much more difficult to accurately evaluate such integrals. This challenge has played an important role in the history of mathematics. Consider the problem of computing the arc length of an ellipse. This requires us to integrate the reciprocal square root of a cubic polynomial $f(t)$. Such integrals are called *elliptic integrals*, and they represent periods of elliptic curves. Furthermore, in an 1841 paper, Abel introduced *abelian integrals*, where $g(t)$ is an algebraic function in one variable t. How to evaluate such an integral? This question leads us to Riemann surfaces and then to their Jacobians. And, voilà, we arrived at the theory of *abelian varieties*.

This chapter presents two current paradigms for accurately computing integrals like (14.3). The first method rests on the theory of D-modules, that is, on the algebraic study of linear differential equations with polynomial coefficients. Our volume is found as a special value of a parametric volume function that is encoded by means of its *Picard–Fuchs differential equation*. This method, which tends to appeal to algebraic geometers, was introduced by Lairez, Mezzarobba, and Safey El Din in [116].

The second approach is due to Lasserre, Henrion, Savorgnan, Tacchi, and Weisser [86, 169, 170]. At first glance, the approach in Section 14.3 might appeal more to readers from analysis and optimization. But, we hope that algebraic geometers will find it interesting as well, as it entails plenty of deep algebraic structure. The idea is to consider all moments $m_\alpha = \int_S \mathbf{x}^\alpha d\mathbf{x}$ of our semialgebraic set S, where again $d\mathbf{x}$ is the Lebesgue measure, and to use relations among the m_α to infer an accurate approximation of $m_0 = \text{vol}(S)$. That numerical inference rests on semidefinite programming [133, Chapter 12].

14.2 D-Modules

In calculus, we learn about definite integrals in order to determine the area under a graph. Likewise, in multivariable calculus, we examine the volume enclosed by a surface. Here, we are interested in areas and volumes of semialgebraic sets. When these sets depend on one or more parameters, their volumes are holonomic functions of the parameters. We explain what this means and how it can be used for accurate evaluation of volume functions. We present the method of [116], following the exposition given in [159].

Definition 14.2 The *Weyl algebra* (cf. [155, 159]) is the quotient of the free $\mathbb{C}$-algebra generated by $2n$ variables $x_1, \ldots, x_n, \partial_1, \ldots, \partial_n$ by the ideal generated by $\partial_i x_i - x_i \partial_i - 1$ for $i = 1, \ldots, n$. We denote the Weyl algebra by

$$D = \mathbb{C}\langle x_1, \ldots, x_n, \partial_1, \ldots, \partial_n \rangle.$$

In the definition of the Weyl algebra, the x_i play the role of the usual variables in calculus. The ∂_i, on the other hand, play the role of partial derivatives. Indeed, if $g : \mathbb{R}^n \to \mathbb{R}$ is a differentiable function, then the product rule yields $\partial_i(x_i \cdot g) = g + x_i(\partial_i g)$. This is precisely the algebraic relation that we quotient out.

Suppose that M is a D-module. We denote the action of D on M by

$$P \bullet f \qquad \text{for } P \in D, f \in M.$$

Definition 14.3 Let M be a D-module. The *annihilator* or *annihilating D-ideal* of an element $f \in M$ is the D-ideal

$$\text{Ann}_D(f) := \{ P \in D \mid P \bullet f = 0 \}.$$

In applications, the D-module M is usually a space of infinitely differentiable functions on a subset of $\mathbb{R}^n$ or $\mathbb{C}^n$. Such D-modules are torsion-free. In the following, we mainly focus on annihilators that are holonomic D-ideals.

Definition 14.4 Let M be a D-module and $f \in M$. For instance, f could be among the functions mentioned above. We say that f is *holonomic* if, for each $k \in \{1, \ldots, n\}$, there is a differential operator

$$P_k = \sum_{i=0}^{m_k} p_{i,k}(\mathbf{x}) \, \partial_k^i \in (D \cap \mathbb{C}[x_1, \ldots, x_n, \partial_k]) \backslash \{0\} \qquad (14.4)$$

that annihilates f; i.e., $P_k \bullet f = 0$. Here, the $p_{i,k}(\mathbf{x})$ are polynomials in $\mathbf{x} = (x_1, \ldots, x_n)$. If this holds, then we say that $\text{Ann}_D(f)$ is a holonomic D-ideal.

The term "holonomic function" is due to Zeilberger [179]. Many interesting functions are holonomic. For instance, holonomic functions in one variable are solutions to ordinary linear differential equations with rational function coefficients. This includes algebraic functions, some elementary trigonometric functions, hypergeometric functions, Bessel functions, period integrals, and many more.

Proposition 14.5 *Every rational function is holonomic.*

Proof Let r be a rational function in $\mathbf{x} = (x_1, \ldots, x_n)$. It is annihilated by the operator $r(\mathbf{x})\partial_i - \partial r/\partial x_i$ for $i = 1, 2, \ldots, n$. By clearing denominators in this operator, we obtain a non-zero differential operator $P_i \in \mathbb{C}[\mathbf{x}]\langle \partial_i \rangle$ of order $m_i = 1$ that annihilates r. $\qquad \square$

To find holonomic annihilating D-ideals, it is helpful to use computer algebra.

Example 14.6 Let $r \in \mathbb{Q}(x_1, \ldots, x_n)$ be a rational function. We can compute its annihilator in Macaulay2 [73] with the built-in command RatAnn. For instance, if $n = 2$ and $r = \frac{x_1}{x_2}$, then we can use the following code

```
needsPackage "Dmodules"
D = QQ[x1,x2,d1,d2, WeylAlgebra => {x1=>d1,x2=>d2}];
rnum = x1;   rden = x2;
I = RatAnn(rnum,rden)
```

This code fragment shows that $r = x_1/x_2$ has $\mathrm{Ann}_D(r) = D\{ \partial_1^2, x_1\partial_1 - 1, x_2\partial_1\partial_2 + \partial_1 \}$. $\diamond$

Now, let M be a D-module, $f \in M$, and $\mathrm{Ann}_D(f)$ a holonomic D-ideal. As in (14.4), let $m_1, \ldots, m_n$ denote the orders of the differential operators $P_1, \ldots, P_n$. Thus, P_k is an operator in ∂_k of order m_k whose coefficients are polynomials in $\mathbf{x}$. We fix initial conditions for f by specifying the following $m_1 m_2 \cdots m_n$ numerical values for a general point $\mathbf{x}_0 \in \mathbb{C}^n$:

$$(\partial_1^{i_1} \cdots \partial_n^{i_n} \bullet f)|_{\mathbf{x}=\mathbf{x}_0} \quad \text{where } 0 \leq i_k < m_k \text{ for } k = 1, \ldots, n. \tag{14.5}$$

The operators P_i together with the initial conditions in (14.5) specify the function f.

Suppose $f(\mathbf{x})$ is an algebraic function. This means that there is a polynomial $F \in \mathbb{C}[y, x_1, \ldots, x_n]$ such that f satisfies the equation $F(f, \mathbf{x}) = 0$. Every algebraic function $f(\mathbf{x})$ in n variables is holonomic. Koutschan [112] developed practical algorithms for manipulating holonomic functions. These are implemented in his Mathematica [102] package HolonomicFunctions [112]. Using the polynomial F as its input, this package can compute a holonomic representation of f. The output is a linear differential operator of lowest degree annihilating f. We now show this with an explicit example.

Example 14.7 Let $n = 1$ and consider the algebraic function $y = f(x)$ that is defined by

$$q(x, y) = x^4 + y^4 + \frac{1}{100}xy - 1. \tag{14.6}$$

Its annihilator in D can be computed as follows:

```
<< RISC`HolonomicFunctions`
q = y^4 + x^4 + x*y/100 - 1;
ann = Annihilator[Root[q, y, 1], Der[x]]
```

This Mathematica code determines an operator P of lowest order in $\mathrm{Ann}_D(f)$. We find

$$P = g_0(x) + g_1(x)\,\partial + g_2(x)\,\partial^2 + g_3(x)\,\partial^3,$$

with coefficients

$$g_3(x) = (2x^4 + 1)^2(25600000000x^{12} - 76800000000x^8 + 76799999973x^4 - 25600000000),$$
$$g_2(x) = 6x^3(2x^4 + 1)(51200000000x^{12} + 76800000000x^8 - 307199999946x^4 + 179199999973),$$
$$g_1(x) = 3x^2(102400000000x^{16} + 204800000000x^{12} + 2892799999572x^8$$
$$-3507199999444x^4 + 307199999953),$$
$$g_0(x) = -3x(102400000000x^{16} + 204800000000x^{12} + 1459199999796x^8$$
$$-1049599999828x^4 + 51199999993).$$

This operator is an encoding of the algebraic function $y = f(x)$ as a holonomic function. ⋄

As discussed above, we specify a holonomic function f in n variables by a holonomic system of linear PDEs together with a list of initial conditions. Initial conditions such as (14.5) are designed to determine the function uniquely inside the linear space $\mathrm{Sol}(I)$, where $I \subseteq \mathrm{Ann}_D(f)$. For instance, in Example 14.7, we need three initial conditions to specify the function $f(\mathbf{x})$ uniquely inside the 3-dimensional solution space to our operator P. We could fix the values at three distinct points, or we could fix the value and the first two derivatives at one special point. We generalize the canonical representation (14.5) as follows.

Definition 14.8 A *holonomic representation* of a function f is a holonomic D-ideal I contained in the annihilator $\mathrm{Ann}_D(f)$ together with a list of linear conditions that specify f uniquely inside the finite-dimensional solution space of holomorphic solutions. Think of the initial conditions for solutions to differential equations.

The next example shows the role of holonomic functions for metric algebraic geometry.

Example 14.9 (The area of a TV screen) Consider the quartic polynomial $q(x, y)$ from (14.6). We are interested in the semialgebraic set $S = \{(x, y) \in \mathbb{R}^2 \mid q(x, y) \leq 0\}$. This convex set is a slight modification of a set known in the optimization literature as "the TV screen". Our aim is to compute the area of the semialgebraic convex set S as accurately as possible. One can get a rough idea of the area of S by sampling. This is illustrated in Figure 14.2. From the polynomial $q(x, y)$ we read off that S is contained in the square defined by $-1.2 \leq x, y \leq 1.2$. We sampled 10000 points uniformly from that square, and for each sample, we checked the sign of q. This is a simple instance of the Metropolis–Hastings sampling algorithm that we discuss in Chapter 15. Points inside S are drawn in blue and points outside S are drawn in pink. By multiplying the area $(2.4)^2 = 5.76$ of the square with the fraction of the number of blue points among the samples, we learn that the area of the TV screen is approximately 3.7077.

We now compute the area more accurately using D-modules. Let $\mathrm{pr}: S \to \mathbb{R}$ be the projection on the x-coordinate, and consider the length of the fibers of this projection:

$$v(x) = \ell(\mathrm{pr}^{-1}(x) \cap S).$$

This length is a holonomic function in the parameter x. Namely, it satisfies the third-order differential operator P we displayed in Example 14.7. The real roots of the resultant

$$\mathrm{Res}_y(q, \partial q/\partial y) = 25600000000x^{12} - 76800000000x^8$$
$$+ 76799999973x^4 - 25600000000 \tag{14.7}$$

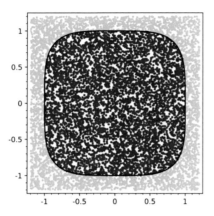

Fig. 14.2: The TV screen is the convex region consisting of the blue points.

are the two branch points $x_0 < x_1$ of the map pr. These values of x_0 and x_1 can be written in radicals over $\mathbb{Q}$. However, here we take an accurate floating point representation:

$$x_1 = -x_0 = 1.0002544658502588454784547666435667500801962761589 7635\ldots$$

The desired area equals

$$\text{vol}(S) = w(x_1),$$

where $w(x) = \int_{x_0}^{x} v(t)dt$ is holonomic (see Proposition 14.15 below). One operator that annihilates w is $P\partial$, where $P \in \text{Ann}_D(v)$ is the third-order operator above. To get a holonomic representation of w, we also need some initial conditions. Clearly, $w(x_0) = 0$. Further initial conditions on w' are derived by evaluating v at other points. By plugging values for x into (14.6) and solving for y, we find $w'(0) = 2$ and $w'(\pm 1) = 1/\sqrt[3]{100}$. Thus, we now have four linear constraints on our function w, albeit at different points.

Our goal is to determine a unique function $w \in \text{Sol}(P\partial)$ by incorporating these four initial conditions, and then to evaluate that holonomic function w at the branch point x_1. To this end, we proceed as follows. We consider any point $x_{\text{ord}} \in \mathbb{R}$ at which the differential operator $P\partial$ is not singular, i.e., the coefficient of the highest order differentiation does not vanish at x_{ord}. Using the command `local_basis_expansion` that is built into the SAGE [171] package `ore_algebra`, we compute a basis of local series solutions to $P\partial$ at the point x_{ord}. Since that point is nonsingular, that basis has the following form:

$$s_{x_{\text{ord}},0}(x) = 1 \qquad + O((x - x_{\text{ord}})^4),\quad s_{x_{\text{ord}},1}(x) = (x - x_{\text{ord}}) + O((x - x_{\text{ord}})^4),$$
$$s_{x_{\text{ord}},2}(x) = (x - x_{\text{ord}})^2 + O((x - x_{\text{ord}})^4),\quad s_{x_{\text{ord}},3}(x) = (x - x_{\text{ord}})^3 + O((x - x_{\text{ord}})^4).$$

Locally at x_{ord}, our solution is given by a unique choice of four coefficients $c_{x_{\text{ord}},i}$, namely

$$w(x) = c_{x_{\text{ord}},0} \cdot s_{x_{\text{ord}},0}(x) + c_{x_{\text{ord}},1} \cdot s_{x_{\text{ord}},1}(x) + c_{x_{\text{ord}},2} \cdot s_{x_{\text{ord}},2}(x) + c_{x_{\text{ord}},3} \cdot s_{x_{\text{ord}},3}(x).$$

At a regular singular point x_{rs}, complex powers of x and $\log(x)$ can appear in the local basis at x_{rs}. Any initial condition at a point determines a linear constraint on its coefficients. For instance, $w'(0) = 2$ implies $c_{0,1} = 2$, and similarly for our initial conditions at $-1, 1$, and x_0. One challenge is that the initial conditions pertain to different points. To address this, we calculate transition matrices that relate the basis above of series solutions at one point to the basis at another point. These are invertible 4×4 matrices.

With the method described above, we find the basis of series solutions at x_1, along with a system of four linear constraints on the four coefficients $c_{x_1,i}$. These constraints are derived from the initial conditions at 0, ± 1, and x_0, using the 4×4 transition matrices. By solving these linear equations, we compute the desired function value up to any desired precision:

$$w(x_1) = 3.70815994474216228834822556114586653712430658199139347 0943....$$

In conclusion, this real number is the area of the TV screen S in Figure 14.2 that is defined by the polynomial $q(x, y)$ in (14.6).   ◇

Before we discuss the approach from Example 14.9 in a general setting, let us first study the structure of the space of holonomic functions. They turn out to have remarkable closure properties. In the following, let f and g be functions in $\mathbf{x} = (x_1, \ldots, x_n)$. For the proof of the following result, see [159, Proposition 2.3].

Proposition 14.10 *Let $f(\mathbf{x})$ be holonomic and $g(\mathbf{x})$ algebraic. Then $f(g(\mathbf{x}))$ is holonomic.*

While algebraic transformations of holonomic functions are again holonomic by Proposition 14.10, this is not true for rational functions of holonomic functions. To see this, we consider holonomic functions in one variable $f(x)$. A necessary condition for a meromorphic function $f(x)$ to be holonomic is that it has only finitely many poles in $\mathbb{C}$.

For a concrete example, we start with the holonomic function $\sin(x)$. This is annihilated by the operator $\partial^2 + 1$. Its reciprocal $f(x) = \frac{1}{\sin(x)}$ has infinitely many poles, so is not holonomic. Hence, the class of holonomic functions is not closed under division. It is also not closed under composition of functions since both $\frac{1}{x}$ and $\sin(x)$ are holonomic.

Proposition 14.11 *If f, g are holonomic functions, then both $f + g$ and $f \cdot g$ are holonomic.*

Proof For each index $i \in \{1, 2, \ldots, n\}$, there exist non-zero operators P_i and Q_i in $\mathbb{C}[\mathbf{x}]\langle \partial_i \rangle$ which satisfy $P_i \bullet f = Q_i \bullet g = 0$. Set $n_i = \text{order}(P_i)$ and $m_i = \text{order}(Q_i)$. Then, the $\mathbb{C}(\mathbf{x})$-linear span of the set $\{\partial_i^k \bullet f\}_{k=0,\ldots,n_i}$ has dimension $\leq n_i$. Similarly, the $\mathbb{C}(\mathbf{x})$-linear span of the set $\{\partial_i^k \bullet g\}_{k=0,\ldots,m_i}$ has dimension $\leq m_i$.

Now consider $\partial_i^k \bullet (f + g)$, which we can write as $\partial_i^k \bullet f + \partial_i^k \bullet g$. Therefore, the $\mathbb{C}(\mathbf{x})$-linear span of $\{\partial_i^k \bullet (f + g)\}_{k=0,\ldots,n_i+m_i}$ has dimension $\leq n_i + m_i$. Hence, there exists a non-zero operator $S_i \in \mathbb{C}[\mathbf{x}]\langle \partial_i \rangle$ such that $S_i \bullet (f + g) = 0$. Since this holds for all indices i, we conclude that the sum $f + g$ is holonomic.

A similar proof works for the product $f \cdot g$. For $i \in \{1, 2, \ldots, n\}$, we consider the set $\{\partial_i^k \bullet (f \cdot g)\}_{k=0,1,\ldots,n_i m_i}$. By applying the product rule, we find that the $n_i m_i + 1$ generators are linearly dependent over the rational function field $\mathbb{C}(\mathbf{x})$. Hence, there is a non-zero operator $T_i \in \mathbb{C}[\mathbf{x}]\langle \partial_i \rangle$ such that $T_i \bullet (f \cdot g) = 0$. Hence, $f \cdot g$ is holonomic.   □

The proof above gives a linear algebra method for computing an annihilating D-ideal for the sum $f + g$ (resp. the product $f \cdot g$), starting from such D-ideals for f and g. The following example illustrates Proposition 14.11.

Example 14.12 This example is similar to one in [179, Section 4.1]. Consider the univariate functions $f(x) = \exp(x)$ and $g(x) = \exp(-x^2)$. Their canonical holonomic representations are $I_f = \langle \partial - 1 \rangle$ with $f(0) = 1$ and $I_g = \langle \partial + 2x \rangle$ with $g(0) = 1$. We are interested in the function $h = f + g$. Its first two derivatives are obtained as follows:

$$\begin{bmatrix} h \\ \partial \bullet h \\ \partial^2 \bullet h \end{bmatrix} = \begin{bmatrix} 1 & 1 \\ 1 & -2x \\ 1 & 4x^2 - 2 \end{bmatrix} \cdot \begin{bmatrix} f \\ g \end{bmatrix}.$$

By computing the left kernel of this 3×2-matrix, we find that $h = f + g$ is annihilated by

$$I_h = \langle (2x + 1)\partial^2 + (4x^2 - 3)\partial - 4x^2 - 2x + 2 \rangle, \quad \text{with } h(0) = 2, \ h'(0) = 1.$$

For the product $j = f \cdot g$ we have $j' = f'g + fg' = f \cdot g + f \cdot (-2xg) = (1 - 2x)j$, so the canonical holonomic representation of j is the D-ideal $I_j = \langle \partial + 2x - 1 \rangle$ with $j(0) = 1$. ◇

Proposition 14.13 *Let f be holonomic in n variables and $m < n$. The restriction of f to the coordinate subspace $\{x_{m+1} = \cdots = x_n = 0\}$ is a holonomic function in $x_1, \ldots, x_m$.*

Proof For $i \in \{m + 1, \ldots, n\}$, we consider the right ideal $x_i D$ in the Weyl algebra D. This ideal is a left module over

$$D_m = \mathbb{C}\langle x_1, \ldots, x_m, \partial_1, \ldots, \partial_m \rangle.$$

The sum of these ideals with $\text{Ann}_D(f)$ is hence a left D_m-module. Its intersection with D_m is called the *restriction ideal*:

$$(\text{Ann}_D(f) + x_{m+1}D + \cdots + x_n D) \cap D_m. \tag{14.8}$$

By [155, Prop. 5.2.4], this D_m-ideal is holonomic and annihilates $f(x_1, \ldots, x_m, 0, \ldots, 0)$.□

Proposition 14.14 *The partial derivatives of a holonomic function are holonomic.*

Proof Let f be holonomic and $P_i \in \mathbb{C}[x]\langle \partial_i \rangle \backslash \{0\}$ with $P_i \bullet f = 0$ for all i. We can write P_i as $P_i = \widetilde{P}_i \partial_i + a_i(x)$, where $a_i \in \mathbb{C}[x]$. If $a_i = 0$, then $\widetilde{P}_i \bullet \frac{\partial f}{\partial x_i} = 0$ and we are done. Assume $a_i \neq 0$. Since both a_i and f are holonomic, by Proposition 14.11, there is a non-zero linear operator $Q_i \in \mathbb{C}[x]\langle \partial_i \rangle$ such that $Q_i \bullet (a_i \cdot f) = 0$. Then $Q_i \widetilde{P}_i$ annihilates $\partial f/\partial x_i$.   □

A key fact about D-modules (see [155, Section 5.5]) is that integration is dual, in the sense of the Fourier transform, to restriction. Here is the dual to Proposition 14.13:

Proposition 14.15 *Let $f : \mathbb{R}^n \to \mathbb{C}$ be a holonomic function. Then the definite integral*

$$F(x_1, \ldots, x_{n-1}) = \int_a^b f(x_1, \ldots, x_{n-1}, x_n) \, dx_n$$

is a holonomic function in $n - 1$ variables, assuming the integral converges.

By dualizing (14.8), we obtain the following D_m-ideal, known as the *integration ideal*:

$$\left(\mathrm{Ann}_D(f) + \partial_{m+1}D + \cdots + \partial_n D \right) \cap D_m \qquad \text{for } m < n.$$

The expression is Fourier dual to the restriction ideal (14.8). This exchanges x_i and ∂_i. If $m = n - 1$, then the integration ideal annihilates the holonomic function F above.

Equipped with our tools for holonomic functions, we now return to volumes of semialgebraic sets, using the method of Lairez, Mezzarobba, and Safey El Din [116]. They compute this volume by deriving a differential operator that encodes periods of an integral [115]:

Definition 14.16 Let $R(t, x_1, \ldots, x_n)$ be a rational function and consider the formal integral

$$\oint R(t, x_1, \ldots, x_n) \, dx_1 \cdots dx_n. \tag{14.9}$$

Fix an open subset Ω of either $\mathbb{R}$ or $\mathbb{C}$. An analytic function $\phi \colon \Omega \to \mathbb{C}$ is a *period* of the integral (14.9) if, for any $s \in \Omega$, there exists a neighborhood $\Omega' \subseteq \Omega$ of s and an n-cycle $\gamma \subset \mathbb{C}^n$ with the following property: For all $t \in \Omega'$, the cycle γ is disjoint from the poles of $R_t := R(t, \bullet)$ and

$$\phi(t) = \int_\gamma R(t, x_1, \ldots, x_n) \, dx_1 \cdots dx_n. \tag{14.10}$$

In the following, we make some assumptions. Let $S = \{f \le 0\} \subset \mathbb{R}^n$ be a compact semialgebraic set, given by $f \in \mathbb{Q}[x_1, \ldots, x_n]$. Furthermore, let $\mathrm{pr} \colon \mathbb{R}^n \to \mathbb{R}$ denote the projection on the first coordinate. The set of *branch points* of pr restricted to the zero locus of f is the following subset of the real line, which is assumed to be finite:

$$\Sigma_f = \left\{ p \in \mathbb{R} \mid \exists \mathbf{y} = (y_2, \ldots, y_n) \in \mathbb{R}^{n-1} : f(p, \mathbf{y}) = 0 \text{ and } \frac{\partial f}{\partial x_i}(p, \mathbf{y}) = 0, \, i = 2, \ldots, n \right\}.$$

The polynomial in the unknown p that defines Σ_f is obtained by eliminating $x_2, \ldots, x_n$. It can be represented as a multivariate resultant, generalizing the Sylvester resultant in (14.7).

Fix an open interval I in $\mathbb{R}$ with $I \cap \Sigma_f = \emptyset$. For any $x_1 \in I$, the set $S_{x_1} := \mathrm{pr}^{-1}(x_1) \cap S$ is compact and semialgebraic in $(n - 1)$-space. We are interested in its volume. By [116, Theorem 9], the function $v \colon I \to \mathbb{R}$, $x_1 \mapsto \mathrm{vol}_{n-1}(S_{x_1})$ is a period of the rational integral

$$\frac{1}{2\pi i} \oint \frac{x_2}{f(x_1, x_2, \ldots, x_n)} \frac{\partial f(x_1, x_2, \ldots, x_n)}{\partial x_2} \, dx_2 \cdots dx_n. \tag{14.11}$$

Let $e_1 < e_2 < \cdots < e_K$ be the branch points in Σ_f and set $e_0 = -\infty$ and $e_{K+1} = \infty$. This specifies the pairwise disjoint open intervals $I_k = (e_k, e_{k+1})$. They satisfy $\mathbb{R} \backslash \Sigma_f = \bigcup_{k=0}^K I_k$. Fix the holonomic functions $w_k(t) = \int_{e_k}^t v(x_1) dx_1$. The volume of S then is obtained as

$$\mathrm{vol}(S) = \int_{e_1}^{e_K} v(x_1) dx_1 = \sum_{k=1}^{K-1} w_k(e_{k+1}).$$

How does one compute such a sum? As a period of the rational integral (14.11), the volume v is a holonomic function on each interval I_k. A key step is to find an operator P in $D_1 = \mathbb{C}\langle x_1, \partial_1 \rangle$ that annihilates $v|_{I_k}$ for all k. The product $P\partial$ annihilates the functions $w_k(x_1)$ for all k. By imposing sufficiently many initial conditions, we can reconstruct the functions w_k from the operator $P\partial$. Initial conditions that come for free are $w_k(e_k) = 0$ for all k. Finally, for evaluating the functions w_k we use methods for numerical integration.

The differential operator P is known as the *Picard–Fuchs equation* of the period in question. The following software packages can be used to compute Picard–Fuchs equations:

- `HolonomicFunctions` by C. Koutschan in `Mathematica` [102],
- `ore_algebra` by M. Kauers in `SAGE` [171],
- `periods` by P. Lairez in `MAGMA` [24],
- `Ore_Algebra` by F. Chyzak in `Maple` [127].

We now apply this to compute volumes. Starting from the polynomial f, we compute the Picard–Fuchs operator $P \in D_1$ along with suitable initial conditions. For each interval I_k we perform the following steps, here described for the `ore_algebra` package in `SAGE`:

(i) Using the command `local_basis_expansion`, compute a local basis of series solutions for the linear differential operator $P\partial$ at various points in $[e_k, e_{k+1})$.
(ii) Using the command `op.numerical_transition_matrix`, numerically compute a transition matrix for the series solution basis from one point to another one.
(iii) From the initial conditions, construct linear relations between the coefficients in the local basis extensions. Using step (ii), transfer them to the branch point e_{k+1}.
(iv) Plug in to the local basis extension at e_{k+1} and thus evaluate the volume of $S \cap \mathrm{pr}^{-1}(I_k)$.

Using this, we compute the volume of a convex body in 3-space, shown in Figure 14.3.

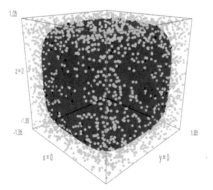

Fig. 14.3: The quartic bounds the convex region consisting of the gray points.

Example 14.17 (Quartic surface) Fix the quartic polynomial

$$f(x, y, z) = x^4 + y^4 + z^4 + \frac{x^3 y}{20} - \frac{xyz}{20} - \frac{yz}{100} + \frac{z^2}{50} - 1, \qquad (14.12)$$

and consider the set $S = \{(x, y, z) \in \mathbb{R}^3 \mid f(x, y, z) \leq 0\}$. Our aim is to compute vol(S).

As in Example 14.9, we get a rough idea by sampling. This is illustrated in Figure 14.3. Our set S is compact, convex, and contained in the cube $-1.05 \leq x, y, z \leq 1.05$. We sampled 10000 points uniformly from that cube. For each sample, we checked the sign of $f(x, y, z)$. By multiplying the volume $(2.1)^3 = 9.261$ of the cube by the fraction of gray points among the sampled points, we found that vol$(S) \approx 6.4771$. In order to gain a higher precision, we now compute the volume of our semialgebraic set S by means of D-modules.

Let pr: $\mathbb{R}^3 \to \mathbb{R}$ be the projection onto the x-coordinate. Let $v(x) = \text{vol}_2 \left(\text{pr}^{-1}(x) \cap S \right)$ denote the area of the fiber over any point x in $\mathbb{R}$. We write $e_1 < e_2$ for the two branch points of the map pr restricted to the quartic surface $\{f = 0\}$. They can be computed with resultants. The projection has 36 complex branch points. The first two of them are real and therefore are the branch points of pr. We obtain $e_1 \approx -1.0023512$ and $e_2 \approx 1.0024985$. By [116, Theorem 9], the area function $v(x)$ is a period of the rational integral

$$\frac{1}{2\pi i} \oint \frac{y}{f(x, y, z)} \frac{\partial f(x, y, z)}{\partial y} \, dy dz.$$

We set $w(t) = \int_{e_1}^{t} v(x) \, dx$. The desired 3-dimensional volume equals vol$(S) = w(e_2)$. Using the function periods in MAGMA [24], we compute a differential operator P of order eight that annihilates $v(x)$. Again, $P\partial$ then annihilates $w(x)$. One initial condition is $w(e_1) = 0$. We obtain eight further initial conditions $w'(x) = \text{vol}_2(S_x)$ for points $x \in (e_1, e_2)$ by running the same algorithm for the 2-dimensional semialgebraic slices $S_x = \text{pr}^{-1}(x) \cap S$. In other words, we make eight subroutine calls to an area measurement as in Example 14.9. From these nine initial conditions, we derive linear relations of the coefficients in the local basis expansion at e_2. These computations are run in SAGE [171] as described in steps (i), (ii), (iii) and (iv) above. We find the approximate volume to be

vol$(S) \approx$ 6.43883248057289354474073389596995618895842088923511697626632892312882 6

915527388764216209149558398903829431137608893452690352556009760102417 1

190804769405534826558114212766135380613959757935305271022089419155701

521586470170874002194384529140686856227759541715097113399134734059 61

76328922060720855163323979691633837600707387601073182477520615047 14

3672504609009234090663777322733903968222962352149636232866131175 57

9306875441483607212256810534811787600582647388671058103268189 11

578448323758536767168707442532146029753762594261578920477859.

All digits in this number are guaranteed to be accurate. For details see [116, Section 4]. ◇

14.3 SDP Hierarchies

We now present our second method for computing volumes. It is based on a hierarchy of semidefinite programs (SDP). This is due to Lasserre and his collaborators. See the articles [86, 169, 170] and references therein.

As before, the goal is to compute the volume of a compact semialgebraic set $S \subset \mathbb{R}^n$ as accurately as possible. We consider an inclusion of semialgebraic sets $S \subset B \subset \mathbb{R}^n$, where B is compact and serves as a bounding box, like $B = [-1, 1]^n$.

We assume that the moments of the Lebesgue measure on B are known or easy to compute, i.e., we have access to the values of the integrals

$$b_\alpha = \int_B \mathbf{x}^\alpha d\mathbf{x} = \int_B x_1^{\alpha_1} x_2^{\alpha_2} \cdots x_n^{\alpha_n} \, dx_1 dx_2 \cdots dx_n \qquad \text{for } \alpha \in \mathbb{N}^n.$$

However, the moments m_α of the Lebesgue measure on S are unknown:

$$m_\alpha = \int_S \mathbf{x}^\alpha d\mathbf{x} = \int_S x_1^{\alpha_1} x_2^{\alpha_2} \cdots x_n^{\alpha_n} \, dx_1 dx_2 \cdots dx_n \qquad \text{for } \alpha \in \mathbb{N}^n. \tag{14.13}$$

These moments will be our decision variables. Our aim is to compute $m_0 = \mathrm{vol}(S)$. The idea is to use the following infinite-dimensional linear program:

> Maximize the integral $\int_S d\mu$, where μ and $\hat{\mu}$ range over measures on $\mathbb{R}^n$, where μ is supported on S, $\hat{\mu}$ is supported on B, and the sum $\mu + \hat{\mu}$ is the Lebesgue measure on B.

The unique optimal solution $(\mu^*, \hat{\mu}^*)$ to this linear program can be characterized as follows: μ^* is the Lebesgue measure on S, $\hat{\mu}^*$ is the Lebesgue measure on $B \backslash S$, and the optimal value is $\mathrm{vol}(S) = \int_S d\mu^*$. This is described in [170, Equation (1)]. The linear programming (LP) dual is given in [170, Equation (2)].

We state our linear program in terms of the moment sequences $\mathbf{m} = (m_\alpha)$ and $\hat{\mathbf{m}} = (\hat{m}_\alpha)$ of the two unknown measures μ and $\hat{\mu}$. Namely, we paraphrase our problem as follows:

> Maximize m_0 subject to $m_\alpha + \hat{m}_\alpha = b_\alpha$ for all $\alpha \in \mathbb{N}^n$,
> where $\mathbf{m}$ and $\hat{\mathbf{m}}$ are the moment sequences of μ and $\hat{\mu}$, $\qquad$ (14.14)
> respectively, with μ supported on S and $\hat{\mu}$ supported on B.

We arrive at the moment problem, which is the question of how to characterize moments of measures. This problem has a long history in mathematics, and an exact characterization is very difficult. However, in recent years, it has been realized that there are effective necessary conditions. These involve semidefinite programming formulations in finite dimensions, which are built via the *localizing matrices* we now define.

In the following, we assume for convenience that S is defined by a single inequality:

$$S = \{\mathbf{x} \in \mathbb{R}^n \mid f(\mathbf{x}) \geq 0\}, \quad \text{where} \quad f(\mathbf{x}) = \sum_\alpha c_\alpha \mathbf{x}^\alpha. \tag{14.15}$$

The theory works for all semialgebraic sets (14.2). Fix an integer d that exceeds the degree of f. We shall construct three symmetric matrices of format $\binom{n+d}{d} \times \binom{n+d}{d}$ whose entries are linear in the decision variables. The rows and columns of our matrices are indexed by elements $\alpha \in \mathbb{N}^n$ with $|\alpha| = \alpha_1 + \cdots + \alpha_n$ at most d. These correspond to monomials $\mathbf{x}^\alpha$ of degree $\leq d$. Our first matrix $M_d(\mathbf{m})$ has the entry $m_{\alpha+\beta}$ in row α and column β. Our second matrix $M_d(\hat{\mathbf{m}})$ has the entry $\hat{m}_{\alpha+\beta}$ in row α and column β. Finally, suppose that the polynomial f defining S has coefficients c_α. Then, our third matrix $M_d(f\mathbf{m})$ has the entry $\sum_\gamma c_\gamma m_{\alpha+\beta+\gamma}$ in row α and column β. We consider the following semidefinite program:

$$\begin{aligned} &\textit{Maximize } m_0 \textit{ subject to } m_\alpha + \hat{m}_\alpha = b_\alpha \\ &\textit{for all } \alpha \in \mathbb{N}^n \textit{ with } |\alpha| \le d, \textit{ where the} \\ &\textit{symmetric matrices } M_d(\mathbf{m}), M_d(\hat{\mathbf{m}}) \\ &\textit{and } M_d(f\mathbf{m}) \textit{ are positive semidefinite.} \end{aligned} \qquad (14.16)$$

The third matrix can be replaced by $M_{d'}(f\mathbf{m})$ where $d' = d - \lceil \deg(f)/2 \rceil$. The objective function value depends on d, and it decreases as d increases. The limit for $d \to \infty$ equals the volume of S. Indeed, this sequence of SDP problems is an approximation to (14.14). The convergence property was proved in [86].

The remainder of this section shows how to solve (14.16) in practice. It is based on [86, 169, 170]. We discuss an implementation in Mathematica [102]. This material was developed by Chiara Meroni. We are very grateful to her for allowing us to include it here.

Our point of departure is the following question: given a sequence of real numbers $\mathbf{m} = (m_\alpha)_\alpha$, does there exist a measure μ_S supported on the set S such that (14.13) holds? Given $d \in \mathbb{N}$, let $\mathbb{N}_d^n$ be the set of $\alpha \in \mathbb{N}^n$ such that $|\alpha| = \alpha_1 + \cdots + \alpha_n \le d$. We also set

$$r = \left\lceil \frac{\deg f}{2} \right\rceil.$$

The *moment matrix* and the *localizing matrix* for our sequence of moments $\mathbf{m}$ are

$$M_d(\mathbf{m}) = \Big(m_{\alpha+\beta} \Big)_{\alpha,\beta \in \mathbb{N}_d^n} \quad \text{and} \quad M_{d-r}(f\mathbf{m}) = \Big(\sum_\gamma c_\gamma m_{\alpha+\beta+\gamma} \Big)_{\alpha,\beta \in \mathbb{N}_{d-r}^n}. \qquad (14.17)$$

The moment matrix has size $\binom{n+d}{d} \times \binom{n+d}{d}$. The localizing matrix has size $\binom{n+d-r}{d-r} \times \binom{n+d-r}{d-r}$.

Proposition 14.18 *A necessary condition for a sequence $\mathbf{m} = (m_\alpha)_\alpha$ to have a representing measure supported on S is that for every $d \in \mathbb{N}$ the following matrix inequalities hold:*

$$M_d(\mathbf{m}) \succcurlyeq 0 \quad \text{and} \quad M_{d-r}(f\mathbf{m}) \succcurlyeq 0.$$

This result is a formulation of Putinar's Positivstellensatz [86, Theorem 2.2]. The positive semi-definiteness of the moment matrix is a necessary condition for $\mathbf{m}$ to have a representing measure; the inequality with the localizing matrix forces the support of the representing measure to be contained in $S = \{f(\mathbf{x}) \ge 0\}$.

Example 14.19 Consider the disc $S = \{(x, y) \in \mathbb{R}^2 \mid f = 1 - x^2 - y^2 \ge 0\}$. Its moments are

$$m_{(\alpha_1, \alpha_2)} = ((-1)^{\alpha_1} + 1)\,((-1)^{\alpha_2} + 1)\, \frac{\Gamma\left(\frac{\alpha_1+1}{2}\right)\Gamma\left(\frac{\alpha_2+1}{2}\right)}{4\Gamma\left(\frac{1}{2}(\alpha_1 + \alpha_2 + 4)\right)},$$

where Γ is the Gamma function. For $d = 3$, the moment and localizing matrices (14.17) are

$$M_3(\mathbf{m}) = \begin{bmatrix} \pi & 0 & \frac{\pi}{4} & 0 & 0 & 0 & 0 & \frac{\pi}{4} & 0 & 0 \\ 0 & \frac{\pi}{4} & 0 & \frac{\pi}{8} & 0 & 0 & 0 & 0 & \frac{\pi}{24} & 0 \\ \frac{\pi}{4} & 0 & \frac{\pi}{8} & 0 & 0 & 0 & 0 & \frac{\pi}{24} & 0 & 0 \\ 0 & \frac{\pi}{8} & 0 & \frac{5\pi}{64} & 0 & 0 & 0 & 0 & \frac{\pi}{64} & 0 \\ 0 & 0 & 0 & 0 & \frac{\pi}{4} & 0 & \frac{\pi}{24} & 0 & 0 & \frac{\pi}{8} \\ 0 & 0 & 0 & 0 & 0 & \frac{\pi}{24} & 0 & 0 & 0 & 0 \\ 0 & 0 & 0 & 0 & \frac{\pi}{24} & 0 & \frac{\pi}{64} & 0 & 0 & \frac{\pi}{64} \\ \frac{\pi}{4} & 0 & \frac{\pi}{24} & 0 & 0 & 0 & 0 & \frac{\pi}{8} & 0 & 0 \\ 0 & \frac{\pi}{24} & 0 & \frac{\pi}{64} & 0 & 0 & 0 & 0 & \frac{\pi}{64} & 0 \\ 0 & 0 & 0 & 0 & \frac{\pi}{8} & 0 & \frac{\pi}{64} & 0 & 0 & \frac{5\pi}{64} \end{bmatrix} \quad \text{and} \quad M_2(f\mathbf{m}) = \begin{bmatrix} \frac{\pi}{2} & 0 & \frac{\pi}{12} & 0 & 0 & \frac{\pi}{12} \\ 0 & \frac{\pi}{12} & 0 & 0 & 0 & 0 \\ \frac{\pi}{12} & 0 & \frac{\pi}{32} & 0 & 0 & \frac{\pi}{96} \\ 0 & 0 & 0 & \frac{\pi}{12} & 0 & 0 \\ 0 & 0 & 0 & 0 & \frac{\pi}{96} & 0 \\ \frac{\pi}{12} & 0 & \frac{\pi}{96} & 0 & 0 & \frac{\pi}{32} \end{bmatrix}.$$

These two matrices are symmetric and positive definite.　◇

We consider the infinite-dimensional *linear program* (LP) on measures whose optimal value is the volume of $S \subset B$. The program was stated above. We use the formulation in [86, Equation 3.1] and [170, Equation 1]:

$$P: \quad \max_{\mu_S, \mu_{B\backslash S}} \int d\mu_S, \quad \text{s.t.} \quad \mu_S + \mu_{B\backslash S} = \mu_B^*. \tag{14.18}$$

Here, μ_S is a positive finite Borel measure supported on S, and μ_B^* is the Lebesgue measure on B. The adjective "infinite-dimensional" refers to the fact that we are optimizing over a set of measures, which is uncountable. Based on the theory of dual Banach spaces, one can talk about dual convex bodies, and construct a duality theory for infinite-dimensional LPs. In our case, the dual to the space of positive finite Borel measures is the set of positive continuous functions. This observation leads to the definition of an LP that is dual to P:

$$P^*: \quad \inf_{\gamma} \int \gamma \, d\mu_B^*, \quad \text{s.t.} \quad \gamma \geq \mathbf{1}_S. \tag{14.19}$$

The decision variable γ is a positive continuous function on B and $\mathbf{1}_S$ is the indicator function of S. There is no duality gap: the optimal values of (14.18) and (14.19) coincide. Note that the optimal value of P^* is an infimum and not a minimum since we approximate the *discontinuous* indicator function $\mathbf{1}_S$ using continuous functions.

The infinite-dimensional LP (14.18) can be approximated by a hierarchy of finite-dimensional *semidefinite programs (SDPs)* [117]. The optimal values of the hierarchy converge monotonically to the optimal value of the LP [86, Theorem 3.2]. There is again a primal and dual version of the SDPs. In our setting, the primal is

$$P_d: \quad \max_{\mathbf{m}, \widehat{\mathbf{m}}} m_0, \quad \text{s.t.} \quad \mathbf{m} + \widehat{\mathbf{m}} = \mathbf{b}, \ M_d(\mathbf{m}) \succcurlyeq 0, \ M_d(\widehat{\mathbf{m}}) \succcurlyeq 0, \ M_{d-r}(f\mathbf{m}) \succcurlyeq 0. \tag{14.20}$$

Here $\mathbf{m} = (m_\alpha)$, $\widehat{\mathbf{m}} = (\widehat{m}_\alpha)$, and $\mathbf{b} = (b_\alpha)$ contains the moments of B indexed by $\alpha \in \mathbb{N}_{2d}^n$. This formulation is [170, Equation 3]. The optimal value of P_d is an upper bound for $\text{vol}(S)$ since we are optimizing over a larger set. The dual SDP is [86, Equation 3.6], which is formulated using sums of squares of polynomials. The authors of [86, 169, 170] implemented the SDPs using GloptiPoly [85]. The next examples are computed with Mathematica [102]. We are going to include the linear condition $\mathbf{m} + \widehat{\mathbf{m}} = \mathbf{b}$ inside the condition on the moment matrix of $\widehat{\mathbf{m}}$, by imposing directly that $M_d(\mathbf{b} - \mathbf{m}) \succcurlyeq 0$.

Example 14.20 (TV screen) Consider $S = \{(x, y) \in [-1.2, 1.2]^2 \mid f(x, y) \geq 0\} \subset \mathbb{R}^2$ where $f = -q$ is the quartic in (14.6). This convex set is shown in Figures 14.2 and 14.4. Recall from Example 14.9 that $\text{vol}(S) = 3.7081599447\ldots$.

Let us now try the SDP formulation above, with $d = 10$. The moment matrices $M_{10}(\mathbf{m})$ and $M_{10}(\mathbf{b} - \mathbf{m})$ have format 66×66. For instance, the second matrix looks like

$$M_{10}(\mathbf{b} - \mathbf{m}) = \begin{bmatrix} 4 - m_{(0,0)} & -m_{(0,1)} & \frac{4}{3} - m_{(0,2)} & -m_{(0,3)} & \cdots \\ -m_{(0,1)} & \frac{4}{3} - m_{(0,2)} & -m_{(0,3)} & \frac{4}{5} - m_{(0,4)} & \cdots \\ \frac{4}{3} - m_{(0,2)} & -m_{(0,3)} & \frac{4}{5} - m_{(0,4)} & -m_{(0,5)} & \cdots \\ -m_{(0,3)} & \frac{4}{5} - m_{(0,4)} & -m_{(0,5)} & \frac{4}{7} - m_{(0,6)} & \cdots \\ \vdots & \vdots & \vdots & \vdots & \ddots \end{bmatrix}.$$

The localizing matrix $M_8(f_1 \mathbf{m})$ has format 45×45. Its (α, β) entry equals

$$m_{\alpha+\beta} - m_{(4,0)+\alpha+\beta} - m_{(0,4)+\alpha+\beta} - \frac{1}{100} m_{(1,1)+\alpha+\beta}.$$

The optimal value of the semidefinite program P_{10} is $4.4644647361\ldots$, the optimal value of P_{14} is $4.3679560947\ldots$, and for P_{18} we get $4.3241824171\ldots$. These numbers are upper bounds for the actual volume, as predicted. However, these bounds are far from the truth.◇

Example 14.21 (Elliptope) Consider $f(x, y, z) = 1 - x^2 - y^2 - z^2 + 2xyz$. This defines the elliptope $S = \{(x, y, z) \in [-1, 1]^3 \mid f(x, y, z) \geq 0\} \subset \mathbb{R}^3$, shown in Figures 14.1 and 14.4. We already know from Example 14.1 that $\text{vol}(S) = \frac{\pi^2}{2} = 4.934802202\ldots$. The upper bounds computed from the SDP (14.20) for $d = 4, 8, 12$ are respectively $7.3254012963\ldots$, $6.6182632506\ldots$, and $6.303035372\ldots$. These numbers are still pretty bad.   ◇

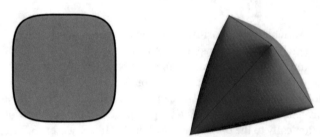

Fig. 14.4: Left: the TV screen from Example 14.20. Right: the elliptope from Example 14.21.

Examples 14.20 and 14.21 suggest that the convergence of the SDP approximation is quite slow. To improve this, one uses the method of *Stokes constraints*. This was introduced in [169, 170]. We shall now explain it. In the linear program P^* (14.19) and in the SDP hierarchy (14.20), we aim to approximate a piecewise-differentiable function $\mathbf{1}_S$ with continuous functions (respectively, polynomials). This produces the well-known *Gibbs effect*, creating many oscillations near the boundary of S in the polynomial solutions of the SDP.

To remedy this, we add linear constraints that do not modify the infinite-dimensional LP problem but add more information to the finite-dimensional SDP. One concrete way to do this uses Stokes' theorem and the fact that f vanishes on the boundary ∂S of $S = \{f \geq 0\}$.

Let U be an open region in $\mathbb{R}^n$ such that the Euclidean closure of U is our semialgebraic set S. Since ∂S is the zero set of a polynomial, it is smooth almost everywhere. The classical theorem of Stokes applies. This theorem states that $\int_{\partial S} \omega = \int_S d\omega$ for any $(n-1)$-differential form ω on $\mathbb{R}^n$. One consequence of Stokes' theorem is *Gauss' formula*

$$\int_{\partial S} V(\mathbf{x}) \cdot N(\mathbf{x}) \, d\mathbf{x} = \int_S \text{div } V(\mathbf{x}) \, d\mathbf{x}.$$

Here, $V(\mathbf{x})$ is a vector field, div denotes the divergence operator, $N(\mathbf{x})$ is the unit normal field of the algebraic variety ∂S that points outwards, and we are integrating against the standard measures on the respective domains. If the vector field is a scalar field times a constant vector $\mathbf{c} \in \mathbb{R}^n$, say $V(\mathbf{x}) = v(\mathbf{x})\mathbf{c}$, and if $\nabla v(\mathbf{x})$ denotes the gradient of v at $\mathbf{x}$, then

$$\mathbf{c} \cdot \int_{\partial S} v(\mathbf{x}) N(\mathbf{x}) \, d\mathbf{x} = \int_S \text{div}(v(\mathbf{x})\mathbf{c}) \, d\mathbf{x} = \mathbf{c} \cdot \int_S \nabla v(\mathbf{x}) \, d\mathbf{x}.$$

This formula holds because $\text{div}(v(\mathbf{x})\mathbf{c}) = \nabla v(\mathbf{x}) \cdot \mathbf{c} + v(\mathbf{x}) \cdot \text{div}(\mathbf{c})$ and the divergence of a constant vector is zero. Since this identity holds for every $\mathbf{c} \in \mathbb{R}^n$, we have

$$\int_{\partial S} v(\mathbf{x}) \cdot N(\mathbf{x}) \, d\mathbf{x} = \int_S \nabla v(\mathbf{x}) \, d\mathbf{x}. \tag{14.21}$$

If $v = 0$ on ∂S, then the left-hand side of (14.21) is zero. This can be expressed in terms of measures and distributions, and added to (14.18) and (14.19) as in [170, Equation 17 and Remark 3]. In our SDP hierarchy, the Stokes constraints are written as follows. Let

$$v(\mathbf{x}) = f(\mathbf{x}) \mathbf{x}^\alpha$$

for any multiindex $\alpha \in \mathbb{N}^n$ with $|\alpha| \leq d+1 - \deg f$. To remedy the Gibbs effect, we require

$$\nabla \big(f(\mathbf{x}) \mathbf{x}^\alpha\big)\big|_{\mathbf{x}^\beta = m_\beta} = 0.$$

This yields n new linear conditions for each α as above.

Example 14.22 For the SDP in Example 14.20, the Stokes constraints for a given α are:

$$\alpha_1 m_{\alpha+(-1,0)} - (\alpha_1 + 4) m_{\alpha+(3,0)} - \alpha_1 m_{\alpha+(-1,4)} - \tfrac{\alpha_1+1}{100} m_{\alpha+(0,1)} = 0,$$

$$\alpha_2 m_{\alpha+(0,-1)} - \alpha_2 m_{\alpha+(4,-1)} - (\alpha_2 + 4) m_{\alpha+(0,3)} - \tfrac{\alpha_2+1}{100} m_{\alpha+(1,0)} = 0.$$

For the SDP in Example 14.21, the Stokes constraints are:

$$\alpha_1 m_{\alpha+(-1,0,0)} - (\alpha_1 + 2) m_{\alpha+(1,0,0)} - \alpha_1 m_{\alpha+(-1,2,0)} - \alpha_1 m_{\alpha+(-1,0,2)} + 2(\alpha_1 + 1) m_{\alpha+(0,1,1)} = 0,$$

$$\alpha_2 m_{\alpha+(0,-1,0)} - \alpha_2 m_{\alpha+(2,-1,0)} - (\alpha_2 + 2) m_{\alpha+(0,1,0)} - \alpha_2 m_{\alpha+(0,-1,2)} + 2(\alpha_2 + 1) m_{\alpha+(1,0,1)} = 0,$$

$$\alpha_3 m_{\alpha+(0,0,-1)} - \alpha_3 m_{\alpha+(2,0,-1)} - \alpha_3 m_{\alpha+(0,2,-1)} - (\alpha_3 + 2) m_{\alpha+(0,0,1)} + 2(\alpha_3 + 1) m_{\alpha+(1,1,0)} = 0.$$

Table 14.1 compares the optimal values in (14.20) with and without Stokes constraints. ◇

As Table 14.1 shows, the convergence with Stokes constraints is much faster than without them. The intuition is that now, with the (dual) Stokes constraints added to P^* (14.19), the

S	volume	d	without Stokes		with Stokes	
			max P_d	time	max P_d	time
	3.708159...	10	4.464464...	0.621093	3.709994...	0.482376
		15	4.367956...	3.545369	3.708191...	3.738137
		20	4.324182...	14.906281	3.708163...	20.592531
	4.934802...	4	7.325401...	0.124392	5.612716...	0.077315
		8	6.618263...	7.222441	4.976796...	7.178571
		12	6.303035...	696.886298	4.937648...	1105.619231

Table 14.1: The optimal values of (14.20) with and without Stokes constraints for Examples 14.20 and 14.21. The column "max P_d" displays the optimal value, whereas the column "time" gives the time, in seconds, for running the command `SemidefiniteOptimization` in `Mathematica` [102].

function we approximate is not just the indicator function of S. A detailed explanation, for a variant of the Stokes constraints, is given in [169]. The authors prove that, when adding these constraints, the optimal solution of the new P^* becomes a minimum. This eliminates any kind of Gibbs effect, and guarantees faster convergence. In [169], the authors mention that, from numerical experiments, it is reasonable to expect that the original Stokes constraints and the new Stokes constraints are equivalent, but there is no formal proof yet.

In this chapter, we considered semialgebraic sets (14.1) that are defined by only one polynomial inequality. This restriction was imposed to simplify the exposition. Two methods for computing their volumes were presented in Sections 14.2 and 14.3. We did not offer a comparison between them: this is a topic for future research. Both methods apply to arbitrary semialgebraic sets. These can be written as finite unions of basic semialgebraic sets (14.2). In conclusion, computing volumes is important for metric algebraic geometry.

Chapter 15
Sampling

In this book, we studied the metric geometry of algebraic varieties from an applied and computational perspective. We now add probability theory into this mix. This chapter is about sampling from a real variety X. Asmussen and Glynn [10] underline the importance of sampling as follows: *"Sampling-based computational methods are a fundamental part of the numerical toolset across an enormous number of different applied domains"*.

Section 15.1 connects us to topological data analysis. Here we explain how sampling can be used to compute topological information about X. In Section 15.2, we discuss algorithms for sampling with density guarantees. The workhorse behind this is the ability to rapidly solve polynomial equations associated with X, namely intersecting with linear subspaces and finding ED critical points. In Section 15.3, we introduce Markov kernels on the variety X. These specify Markov chains that take steps on the variety X. While Markov chains on discrete state spaces are familiar in combinatorics and algebra, now the state space X is continuous. This requires some analysis that may be unfamiliar to our readers.

Markov chains on X can be modified with the Metropolis–Hastings algorithm (Algorithm 6), in order to reach a desired stationary distribution. For instance, if X is compact, we could be interested in sampling from the uniform distribution. We will explain how to do this in Section 15.4. The name Chow in the section title is an allusion to Chow forms. These encode a variety X through its linear sections. In our probabilistic setting, distributions on X are represented by distributions on the Grassmannian. This can be viewed as an interpretation of the Chow form for random algebraic geometry. For the resulting class of sampling algorithms, we propose the acronym CMCMC, short for Chow Markov Chain Monte Carlo.

15.1 Homology from Finite Samples

Let X be a real algebraic variety in $\mathbb{R}^n$. Our aim is to compute a finite subset $S \subset X$, which we call a *sample*. A sample on a curve is shown in Figure 15.1. The sample S yields a discrete approximation of X that can be used to explore properties of the variety.

© The Author(s) 2024
P. Breiding et al., *Metric Algebraic Geometry*, Oberwolfach Seminars 53,
https://doi.org/10.1007/978-3-031-51462-3_15

For instance, if $g : X \to \mathbb{R}$ is a function, we can find a lower bound for the optimization problem $\max_{\mathbf{x} \in X} g(\mathbf{x})$ by computing $\max_{\mathbf{s} \in S} g(\mathbf{s})$. Or, assuming that X has finite volume, we can estimate the average value $\frac{1}{\text{vol}(X)} \int_{\mathbf{x} \in X} g(\mathbf{x}) d\mathbf{x}$ by the finite average $\frac{1}{|S|} \sum_{\mathbf{s} \in S} g(\mathbf{s})$,

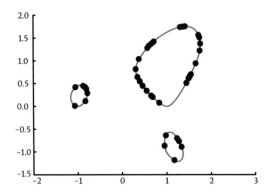

Fig. 15.1: A sample of points on the curve $X = \{x^4 + y^4 - 2x^2 - 2xy^2 - y + 1 = 0\}$.

In what follows, we use the Euclidean distance $d(\mathbf{x}, \mathbf{y}) = \|\mathbf{x} - \mathbf{y}\|$ on the ambient space $\mathbb{R}^n$. This means that we consider our variety X with an *extrinsic metric*. The next definition refers to this extrinsic metric on the variety X.

Definition 15.1 A finite subset S of X is called ε-*dense* in X if, for all $\mathbf{x} \in X$, there exists an $\mathbf{s} \in S$ with $d(\mathbf{x}, \mathbf{s}) < \varepsilon$. If this holds, then we also say that S is an ε-*sample* for X.

In this section, we are computing random samples on an algebraic variety X. In this computation, we can replace X by its smooth locus $\text{Reg}(X)$ and sample from the manifold $\text{Reg}(X)$. To keep the notation simple, we assume throughout this chapter that the variety X is smooth. It is a manifold that is embedded in the Euclidean space $\mathbb{R}^n$.

We first recall a theorem due to Niyogi, Smale, and Weinberger [138]. This gives conditions for when the homology of X can be computed from a finite sample S. The idea is to compute the homology of the union of ε-balls $U = \bigcup_{\mathbf{s} \in S} B_{\varepsilon}(\mathbf{s})$ of an ε-sample S. The homology groups of a union of balls U can be computed from the associated Čech complex. The theorem explains how small ε must be for this to work.

Theorem 15.2 *Let $\varepsilon > 0$, let S be an ε-sample for X, and write $U = \bigcup_{\mathbf{s} \in S} B_{\varepsilon}(\mathbf{s})$ for the union of all ε-balls around all points in S. If $\varepsilon < \sqrt{\frac{3}{20}} \, \tau(X)$, where $\tau(X)$ is the reach, then X is a deformation retract of U. In this case, the homology of X equals the homology of U.*

Proof See [138, Proposition 3.1].   □

Di Rocco, Eklund, and Gäfvert [56] showed that, in Theorem 15.2, the reach $\tau(X)$ can be replaced by the *local reach* $\tau(\mathbf{x})$. For $\mathbf{x} \in X$, the local reach is defined as the quantity

$$\tau(\mathbf{x}) := \sup \{r \geq 0 \mid \text{every } \mathbf{u} \in B_r(\mathbf{x}) \text{ has a unique closest point on } X\}. \tag{15.1}$$

Thus, the local reach $\tau(\mathbf{x})$ is the distance from the point $\mathbf{x} \in X$ to the medial axis $\mathrm{Med}(X)$; cf. Chapter 7. The global reach $\tau(X)$ is bounded above by the local reach at each point; in fact, we have $\tau(X) = \inf_{\mathbf{x} \in X} \tau(\mathbf{x})$. Thus, using the local reach instead of the global reach for upper bounding ε can make a significant difference. The result from [56] is: *If $\varepsilon < \frac{4}{5}\tau(\mathbf{s})$ for all points $\mathbf{s}$ in an ε-sample $S \subset X$, then X is a deformation retract of $U = \bigcup_{\mathbf{s} \in S} B_\varepsilon(\mathbf{s})$.*

If we only seek homology for small dimensions, then we can relax the conditions on ε. We first need the notion of k-*bottlenecks*. For a finite subset $B \subset X$, let $\Gamma(B)$ denote the subset of points in $\mathbb{R}^n$ that are equidistant to all points in B. This is an affine-linear space.

Definition 15.3 Let $B = \{\mathbf{x}_1, \ldots, \mathbf{x}_k\} \subset X$ for $k \geq 2$. Let $L_i := \mathbf{x}_i + N_{\mathbf{x}_i} X$ denote the affine normal space of X at $\mathbf{x}_i$. If $L_1 \cap \cdots \cap L_k \cap \Gamma(B) \cap \mathrm{conv}(B) \neq \emptyset$, we say that B is a k-*bottleneck* of X. The *width* of B is

$$\ell(B) := \inf_{\mathbf{u} \in \Gamma(B)} d(\mathbf{x}_1, \mathbf{u}).$$

Intuitively, k-bottlenecks identify data points $\mathbf{u} \in \mathbb{R}^n$ that have precisely k nearest points on X. The convexity condition implies that $\mathbf{u}$ is in the "middle" of the $\mathbf{x}_i$, and not outside. For instance, the 2-bottlenecks of X are precisely the bottlenecks studied in Section 7.1.

The *weak feature size* $\mathrm{wfs}(X)$ of the variety X is the smallest width of a k-bottleneck:

$$\mathrm{wfs}(X) := \min_{2 \leq k \leq \mathrm{EDdegree}(X)} \inf \{ \ell(B) \mid B \text{ is a } k\text{-bottleneck of } X \}.$$

The weak feature size is always greater than or equal to the reach. The reason is that the exponential map φ_ε in (6.12) cannot be injective when $\varepsilon > \ell(B)$ for a k-bottleneck $B = \{\mathbf{x}_1, \ldots, \mathbf{x}_k\}$. Indeed, if this holds, then the point $\mathbf{u} \in \Gamma(B)$ with $\ell(B) = d(\mathbf{x}_1, \mathbf{u})$ has k points $(\mathbf{x}_1, \mathbf{v}_1), \ldots, (\mathbf{x}_k, \mathbf{v}_k)$ in its fiber under φ_ε.

For the next theorem, we fix $\varepsilon > 0$ and a finite set $S \subset \mathbb{R}^n$. We define a two-dimensional simplicial complex C with vertex set S as follows. For $\mathbf{x}, \mathbf{y} \in S$, the pair $\{\mathbf{x}, \mathbf{y}\}$ is an edge of C if and only if at least one of the following two metric conditions is satisfied:

1. $d(\mathbf{x}, \mathbf{y}) \leq 2\varepsilon$, or
2. $d(\mathbf{x}, \mathbf{y}) \leq \sqrt{8}\varepsilon$ and there is $\mathbf{z} \in S$ with $d(\mathbf{x}, \mathbf{z}) \leq 2\varepsilon$ and $d(\mathbf{y}, \mathbf{z}) \leq 2\varepsilon$.

We add the triangle with vertices $\mathbf{x}, \mathbf{y}, \mathbf{z} \in S$ to C if and only if there is an edge between $\mathbf{x}$ and $\mathbf{y}$, between $\mathbf{x}$ and $\mathbf{z}$, and between $\mathbf{y}$ and $\mathbf{z}$. The result is called *Vietoris–Rips complex* (at scale ε). The following was proved in [56].

Theorem 15.4 *Let $S \subset X$ be an ε-sample, where $\varepsilon < \mathrm{wfs}(X)$. Let C be the Vietoris–Rips complex above. Then, $H_0(X) \cong H_0(C)$ and $H_1(X) \cong H_1(C)$.*

Remark 15.5 Theorem 15.4 is based on [46, Theorem 1]. Here, Chazal and Lieutier proved that, for $\varepsilon < \mathrm{wfs}(X)$, the tubular neighborhood $\bigcup_{\mathbf{x} \in X} B_\varepsilon(\mathbf{x})$ is homotopy equivalent to X.

Topological data analysis is concerned with the homology of a data space X. One can try to compute this from an ε-sample. The difficulty is to determine ε a priori. The input is some representation of X. This could be a set of polynomial equations that define X. The difficulty lies in the fact that, in order to choose ε reliably, we need to compute some invariants of X, such as the reach $\tau(X)$, the local reach $\tau(\mathbf{s})$ at various points $\mathbf{s} \in X$, the k-bottlenecks and their widths, or the weak feature size $\mathrm{wfs}(X)$. All of these are computations in metric algebraic geometry, based on what was explained in previous chapters.

15.2 Sampling with Density Guarantees

This section presents two sampling algorithms that are guaranteed to compute an ε-sample in a box. The input is a variety X in $\mathbb{R}^n$. Both algorithms compute a sample of $X \cap R$, where

$$R = [a_1, b_1] \times \cdots \times [a_n, b_n] \subset \mathbb{R}^n.$$

If X is compact, then it is desirable to use a box R that contains the variety X. To compute such an R, we first sample a point $\mathbf{u} \in \mathbb{R}^n$ at random. Then, we compute the ED critical points on X with respect to $\mathbf{u}$; i.e., the critical points of the Euclidean distance function $X \to \mathbb{R}, \mathbf{x} \mapsto d(\mathbf{x}, \mathbf{u})$. From this, we infer $r := \max_{\mathbf{x} \in X} d(\mathbf{x}, \mathbf{u})$ and set R to be the box with center $\mathbf{u}$ and side length $2r$. Alternatively, we can maximize and minimize the n coordinate functions over X. This will furnish a box with optimal values for $a_1, b_1, \ldots, a_n, b_n$.

The first sampling algorithm we present is due to Dufresne, Edwards, Harrington, and Hauenstein [65]. Fix a box R. The basic idea for sampling from $X \cap R$ is to sample points $\mathbf{u} \in \mathbb{R}^n$ and then to collect the ED critical points with respect to $\mathbf{u}$. The complexity of this approach therefore depends on the Euclidean distance degree of the variety X.

The algorithm in [65] works recursively by dividing the edges of the box R in half, thus splitting R into 2^n subboxes. In addition, one implements a database $\mathcal{D}$ that contains information about all the regions in R that already have been covered by at least one sample point. Algorithm 4 provides a complete description. Its correctness is proved in [65].

Theorem 15.6 (Theorem 4.4 in [65]) *Algorithm 4 terminates with an ε-sample of $X \cap R$.*

The second algorithm we present is due to Di Rocco, Eklund, and Gäfvert [56]. Their algorithm is also based on computing ED critical points on X. But, in addition, it also adds linear slices to the sampling. We need a few definitions. We set $d := \dim X$. For every $1 \leq k \leq d$, denote by $\mathcal{T}_k$ the set of subsets of $\{1, \ldots, n\}$ with k elements. Given $T = \{t_1, \ldots, t_k\} \in \mathcal{T}_K$, we let $V_T \subseteq \mathbb{R}^n$ be the k-dimensional coordinate plane spanned by $\mathbf{e}_{t_1}, \ldots, \mathbf{e}_{t_k}$. For $\delta > 0$, we consider the grid

$$G_T(\delta) := \{\delta \cdot (a_1 \cdot \mathbf{e}_{t_1} + \cdots + a_k \cdot \mathbf{e}_{t_k}) \mid a_1, \ldots, a_k \in \mathbb{Z}\} \cong \delta \cdot \mathbb{Z}^k.$$

Let $\pi_T : \mathbb{R}^n \to V_T$ be the projection. The affine-linear spaces $\pi_T^{-1}(g)$ for $g \in G_T(\delta)$ specify the faces of a cubical tessellation with side length δ.

Let $B(X)$ be the width of the smallest bottleneck of X. The method from [56] is presented in Algorithm 5. It takes as input a number $0 < \delta < \frac{1}{\sqrt{n}} \min\{\varepsilon, 2B(X)\}$ for the grid size. Then, the sample is given by

$$S_\delta := \bigcup_{T \in \mathcal{T}_d} \bigcup_{g \in G_T(\delta)} X \cap \pi_T^{-1}(g). \tag{15.2}$$

Algorithm 4: Finding ε-samples by ED optimization in subdivided boxes [65].

Input: A real algebraic variety $X \subset \mathbb{R}^n$, $\varepsilon > 0$, and a box $R = [a_1, b_1] \times \cdots \times [a_n, b_n]$.
Output: An ε-sample $S \subset X \cap R$.

1 Initialize $S = \emptyset$ and $\mathcal{D} = \emptyset$. The set S will contain the sample points. The set $\mathcal{D}$ serves as a
    database containing balls in $\mathbb{R}^n$ that have already been covered in the process of the algorithm.
2 **for** *each subbox R' of R that is not yet covered* **do**
3   Compute the midpoint $\mathbf{u}$ of R'.
4   Compute the real ED critical points $E \subset X$ with respect to $\mathbf{u}$ and compute
        $r := \min_{\mathbf{x} \in X} \|\mathbf{x} - \mathbf{u}\|$.
5   Add the points in E to S.
6   Add $B_r(\mathbf{u})$ to $\mathcal{D}$ (this open ball does not contain any point in X, so we do not need to
        consider this region any further and can label it as being covered).
7   Add $B_\varepsilon(\mathbf{y})$ for $\mathbf{y} \in E$ to $\mathcal{D}$.
8   **if** *the union of balls in $\mathcal{D}$ cover R'* **then**
9   Label R' and all of its subboxes as covered.
10   **else**
11   Split R' into 2^n smaller subboxes.
12   **end**
13 **end**
14 **if** *all subboxes of R are labeled as covered* **then**
15   **return** S.

Algorithm 5: Find an ε-dense sample with bottlenecks, grids, and ED methods [56].

Input: A real algebraic variety $X \subset \mathbb{R}^n$, a real number $\delta > 0$ with $\delta < \frac{1}{\sqrt{n}} \min\{\varepsilon, 2B(X)\}$, and
    a box R.
Output: An ε-dense sample $S \subset X \cap R$.

1 Initialize $S_\delta = \emptyset$ and $S'_\delta = \emptyset$.
2 Set $d := \dim X$.
3 **for** $T \in \mathcal{T}_d$ and $g \in G_T(\delta)$ **do**
4   Compute $X \cap \pi_T^{-1}(g)$.
5   Add the points in $X \cap \pi_T^{-1}(g)$ to S_δ.
6 **end**
7 **if** $d > 1$ **then**
8   **for** $1 \leq k < d$ **do**
9   **for** $T \in \mathcal{T}_k$ and $g \in G_T(\delta)$ **do**
10   Sample a random point $\mathbf{u} \in \mathbb{R}^n$.
11   Compute $E(T, g, \mathbf{u})$, which are the ED critical points on $X \cap \pi_T^{-1}(g)$ with respect
            to $\mathbf{u}$.
12   Add the points in $E(T, g, \mathbf{u})$ to S'_δ.
13   **end**
14   **end**
15 Set $S := S_\delta \cup S'_\delta$.
16 **return** S.

In words, the sample S_δ consists of the points that are obtained by intersecting the variety X
with the collection of affine-linear spaces $\pi_T^{-1}(g)$ that are indexed by $T \in \mathcal{T}_d$ and $g \in G_T(\delta)$.
The dimension of $\pi_T^{-1}(g)$ equals the codimension of X. To ensure transversal intersections,
we can always modify $G_T(\delta)$ by a random translation. If $d = \dim X > 1$, then we compute

an additional sample. First, we pick a random point $\mathbf{u} \in \mathbb{R}^n$. Denote by $E(T, g, \mathbf{u})$ the ED critical points on $X \cap \pi_T^{-1}(g)$ with respect to $\mathbf{u}$. The additional sample is the set

$$S'_\delta = \bigcup_{k=1}^{d-1} \bigcup_{T \in \mathcal{T}_k} \bigcup_{g \in G_T(\delta)} E(T, g, \mathbf{u}). \tag{15.3}$$

The motivation for this extra sample is that $E(T, g, \mathbf{u})$ contains a point on every connected component of the variety $X \cap \pi_T^{-1}(g)$. The algorithm is summarized in Algorithm 5. It requires us to compute bottlenecks of X in order to specify the input. In the algorithm itself, ED critical points must be found many times. Here, our Chapter 2 becomes relevant.

Theorem 15.7 (Theorem 4.6 in [56]) *Algorithm 5 outputs an ε-sample of $X \cap R$.*

We conclude this section with a third alternative for obtaining ε-samples. This comes from the work of Niyogi, Smale, and Weinberger [138], which we already saw in Theorem 15.2. Suppose that the variety X is compact, and suppose further that we are able to sample from the uniform distribution on X. The algorithm simply consists of sampling k i.i.d. points $S = \{\mathbf{x}_1, \ldots, \mathbf{x}_k\}$ from the uniform distribution on X.

The result in [138] implies that S is an ε-sample with high probability provided k is large enough. More precisely, for a given $0 < \delta < 1$, they specify a required sample size k such that the probability that S is an ε-sample for X is at least $1 - \delta$. That sample size k depends on the reach and the volume of the variety X. The study and computation of these metric quantities is the main point of this book. We invite the reader to work through the previous chapters while taking a perspective towards random points, sampling, and its role in machine learning (cf. Chapter 10). This requires examining probability distributions on a real algebraic variety. We explain this in the next sections and we present algorithms for sampling from such probability distributions.

15.3 Markov Chains on Varieties

We will review one popular class of methods for sampling from probability distributions, namely *Markov Chain Monte Carlo* (MCMC) methods. When the state space is finite, this amounts to a random walk on a finite graph. Experts in algebraic statistics [167] are familiar with MCMC methods that rest on Markov bases. These are used to carry out Fischer's Exact Test for conditional distributions; see [167, Chapter 9].

In this section, the situation is different because the state space X is continuous. As before, we assume that X is a real algebraic variety in $\mathbb{R}^n$. Now, the description of Markov chains requires a dose of analysis. In the following discussion, $\mathcal{A}$ denotes a σ-algebra on X. Think of $\mathcal{A}$ as the collection of subsets of X that we can assign a measure to. The elements in $\mathcal{A}$ are called *measurable sets*. If $\mu : \mathcal{A} \to \mathbb{R}$ is a measure and $f : X \to \mathbb{R}$ is a measurable function, then we can integrate f against μ. This integral is written as

$$\int_A f(\mathbf{x}) \, \mu(d\mathbf{x}) \qquad \text{for} \quad A \in \mathcal{A}.$$

The Lebesgue measure on X, which is induced from the ambient $\mathbb{R}^n$, is simply denoted by d$\mathbf{x}$. If μ is a probability measure with a probability density ϕ, then $\mu(\mathrm{d}\mathbf{x}) = \phi(\mathbf{x})\mathrm{d}\mathbf{x}$. For a measurable set $A \in \mathcal{A}$, let $\mathrm{vol}(A) = \int_A \mathrm{d}\mathbf{x}$ be the volume of A. Assume that X is compact. Then $\mathrm{vol}(A) < \infty$ and the probability distribution with probability measure $\mu(A) := \mathrm{vol}(A)/\mathrm{vol}(X)$ is called the *uniform distribution* on X.

Example 15.8 ($n = 2$) A sample of i.i.d. (independent and identically distributed) points from the uniform distribution on the plane curve $X = \{x^4 + y^4 - 2x^2 - 2xy^2 - y + 1 = 0\}$ is shown in Figure 15.1. It is important that each connected component is covered.   ◇

We next discuss sampling from X by using MCMC methods. These methods set up a *Markov process* on X. Let us recall some basic definitions from the theory of Markov chains; see, e.g., [131, Chapter 3]. In this context, X is also called the *state space*. A *Markov kernel* is a map $p : X \times \mathcal{A} \to [0, 1]$, such that

1. $p(\mathbf{x}, \cdot)$ is a probability measure for all $\mathbf{x} \in X$;
2. $p(\cdot, A)$ is a measurable function for all $A \in \mathcal{A}$.

A stochastic process $\mathbf{x}_0, \mathbf{x}_1, \mathbf{x}_2, \mathbf{x}_3, \ldots$ on X is a sequence of random points on X.

A *Markov process* is a special type of stochastic process. We fix a point $\mathbf{x} \in X$. A (time-homogeneous) Markov process with starting point $\mathbf{x}_0 = \mathbf{x}$ is the stochastic process where the probability that the next k points lie in certain measurable sets $A_1, \ldots, A_k$ is the following iterated integral over the Markov kernel:

$$\mathrm{Prob}(\mathbf{x}_k \in A_k, \ldots, \mathbf{x}_1 \in A_1 \mid \mathbf{x}_0 = \mathbf{x})$$

$$:= \int_{\mathbf{y}_1 \in A_1} \cdots \int_{\mathbf{y}_{k-1} \in A_{k-1}} p(\mathbf{x}, \mathrm{d}\mathbf{y}_1) p(\mathbf{y}_1, \mathrm{d}\mathbf{y}_2) \cdots p(\mathbf{y}_{k-1}, A_k).$$

A Markov process satisfies the *Markov property*: the probability law of the k-th state only depends on the position of the $(k - 1)$-th state, but not on the earlier states $\mathbf{x}_0, \ldots, \mathbf{x}_{k-2}$.

In the discrete setting [167, Section 1.1], the Markov property is expressed as a conditional probability, where one conditions on $\mathbf{x}_i = \mathbf{z}_i$ and the $\mathbf{z}_i$ are fixed points. In our continuous setting, this is not possible because it involves conditioning on an event of probability zero. Instead, one has a more technical description. Let

$$\mu^k(\mathbf{x}, \cdot) : A \mapsto \mathrm{Prob}(\mathbf{x}_k \in A \mid \mathbf{x}_0 = \mathbf{x}) \tag{15.4}$$

denote the probability law of the k-th state. Then, the Markov property is

$$\mathrm{Prob}(\mathbf{x}_k \in A \mid \mathbf{x}_{k-1}, \ldots, \mathbf{x}_1, \mathbf{x}_0 = \mathbf{x}) = \mathrm{Prob}(\mathbf{x}_1 \in A \mid \mathbf{x}_0 = \mathbf{z}) = p(\mathbf{z}, A), \quad \text{where } \mathbf{z} \sim \mu^k.$$

This is found in the textbook by Meyn and Tweedie [131, Proposition 3.4.3]. The left-hand side is a probability conditioned on the σ-algebra generated by the random variables $\mathbf{x}_i$.

We will not discuss the technical details of the Markov property. Instead, we explain what it means. If the $(k - 1)$-th state $\mathbf{x}_{k-1}$ is at position $\mathbf{z}$, then the probability of passing from $\mathbf{z}$ to a k-th state $\mathbf{x}_k \in A$ is given by the Markov kernel $p(\mathbf{z}, A)$. As in the discrete setting, the probability law of the $\mathbf{x}_k$ state only depends on the position of $\mathbf{x}_{k-1}$, but not on the earlier states. Due to its role in the transition from one state to the next state, the Markov kernel p is also called *transition probability*. Let us now see some examples.

Example 15.9 (Markov Chains in $\mathbb{R}^2$) For our state space, we take $X = \mathbb{R}^2$. We shall present two examples of Markov Chains in $\mathbb{R}^2$ starting at the origin. For the first, we take $p_1(\mathbf{x}, d\mathbf{y}) = (2\pi)^{-1} \cdot \exp(-\frac{1}{2}\|\mathbf{y}\|^2)\,d\mathbf{y}$ as the kernel. The stochastic process arising from this passes from the state $\mathbf{x} \in \mathbb{R}^2$ to the next state by sampling a normal vector $\mathbf{y}$ in $\mathbb{R}^2$ with covariance matrix the identity and mean value $\mathbf{0}$. The next state is then $\mathbf{y}$. In particular, the transition probability is entirely independent of the present state $\mathbf{x}$.

Our second example has the kernel $p_2(\mathbf{x}, d\mathbf{y}) = (2\pi)^{-1} \exp(-\frac{1}{2}\|\mathbf{x} - \mathbf{y}\|^2)\,d\mathbf{y}$. This is more interesting than the previous process. In this Markov chain, passing from a state $\mathbf{x} \in \mathbb{R}^2$ to the next state $\mathbf{y}$ works by sampling a normal vector $\mathbf{u}$ in $\mathbb{R}^2$ with covariance matrix the identity and mean value $\mathbf{0}$. The vector $\mathbf{u}$ is the step, so the next state is $\mathbf{y} = \mathbf{x} + \mathbf{u}$.

We can simulate the first $n = 10$ steps in this Markov chain in Julia [20] as follows.

```
x0 = [0; 0]
states = [];
push!(states, x0);
n = 10
for k in 1:n
    x = states[k];
    y = x + randn(2);
    push!(states, y);
end
```

Figure 15.2 shows one realization of this chain.   ◇

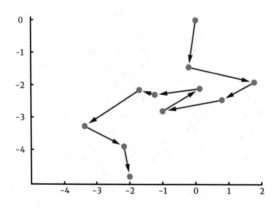

Fig. 15.2: A sample of the first 10 steps of the Markov chain with transition probability at state $\mathbf{x}$ given by sampling a normal vector in $\mathbb{R}^2$ with covariance matrix the identity and mean value $\mathbf{x}$. See Example 15.9.

Our aim is to obtain Markov chains on an arbitrary real variety $X \subset \mathbb{R}^n$. Suppose that X has dimension d and degree at least two. We can create a Markov chain on X using the following simple geometric idea. Suppose that chain is in state $\mathbf{x} \in X$. We sample a random

linear space L of complimentary dimension $n - d$ passing through $\mathbf{x}$. Then, we sample a
point uniformly at random from $X \cap L$. This is the next state in the Markov chain.

Example 15.10 (Markov Chain on a Surface) Fix $n = 3$ and let X be the surface defined
by $z - xy = 0$. Given a point $\mathbf{x}$ on this surface, we sample the line $L = \{A\mathbf{x} = \mathbf{b}\}$ by
sampling A with Gaussian entries and then setting $\mathbf{b} = A\mathbf{x}$. In addition, we only accept
states in the box $R = [-8, 8] \times [-8, 8] \times [-64, 64]$. The first few steps in this chain starting
at $\mathbf{x}_0 = (0, 0, 0)$ are implemented in Julia [20] as follows.

```julia
using HomotopyContinuation
@var x y z
f = System([z - x * y], variables = [x; y; z]);
is_in_R(p) = abs(p[1])<8 && abs(p[2])<8 && abs(p[3])<64;
n = 10;
x0 = [0.0; 0.0; 0.0];
states = [x0];
for k in 1:n
    p = last(states);
    A = randn(2,3); b = A*p;
    L = LinearSubspace(A, b);
    S = solve(f, target_subspace = L);
    points = real_solutions(S);
    filter!(is_in_R, points);
    push!(states, rand(points));
end
```

Figure 15.3 shows a realization. We think of steps taken on the surface itself. The arrows
are drawn curvy in order to highlight the nonlinear state space of this Markov chain.   ◊

Fig. 15.3: A few steps of the Markov chain from Example 15.10 on the surface with equation $z - xy = 0$.

Let us now work towards sampling. We will recall results from the survey [153] by
Roberts and Rosenthal. A probability distribution π on X is called *stationary* for a Markov
kernel p if it satisfies the following linear equation:

$$\int_{\mathbf{x} \in X} p(\mathbf{x}, A) \, \pi(\mathrm{d}\mathbf{x}) = \pi(A) \quad \text{for all } A \in \mathcal{A}. \tag{15.5}$$

The idea of MCMC methods for sampling from a probability distribution π is to set up a Markov process on X with stationary distribution π. The Markov chain starts at $\mathbf{x} \in X$. We want the probability measure $\mu^k(\mathbf{x}, \cdot)$ of the k-th state, as defined in (15.4), to converge to π as $k \to \infty$. Convergence is measured by the total variation distance. The *total variation distance* between two measures μ and ν is

$$d_{\mathrm{TV}}(\mu, \nu) := \sup_{A \in \mathcal{A}} |\mu(A) - \nu(A)|.$$

Thus, we aim to define a Markov process starting at an initial point $\mathbf{x} \in X$ such that the distribution of the k-th state converges to π as $k \to \infty$; i.e., $\lim_{k \to \infty} d_{\mathrm{TV}}(\pi, \mu^k(\mathbf{x}, \cdot)) = 0$. The key properties for achieving this are *irreducibility* and *aperiodicity*.

Definition 15.11 A Markov chain with kernel p is called *irreducible* if, for all states $\mathbf{x} \in X$ and all measurable sets $A \in \mathcal{A}$ with $\mathrm{vol}(A) > 0$, there exists $k \in \mathbb{N}$ such that $\mu^k(\mathbf{x}, A) > 0$.

The interpretation of irreducibility is that all sets with positive volume will eventually be reached by the Markov chain, and here the chain can start at any point $\mathbf{x} \in X$.

Remark 15.12 The expression $\mathrm{vol}(A)$ in Definition 15.11 refers to the Lebesgue measure. One can replace the Lebesgue measure with any other measure ν. In that case, one speaks of ν-irreducible Markov chains.

Definition 15.13 Consider a Markov chain with kernel p and suppose that it has a stationary distribution π. Let $r \geq 2$. We call the chain *periodic* with period r if there exist pairwise disjoint subsets $A_1, \ldots, A_r \in \mathcal{A}$ that satisfy the following two conditions for all indices i:

(a) $\pi(A_i) > 0$ and
(b) $p(\mathbf{x}, A_{i+1 \bmod r}) = 1$ for all $\mathbf{x} \in A_i$.

Otherwise, the chain is called *aperiodic*.

Aperiodicity means that the Markov process does not move periodically between the sets $A_1, \ldots, A_r$. The next lemma covers most of the Markov chains that arise from constructions in algebraic geometry. Any Markov chain defined by a kernel $p(\mathbf{x}, A) = \int_A \phi(\mathbf{x}, \mathbf{y}) \, d\mathbf{y}$ with positive continuous probability density ϕ satisfies the hypothesis of the lemma.

Lemma 15.14 *Consider a Markov chain with kernel p and stationary distribution π that is absolutely continuous with respect to the Lebesgue measure. Suppose that the kernel is positive, i.e., we have $p(\mathbf{x}, A) > 0$ for all $\mathbf{x} \in X$ and $A \in \mathcal{A}$ with $\mathrm{vol}(A) > 0$. Then, the Markov chain on the variety X corresponding to p is irreducible and aperiodic.*

Proof For irreducibility, fix $\mathbf{x} \in X$ and a measurable set $A \in \mathcal{A}$ with $\mathrm{vol}(A) > 0$. Then,

$$\mu^1(\mathbf{x}, A) = \mathrm{Prob}(\mathbf{x}_1 \in A \mid \mathbf{x}_0 = \mathbf{x}) = p(\mathbf{x}, A) > 0.$$

For aperiodicity, assume that there exist pairwise disjoint $A_1, \ldots, A_r \in \mathcal{A}$ with $\pi(A_i) > 0$ such that the Markov chain moves periodically between the A_i. Since π is absolutely continuous with respect to the Lebesgue measure, we also have $\mathrm{vol}(A_i) > 0$ for all i. Fix i and let $\mathbf{x} \in A_i$. Then, $p(\mathbf{x}, A_i) > 0$ and so $p(\mathbf{x}, A_{i+1 \bmod r}) < 1$, which is a contradiction. $\square$

For an irreducible and aperiodic Markov chain with stationary distribution, we have the following convergence result [153, Theorem 4]. This is a continuous analogue of the Perron–Frobenius theorem from linear algebra, which concerns distributions on finite state spaces.

Theorem 15.15 *Consider an irreducible and aperiodic Markov chain with kernel p and stationary distribution π. For almost all $\mathbf{x} \in X$, it converges to π in total variation distance:*

$$\lim_{k \to \infty} d_{\mathrm{TV}}(\pi, \mu^k(\mathbf{x}, \cdot)) = 0.$$

Theorem 15.15 says that irreducible and aperiodic Markov chains converge to stationary distributions. See Sullivant's book [167, Theorem 9.2.3] for the discrete case. The speed of convergence is called *mixing time* in the literature. For instance, a basic result on mixing times of Markov chains with continuous state space is [153, Theorem 8]. Roberts and Rosenthal [153] attribute this to Doeblin, Doob, and also to Markov.

For Markov chains arising in the context of algebraic varieties, for instance, the one in Example 15.10 defined by linear sections, it is interesting to study and describe their stationary distributions. However, it is not even clear whether they exist. In the next section, we turn the perspective around. Our starting point will be a stationary distribution π, and we will set up a Markov chain whose stationary distribution is π.

15.4 Chow goes to Monte Carlo

The algebraic geometer Wei-Liang Chow obtained his doctoral degree from Leipzig University in 1936. He is famous for fundamental contributions to the intersection theory of algebraic varieties. The Chow form encodes a variety of dimension d by the linear subspaces of codimension $d + 1$ that intersect it. We learn about Chow varieties and Chow polytopes in the book by Gel'fand, Kapranov, and Zelevinsky [71]. Metric algebraic geometry suggests that we take a look at objects named after Chow through the probabilistic lens.

In this section, we study the problem of sampling from a variety X using the Markov Chain Monte Carlo (MCMC) paradigm. This involves intersecting X with random linear spaces, whence our pointer to Chow. This gives rise to a class of algorithms, which we name *Chow–Markov Chain Monte Carlo* (CMCMC).

Our point of departure is Theorem 15.15. This result has the following algorithmic consequence. For sampling from a distribution π, we set up an irreducible and aperiodic Markov chain with stationary distribution π. Then, if we let the Markov chain run long enough, the points in the process will have a probability distribution close to π. This is the idea underlying MCMC. The key task is thus to find and implement such a chain. We shall explain how to find a Markov chain whose stationary distribution is π. One approach to verifying that π is the stationary distribution is to show that the chain is reversible.

Definition 15.16 A Markov chain with kernel p is *reversible* with respect to π if

$$\int_{\mathbf{x} \in B} p(\mathbf{x}, A)\, \pi(\mathrm{d}\mathbf{x}) \;=\; \int_{\mathbf{x} \in A} p(\mathbf{x}, B)\, \pi(\mathrm{d}\mathbf{x}) \quad \text{for all } A, B \in \mathcal{A}.$$

Lemma 15.17 *If a Markov chain with kernel p is reversible with respect to a probability distribution π, then π is a stationary distribution for that Markov chain.*

Proof We have to check the equation in (15.5). Fix $A \in \mathcal{A}$. Then

$$\int_{\mathbf{x} \in X} p(\mathbf{x}, A) \, \pi(\mathrm{d}\mathbf{x}) = \int_{\mathbf{x} \in A} p(\mathbf{x}, X) \, \pi(\mathrm{d}\mathbf{x}) = \pi(A),$$

since $p(\mathbf{x}, \cdot)$ is a probability measure, so that $p(\mathbf{x}, X) = 1$ for all $\mathbf{x} \in X$.   □

Lemma 15.17 implies that the *Metropolis–Hastings algorithm* (see Algorithm 6 below) creates a Markov chain with stationary distribution π. We refer to [153, Proposition 2], and [167, Proposition 9.2.2] for the discrete case.

The Metropolis–Hastings algorithm works for a target distribution π that has a density ϕ. The basic idea is to take another Markov chain whose kernel $p(\mathbf{x}, A)$ has a density $q(\mathbf{x}, \mathbf{y})$; i.e., $p(\mathbf{x}, \mathrm{d}\mathbf{y}) = q(\mathbf{x}, \mathbf{y}) \, \mathrm{d}\mathbf{y}$. The density q is called a *proposal density*. Sampling from the proposal density creates a random proposal point $\mathbf{y}$, which is either accepted or rejected depending on how likely it is that the proposal point $\mathbf{y}$ was sampled from π. Notice that in the algorithm we only need to evaluate $\phi(\mathbf{y})$ and $q(\mathbf{x}, \mathbf{y})$ up to scaling.

Algorithm 6: The Metropolis–Hastings algorithm.

Input: A probability measure π on X with density $\phi(\mathbf{y})$. A Markov kernel $p(\mathbf{x}, A)$ on X with density $q(\mathbf{x}, \mathbf{y})$. A fixed starting point $\mathbf{x} \in X$.
Output: A Markov chain on X with stationary distribution π.

1 Set $\mathbf{x}_0 = \mathbf{x}$.
2 **for** $k = 0, 1, 2, \ldots$ **do**
3 $\quad$ Sample $\mathbf{y} \sim p(\mathbf{x}_k, \cdot)$.
4 $\quad$ **if** $\phi(\mathbf{x}_k) = 0$ *or* $q(\mathbf{x}_k, \mathbf{y}) = 0$ **then**
5 $\quad\quad |$ Set $w(\mathbf{x}_k, \mathbf{y}) = 0$
6 $\quad$ **else**
7 $\quad\quad |$ Compute $w(\mathbf{x}_k, \mathbf{y}) = \min \left\{ 1, \dfrac{\phi(\mathbf{y}) \cdot q(\mathbf{y}, \mathbf{x}_k)}{\phi(\mathbf{x}_k) \cdot q(\mathbf{x}_k, \mathbf{y})} \right\}.$
8 $\quad$ **end**
9 $\quad$ Sample a Bernoulli random variable $\beta \in \{0, 1\}$ with $\mathrm{Prob}\{\beta = 1\} = w(\mathbf{x}_k, \mathbf{y})$.
10 $\quad$ **if** $\beta = 1$ **then**
11 $\quad\quad |$ Set $\mathbf{x}_{k+1} := \mathbf{y}$.
12 $\quad$ **else**
13 $\quad\quad |$ Return to line 3.
14 $\quad$ **end**
15 **end**

Example 15.18 (The Symmetric Metropolis Algorithm) For a symmetric proposal density $q(\mathbf{x}, \mathbf{y}) = q(\mathbf{y}, \mathbf{x})$, the Metropolis–Hastings algorithm (Algorithm 6) is called *Symmetric Metropolis Algorithm*. For instance, the Markov kernel from Example 15.9 with density $q(\mathbf{x}, \mathbf{y}) \propto \exp(-\frac{1}{2} \|\mathbf{x} - \mathbf{y}\|^2)$ is symmetric.   ◇

Example 15.19 (Random Walks) We speak of a *Random–Walk Metropolis–Hastings Algorithm* if the proposal density $q(\mathbf{x}, \mathbf{y})$ is a function of the difference $\mathbf{x} - \mathbf{y}$. An example for this is the density from Example 15.9.   ◇

Example 15.20 (Independence Sampler) We call Algorithm 6 an *independence sampler* if $q(\mathbf{x}, \mathbf{y})$ does not depend on $\mathbf{x}$. In this case, samples from Algorithm 6 are independent. ◇

The Metropolis–Hastings algorithm and its variants in the previous examples can be used for any sample space. We now turn to the setting of algebraic geometry, where our objects are algebraic varieties.

Example 15.21 (Sampling points that are close to a variety) This follows the work of Hauenstein and Kahle in [81]. We will see how Algorithm 6 can be used for sampling points near a variety $X \subset \mathbb{R}^n$. To this end, choose a box $R \subset \mathbb{R}^n$ and fix $\sigma^2 > 0$. Writing $d(\,\cdot\,, \,\cdot\,)$ for the Euclidean distance, we consider the probability measure π on R given by

$$\phi(\mathbf{u}) \propto \exp\left(- \tfrac{1}{2\sigma^2} \, d(\mathbf{u}, X)^2 \right).$$

Sampling from π produces points that are likely to be close to X. Let X be the Trott curve in (2.7) and let the box be $R = [-2, 2] \times [-2, 2]$. We implement the Metropolis–Hasting algorithm in Julia [20] where the proposal density is the standard Gaussian in $q(\mathbf{x}, \mathbf{y}) \propto \exp(-\tfrac{1}{2} \|\mathbf{y}\|^2)$. This is an independence sampler (cf. Example 15.20). First, we set up the polynomial system for computing $d(\mathbf{u}, X)$. We solve it for a general complex point $\mathbf{u}$ that we will use later for running parameter homotopies.

```
using HomotopyContinuation, LinearAlgebra
@var x[1:2] u[1:2] l
f = 144(x[1]^4+x[2]^4) - 225(x[1]^2+x[2]^2) + 350x[1]^2*x[2]^2 + 81;
df = differentiate(f, x);
F = System([f; df - l .* (x-u)], variables = [x; l], parameters = u);
uC = randn(ComplexF64, 2);
SC = solve(F, target_parameters = uC)
```

Next, we implement the densities q and ϕ. The variance is chosen to be $\sigma^2 = 1/100$. We start out with a function that is a membership test for the box R.

```
is_in_R(p) = abs(p[1])<2 && abs(p[2])<2;
q(u, v) = exp(-1/2 * norm(v)^2);
sigma_sq = 1/100;
function phi(u)
    S = solve(F, solutions(SC),
                  start_parameters = uC,
                  target_parameters = u,
                  show_progress = false);
    R = map(s -> s[1:2], real_solutions(S));
    d = minimum([norm(r - u) for r in R]);
    exp(-d^2 / (2*sigma_sq));
end
```

Finally, we run the Metropolis–Hastings algorithm for 10.000 steps:

```
n = 10000;
u0 = randn(2);
states = [u0];
for k in 1:n
    uk = last(states);
    v = randn(2);
    if is_in_R(v)
        a = min(1, phi(v) * q(v, uk) / (phi(uk) * q(uk, v)));
        b = rand();
        if a > b; push!(states, v); end
    end
end
```

The result is shown in Figure 15.4.   ◇

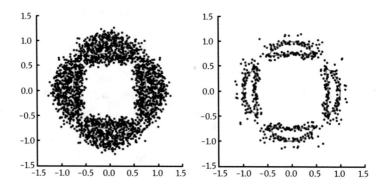

Fig. 15.4: Two samples of points near the Trott curve, generated with the Metropolis–Hastings algorithm, as explained in Example 15.21. The variance parameters are $\sigma^2 = 1/100$ (left) and $\sigma^2 = 1/400$ (right).

One difficulty with Algorithm 6 when sampling from a variety X is to identify a proposal distribution with a density. Take, for instance, the Markov chain from Example 15.10, where the next step $\mathbf{x}_{k+1} \in X$ is computed from the current step $\mathbf{x}_k$ by taking a random linear space L through $\mathbf{x}_k$ and sampling uniformly from the points in $X \cap L$. It is straightforward to describe the generation of this random variable. But it is not clear how to compute its density (or if such a density even exists). In such a scenario, we must prove that the chain is reversible with respect to π. A further complication arises when X is not connected. In that case, the proposals must be chosen in a way that no connected component of X is missed.

We shall present two approaches to Metropolis–Hastings that do not require the knowledge of a proposal density. The first is the algorithm of Breiding and Marigliano [28], which uses an independence sampler. Second, we consider the algorithm of Lelièvre, Stoltz, and Zhang [123], which is based on a random walk. We attach the label CMCMC to both algorithms, since they are based on linear slicings of the variety X.

The CMCMC algorithm in [28] rests on a Markov chain in the affine Grassmannian

$$\mathrm{Gr}_a(c, \mathbb{R}^n) = \{L \subset \mathbb{R}^n \mid L \text{ is an affine linear space of dimension } c\},$$

where $c = \operatorname{codim} X$. Our next theorem describes the density with which we sample a linear space $L \in \operatorname{Gr}_a(c, \mathbb{R}^n)$ (for instance, using Algorithm 6). As before, we assume that the target distribution π has a density $\phi(\mathbf{y})$. For a pair $(A, \mathbf{b}) \in \mathbb{R}^{d \times n} \times \mathbb{R}^d$, we define

$$\bar{\phi}(A, \mathbf{b}) \; := \; \sum_{\mathbf{x} \in X: A\mathbf{x} = \mathbf{b}} \frac{\phi(\mathbf{x})}{\alpha(\mathbf{x})}, \quad \text{where} \quad \alpha(\mathbf{x}) \; := \; \frac{\sqrt{1 + \langle \mathbf{x}, P_{\mathbf{x}}\mathbf{x} \rangle}}{(1 + \|\mathbf{x}\|^2)^{(d+1)/2}} \, \frac{\Gamma\!\left(\frac{d+1}{2}\right)}{\sqrt{\pi}^{\,d+1}}.$$

Here, $d = \dim X$ and $P_{\mathbf{x}}$ is the orthogonal projection onto the normal space $N_{\mathbf{x}}X$. The additional factor α is related to the change of variables when embedding $\mathbb{R}^n$ into the n-dimensional real projective space.

Theorem 15.22 *Let $\varphi(A, \mathbf{b})$ be the probability density for which the entries of the pair $(A, \mathbf{b}) \in \mathbb{R}^{d \times n} \times \mathbb{R}^d$ are i.i.d. standard Gaussian. The following formula defines a probability density on the affine Grassmannian $\operatorname{Gr}_a(c, \mathbb{R}^n)$:*

$$\psi(A, \mathbf{b}) \; := \; \frac{\varphi(A, \mathbf{b}) \cdot \bar{\phi}(A, \mathbf{b})}{\mathbb{E}_{(A,\mathbf{b}) \sim \varphi} \, \bar{\phi}(A, \mathbf{b})}.$$

The random linear space $L := \{\mathbf{x} \in \mathbb{R}^n \mid A\mathbf{x} = \mathbf{b}\} \in \operatorname{Gr}_a(c, \mathbb{R}^n)$ for $(A, b) \sim \psi$ satisfies:

(a) $X \cap L$ is finite with probability one.
(b) If we choose $\mathbf{x}$ in the finite set $X \cap L$ with probability $\phi(\mathbf{x})/(\alpha(\mathbf{x}) \cdot \bar{\phi}(A, \mathbf{b}))$, then the random point $\mathbf{x}$ is distributed according to the target density ϕ on the variety X.

Proof See [28, Theorem 1.1]. □

The algorithm in [123] is similar to the Metropolis–Hastings algorithm. The basic idea is as follows. Suppose we want to sample from a target density $\phi(\mathbf{y}) = \exp(-V(\mathbf{y}))$. Here, $V(\mathbf{y})$ is a smooth function, called the *potential function*. Given a point $\mathbf{x} \in X$, we create a proposal distribution by sampling first in a random tangent direction $\mathbf{v} \in T_{\mathbf{x}}X$ and then computing the intersection of X with the random linear space

$$L \; = \; \mathbf{x} + \mathbf{v} + N_{\mathbf{x}}X \in \operatorname{Gr}_a(c, \mathbb{R}^n).$$

This creates a Markov chain on X. However, it is not clear how to compute the density for this random proposal. The authors of [123] prove directly the reversibility of their Markov chain, so they can apply Lemma 15.17. Algorithm 7 shows a version of their approach.

Remark 15.23 In practice, the random tangent vector $\mathbf{v}$ in line 3 of Algorithm 7 can be sampled as follows. Let $U \in \mathbb{R}^{n \times d}$ be a matrix whose columns form an orthonormal basis of $T_{\mathbf{x}}X$. Such a matrix can be computed with the Gram–Schmidt algorithm. If we sample $\mathbf{u} \in \mathbb{R}^d$ with i.i.d. $N(0, \sigma^2)$-entries, then we can take $\mathbf{v} = U\mathbf{u}$.

The following theorem asserts that Algorithm 7 works correctly. For the proof we refer to [123, Theorem 1]. The second statement follows from Lemma 15.17.

Theorem 15.24 *Algorithm 7 yields a Markov chain on X that is reversible with respect to the desired distribution π. In particular, the Markov chain has π as stationary distribution.*

Algorithm 7: The CMCMC algorithm due to Lelièvre, Stoltz, and Zhang.

Input: A probability measure π on X with a density $\exp(-V(\mathbf{y}))$, where V is a smooth function
on X. A variance parameter $\sigma^2 > 0$. A fixed starting point $\mathbf{x} \in X$.

Output: A Markov chain on X with stationary distribution π.

1 Set $\mathbf{x}_0 = \mathbf{x}$.
2 **for** $k = 0, 1, 2, \ldots$ **do**
3 $\quad$ Draw a random tangent vector $\mathbf{v} \in T_{\mathbf{x}_k} X$ by sampling $\mathbf{v}$ from the multivariate normal
$\quad$ distribution on $T_{\mathbf{x}_k} X$ with mean $\mathbf{0}$ and covariance matrix $\sigma^2 \cdot I$.
4 $\quad$ Set $L = \mathbf{x}_k + \mathbf{v} + N_{\mathbf{x}_k} X$.
5 $\quad$ Sample a point $\mathbf{y} \in X \cap L$ uniformly.
6 $\quad$ Compute $\mathbf{w} \in T_{\mathbf{y}} X$ such that $\mathbf{x}_k \in K := \mathbf{y} + \mathbf{w} + N_{\mathbf{y}} X$.
7 $\quad$ Compute

$$w(\mathbf{x}_k, \mathbf{y}) = \min \left\{ 1, \frac{|X \cap L|}{|X \cap K|} \cdot \exp\left(-(V(\mathbf{y}) - V(\mathbf{x})) \right) \cdot \exp\left(-\frac{1}{2\sigma^2} (\|\mathbf{w}\|^2 - \|\mathbf{v}\|^2) \right) \right\}.$$

8 $\quad$ Sample a Bernoulli random variable $\beta \in \{0, 1\}$ with $\mathrm{Prob}\{\beta = 1\} = w(\mathbf{x}_k, \mathbf{y})$.
9 $\quad$ **if** $\beta = 1$ **then**
10 $\quad\quad$ Set $\mathbf{x}_{k+1} := \mathbf{y}$.
11 $\quad$ **else**
12 $\quad\quad$ Return to line 3.
13 $\quad$ **end**
14 **end**

We have now reached the end of this book. Our presentation featured a wide range of tools and results from computational algebraic geometry that involve metric aspects of varieties. In addressing problems from applications, we are naturally led to cross paths with differential geometry, and – in this final chapter – also with probability theory. We learned that, in order to reliably sample from a variety X, it is essential to have a solid algebraic understanding of tools like ED degree, curvature, reach, medial axes, bottlenecks, etc.

References

1. Hirotachi Abo, Anna Seigal, and Bernd Sturmfels. *Eigenconfigurations of tensors*, pages 1–25. 2017.
2. Daniele Agostini, Taylor Brysiewicz, Claudia Fevola, Lukas Kühne, Bernd Sturmfels, and Simon Telen. Likelihood degenerations. *Adv. Math.*, 414, 2023.
3. Chris Aholt and Luke Oeding. The ideal of the trifocal variety. *Math. of Computation*, 83, 2012.
4. Chris Aholt, Bernd Sturmfels, and Rekha Thomas. A Hilbert scheme in computer vision. *Canadian Journal of Mathematics*, 65, 2011.
5. Yulia Alexandr and Alexander Heaton. Logarithmic Voronoi cells. *Algebr. Stat.*, 12:75–95, 2021.
6. Carlos Améndola, Lukas Gustafsson, Kathlén Kohn, Orlando Marigliano, and Anna Seigal. The maximum likelihood degree of linear spaces of symmetric matrices. *Le Matematiche*, 76(2):535–557, 2021.
7. Carlos Améndola, Kathlén Kohn, Philipp Reichenbach, and Anna Seigal. Invariant theory and scaling algorithms for maximum likelihood estimation. *SIAM Journal on Applied Algebra and Geometry*, 5(2):304–337, 2021.
8. Raman Arora, Amitabh Basu, Poorya Mianjy, and Anirbit Mukherjee. Understanding deep neural networks with rectified linear units. In *International Conference on Learning Representations*, 2018.
9. Sanjeev Arora, Nadav Cohen, and Elad Hazan. On the optimization of deep networks: Implicit acceleration by overparameterization. In *International Conference on Machine Learning*, pages 244–253, 2018.
10. Søren Asmussen and Peter Glynn. *Stochastic Simulation: Algorithms and Analysis*. Springer, 2007.
11. Pierre Baldi and Kurt Hornik. Neural networks and principal component analysis: Learning from examples without local minima. *Neural networks*, 2:53–58, 1989.
12. Jiakang Bao, Yang-Hui He, Edward Hirst, Johannes Hofscheier, Alexander Kasprzyk, and Suvajit Majumder. Hilbert series, machine learning, and applications to physics. *Physics Letters B*, 827, 2022.
13. Alfred Basset. *An elementary treatise on cubic and quartic curves*. Cambridge, Deighton, Bell, 1901.
14. Saugata Basu and Antonio Lerario. Hausdorff approximations and volume of tubes of singular algebraic sets. *Mathematische Annalen*, 387:79–109, 2023.
15. Daniel Bates, Paul Breiding, Tianran Chen, Jonathan Hauenstein, Anton Leykin, and Frank Sottile. Numerical nonlinear algebra. *arXiv:2302.08585*.
16. Adrian Becedas, Kathlén Kohn, and Lorenzo Venturello. Voronoi diagrams of algebraic varieties under polyhedral norms. *Journal of Symbolic Computation*, 120, 2024.
17. Valentina Beorchia, Francesco Galuppi, and Lorenzo Venturello. Eigenschemes of ternary tensors. *SIAM Journal on Applied Algebra and Geometry*, 5(4):620–650, 2021.
18. Edgar Bernal, Jonathan Hauenstein, Dhagash Mehta, Margaret Regan, and Tingting Tang. Machine learning the real discriminant locus. *Journal of Symbolic Computation*, 115:409–426, 2023.
19. David Bernstein, Anatoliy Kušnirenko, and Askold Hovanskiĭ. Newton polyhedra. *Uspehi Mat. Nauk*, 31(3(189)):201–202, 1976.

© The Author(s) 2024
P. Breiding et al., *Metric Algebraic Geometry*, Oberwolfach Seminars 53,
https://doi.org/10.1007/978-3-031-51462-3

20. Jeff Bezanson, Alan Edelman, Stefan Karpinski, and Viral Shah. Julia: A fresh approach to numerical computing. *SIAM Review*, 59:65–98, 2017.

21. Lenore Blum, Felipe Cucker, Michael Shub, and Steve Smale. *Complexity and Real Computation.* Springer, 1998.

22. Tobias Boege, Jane Coons, Chris Eur, Aida Maraj, and Frank Röttger. Reciprocal maximum likelihood degrees of Brownian motion tree models. *Le Matematiche*, 76(2):383–398, 2021.

23. Viktoriia Borovik and Paul Breiding. A short proof for the parameter continuation theorem. *arXiv:2302.14697.*

24. Wieb Bosma, John Cannon, and Catherine Playoust. The Magma algebra system. I. The user language. *J. Symbolic Comput.*, 24(3-4):235–265, 1997.

25. Madeline Brandt and Madeleine Weinstein. Voronoi cells in metric algebraic geometry of plane curves. arXiv:1906.11337.

26. Paul Breiding, Sara Kališnik, Bernd Sturmfels, and Madeleine Weinstein. Learning algebraic varieties from samples. *Rev. Mat. Complut.*, 31(3):545–593, 2018.

27. Paul Breiding, Julia Lindberg, Wern Juin Gabriel Ong, and Linus Sommer. Real circles tangent to 3 conics. *Le Matematiche*, 78:149–175, 2023.

28. Paul Breiding and Orlando Marigliano. Random points on an algebraic manifold. *SIAM Journal on Mathematics of Data Science*, 2(3):683–704, 2020.

29. Paul Breiding, Kristian Ranestad, and Madeleine Weinstein. Critical curvature of algebraic surfaces in three-space. *arXiv:2206.09130.*

30. Paul Breiding, Kemal Rose, and Sascha Timme. Certifying zeros of polynomial systems using interval arithmetic. *ACM Trans. Math. Software*, 49, 2023.

31. Paul Breiding and Sascha Timme. HomotopyContinuation.jl: A package for homotopy continuation in Julia. In *Mathematical Software – ICMS*, pages 458–465. Springer, 2018.

32. Paul Breiding and Nick Vannieuwenhoven. The condition number of Riemannian approximation problems. *SIAM Journal on Optimization*, 31:1049–1077, 2021.

33. Joan Bruna, Kathlén Kohn, and Matthew Trager. Pure and spurious critical points: A geometric study of linear networks. *Internat. Conf. on Learning Representations*, 2020.

34. Peter Bürgisser. Condition of intersecting a projective variety with a varying linear subspace. *SIAM Journal on Applied Algebra and Geometry*, 1:111–125, 2017.

35. Peter Bürgisser and Felipe Cucker. *Condition: The Geometry of Numerical Algorithms*, volume 349 of *Grundlehren der mathematischen Wissenschaften*. Springer, 2013.

36. Peter Bürgisser, Felipe Cucker, and Martin Lotz. The probability that a slightly perturbed numerical analysis problem is difficult. *Math. Comput.*, 77:1559–1583, 2008.

37. Freddy Cachazo, Nick Early, Alfredo Guevara, and Sebastian Mizera. Scattering equations: From projective spaces to tropical Grassmannians. *J. High Energy Phys.*, (6), 2019.

38. Freddy Cachazo, Bruno Umbert, and Yong Zhang. Singular solutions in soft limits. *J. High Energy Phys.*, (5), 2020.

39. Dustin Cartwright and Bernd Sturmfels. The number of eigenvalues of a tensor. *Linear Algebra and its Applications*, 438(2):942–952, 2013.

40. Jean-Dominique Cassini. *De l'Origine et du progrès de l'astronomie et de son usage dans la géographie et dans la navigation.* L'Imprimerie Royale, 1693.

41. Fabrizio Catanese, Serkan Hoşten, Amit Khetan, and Bernd Sturmfels. The maximum likelihood degree. *Amer. J. Math.*, 128(3):671–697, 2006.

42. Eduardo Cattani, Angelica Cueto, Alicia Dickenstein, Sandra Di Rocco, and Bernd Sturmfels. Mixed discriminants. *Mathematische Zeitschrift*, 274, 2011.

43. Türkü Özlüm Çelik, Asgar Jamneshan, Guido Montúfar, Bernd Sturmfels, and Lorenzo Venturello. Optimal transport to a variety. In *Mathematical Aspects of Computer and Information Sciences*, volume 11989 of *Lecture Notes in Comput. Sci.*, pages 364–381. Springer, 2020.

44. Türkü Özlüm Çelik, Asgar Jamneshan, Guido Montúfar, Bernd Sturmfels, and Lorenzo Venturello. Wasserstein distance to independence models. *J. Symbolic Comput.*, 104:855–873, 2021.

45. Michel Chasles. *Aperçu historique sur l'origine et le développement des méthodes en géométrie: particulièrement de celles qui se rapportent à la géométrie moderne, suivi d'un mémoire de géométrie sur deux principes généraux de la science, la dualité et l'homographie.* M. Hayez, 1837.

46. Frédéric Chazal and André Lieutier. The "λ-medial axis". *Graphical Models*, 67(4):304–331, 2005.

47. Heng-Yu Chen, Yang-Hui He, Shailesh Lal, and Suvajit Majumder. Machine learning Lie structures & applications to physics. *Physics Letters B*, 817, 2021.

48. Luca Chiantini, Giorgio Ottaviani, and Nick Vannieuwenhoven. An algorithm for generic and low-rank specific identifiability of complex tensors. *SIAM J. Matrix Anal. Appl.*, 35(4):1265–1287, 2014.

49. Diego Cifuentes, Corey Harris, and Bernd Sturmfels. The geometry of SDP-exactness in quadratic optimization. *Math. Program.*, 182:399–428, 2020.

50. Diego Cifuentes, Kristian Ranestad, Bernd Sturmfels, and Madeleine Weinstein. Voronoi cells of varieties. *J. Symbolic Comput.*, 109:351–366, 2022.

51. Roger Cotes. *Harmonia mensurarum*. Robert Smith, 1722.

52. David Cox, John Little, and Donal O'Shea. *Ideals, Varieties, and Algorithms: An introduction to computational algebraic geometry and commutative algebra*. Undergraduate Texts in Mathematics. Springer, 2015.

53. Vin de Silva and Lek-Heng Lim. Tensor rank and the ill-posedness of the best low-rank approximation problem. *SIAM J. Matrix Anal. Appl.*, 30(3):1084–1127, 2008.

54. Michel Demazure. Sur deux problemes de reconstruction. Technical Report 882, INRIA, 1988.

55. James Demmel. On condition numbers and the distance to the nearest ill-posed problem. *Numer. Math.*, 51(3):251–289, 1987.

56. Sandra Di Rocco, David Eklund, and Oliver Gäfvert. Sampling and homology via bottlenecks. *Math. Comp.*, 91(338):2969–2995, 2022.

57. Sandra Di Rocco, David Eklund, and Madeleine Weinstein. The bottleneck degree of algebraic varieties. *SIAM J. Appl. Algebra Geom.*, 4:227–253, 2020.

58. Manfredo do Carmo. *Riemannian Geometry*. Birkhäuser, 1993.

59. Michael Douglas, Subramanian Lakshminarasimhan, and Yidi Qi. Numerical Calabi-Yau metrics from holomorphic networks. In *Mathematical and Scientific Machine Learning*, pages 223–252, 2022.

60. Jan Draisma, Emil Horobeţ, Giorgio Ottaviani, Bernd Sturmfels, and Rekha Thomas. The Euclidean distance degree of an algebraic variety. *Found. Comput. Math.*, 16:99–149, 2016.

61. Jan Draisma and Jose Rodriguez. Maximum likelihood duality for determinantal varieties. *Int. Math. Res. Not. IMRN*, (20):5648–5666, 2014.

62. Simon Du, Wei Hu, and Jason Lee. Algorithmic regularization in learning deep homogeneous models: Layers are automatically balanced. *Advances in Neural Information Processing Systems*, 31, 2018.

63. Eliana Duarte, Orlando Marigliano, and Bernd Sturmfels. Discrete statistical models with rational maximum likelihood estimator. *Bernoulli*, 27:135–154, 2021.

64. Timothy Duff, Kathlén Kohn, Anton Leykin, and Tomas Pajdla. PLMP – Point-line minimal problems in complete multi-view visibility. In *IEEE/CVF International Conference on Computer Vision*, pages 1675–1684, 2019.

65. Emily Dufresne, Parker Edwards, Heather Harrington, and Jonathan Hauenstein. Sampling real algebraic varieties for topological data analysis. *IEEE International Conference on Machine Learning and Applications*, 2019.

66. Alan Edelman and Eric Kostlan. How many zeros of a random polynomial are real? *Math. Soc. Mathematical Reviews*, 32:1–37, 1995.

67. Chris Eur, Tara Fife, Jose Samper, and Tim Seynnaeve. Reciprocal maximum likelihood degrees of diagonal linear concentration models. *Le Matematiche*, 76(2):447–459, 2021.

68. Shmuel Friedland and Giorgio Ottaviani. The number of singular vector tuples and uniqueness of best rank-one approximation of tensors. *Foundations of Comp. Math.*, 14(6):1209–1242, 2014.

69. William Fulton. *Intersection theory*. Springer, 1998.

70. William Fulton, Steven Kleiman, and Robert MacPherson. About the enumeration of contacts. *Algebraic Geometry – Open Problems*, 997:156–196, 1983.

71. Israel Gel'fand, Mikhail Kapranov, and Andrei Zelevinsky. *Discriminants, Resultants, and Multidimensional Determinants*. Birkhäuser, 1994.

72. Alfred Gray. *Tubes*. Birkhäuser, 2004.

73. Dan Grayson and Michael Stillman. Macaulay2, a software system for research in algebraic geometry. *available at* https://macaulay2.com.
74. Elisenda Grigsby, Kathryn Lindsey, Robert Meyerhoff, and Chenxi Wu. Functional dimension of feedforward ReLU neural networks. *arXiv:2209.04036*.
75. Philipp Grohs and Gitta Kutyniok. *Mathematical Aspects of Deep Learning*. Cambridge University Press, 2022.
76. Richard Hartley. Projective reconstruction and invariants from multiple images. *IEEE Transactions on Pattern Analysis and Machine Intelligence*, 16(10):1036–1041, 1994.
77. Richard Hartley and Fredrik Kahl. Optimal algorithms in multiview geometry. In *Asian Conference on Computer Vision*, pages 13–34. Springer, 2007.
78. Richard Hartley and Frederik Schaffalitzky. Reconstruction from projections using Grassmann tensors. *International Journal of Computer Vision*, 83(3):274–293, 2009.
79. Richard Hartley and Peter Sturm. Triangulation. *Computer Vision and Image Understanding*, 68(2):146–157, 1997.
80. Richard Hartley and Andrew Zisserman. *Multiple View Geometry in Computer Vision*. Cambridge, 2nd edition, 2003.
81. Jonathan Hauenstein and David Kahle. Stochastic exploration of real varieties via variety distributions. *Preprint available at* www.nd.edu/~jhauenst/preprints/khVarietyDistributions.pdf, 2023.
82. Jonathan Hauenstein, Jose Rodriguez, and Bernd Sturmfels. Maximum likelihood for matrices with rank constraints. *J. Algebr. Stat.*, 5:18–38, 2014.
83. Jonathan Hauenstein and Frank Sottile. Algorithm 921: alphaCertified: Certifying solutions to polynomial systems. *ACM Trans. Math. Software*, 38(4), 2012.
84. Kathryn Heal, Avinash Kulkarni, and Emre Can Sertöz. Deep learning Gauss–Manin connections. *Advances in Applied Clifford Algebras*, 32(2):24, 2022.
85. Didier Henrion, Jean Bernard Lasserre, and Johan Löfberg. GloptiPoly 3: moments, optimization and semidefinite programming. *Optimization Methods and Software*, 24:761–779, 2009.
86. Didier Henrion, Jean Bernard Lasserre, and Carlo Savorgnan. Approximate volume and integration for basic semialgebraic sets. *SIAM Rev.*, 51(4):722–743, 2009.
87. Otto Hesse. Die cubische Gleichung, von welcher die Lösung des Problems der Homographie von M. Chasles abhängt. *Journal für die reine und angewandte Mathematik*, 62, 1863.
88. Anders Heyden and Kalle Åström. Algebraic properties of multilinear constraints. *Mathematical Methods in the Applied Sciences*, 20(13):1135–1162, 1997.
89. Nicholas Higham. *Accuracy and stability of numerical algorithms*. SIAM, 1996.
90. Serkan Hoşten, Amit Khetan, and Bernd Sturmfels. Solving the likelihood equations. *Found. Comput. Math.*, 5(4):389–407, 2005.
91. Audun Holme. The geometric and numerical properties of duality in projective algebraic geometry. *Manuscripta mathematica*, 61:145–162, 1988.
92. Emil Horobeţ and Madeleine Weinstein. Offset hypersurfaces and persistent homology of algebraic varieties. *Comput. Aided Geom. Design*, 74, 2019.
93. David Hough. *Explaining and Ameliorating the Ill Condition of Zeros of Polynomials*. PhD thesis, University of California Berkeley, 1977.
94. Ralph Howard. The kinematic formula in Riemannian homogeneous spaces. *Mem. Amer. Math. Soc.*, 106(509), 1993.
95. Petr Hruby, Timothy Duff, Anton Leykin, and Tomas Pajdla. Learning to solve hard minimal problems. In *IEEE/CVF Conference on Computer Vision and Pattern Recognition*, pages 5532–5542, 2022.
96. Kun Huang, Robert Fossum, and Yi Ma. Generalized rank conditions in multiple view geometry with applications to dynamical scenes. In *European Conference on Computer Vision*, pages 201–216. Springer, 2002.
97. Zongyan Huang, Matthew England, David Wilson, James Bridge, James Davenport, and Lawrence Paulson. Using machine learning to improve cylindrical algebraic decomposition. *Mathematics in Computer Science*, 13:461–488, 2019.
98. Birkett Huber and Bernd Sturmfels. A polyhedral method for solving sparse polynomial systems. *Math. Comp.*, 64(212):1541–1555, 1995.

99. June Huh. The maximum likelihood degree of a very affine variety. *Compos. Math.*, 149(8):1245–1266, 2013.

100. June Huh. Varieties with maximum likelihood degree one. *J. Algebr. Stat.*, 5:1–17, 2014.

101. June Huh and Bernd Sturmfels. Likelihood geometry. In *Combinatorial Algebraic Geometry*, volume 2108 of *Lecture Notes in Math.*, pages 63–117. Springer, 2014.

102. Wolfram Research, Inc. Mathematica. Champaign, IL, 2023.

103. Kenji Kawaguchi. Deep learning without poor local minima. *Advances in Neural Information Processing Systems*, 29, 2016.

104. Joe Kileel and Kathlén Kohn. Snapshot of algebraic vision. *arXiv:2210.11443*.

105. Joe Kileel, Matthew Trager, and Joan Bruna. On the expressive power of deep polynomial neural networks. *Advances in Neural Information Processing Systems*, 32, 2019.

106. Steven Kleiman. Tangency and duality. In *Proceedings 1984 Vancouver Conference in Algebraic Geometry*, pages 163–226. Amer. Math. Soc., 1986.

107. Felix Klein. Eine neue Relation zwischen den Singularitäten einer algebraischen Curve. *Math. Ann.*, 10(2):199–209, 1876.

108. Kathlén Kohn, Thomas Merkh, Guido Montúfar, and Matthew Trager. Geometry of linear convolutional networks. *SIAM Journal on Applied Algebra and Geometry*, 6(3):368–406, 2022.

109. Kathlén Kohn, Guido Montúfar, Vahid Shahverdi, and Matthew Trager. Function space and critical points of linear convolutional networks. *arXiv:2304.05752*.

110. Eric Kostlan. On the distribution of roots of random polynomials. In *From Topology to Computation: Proceedings of the Smalefest (Berkeley, CA)*, pages 419–431. Springer, 1993.

111. Eric Kostlan. On the expected number of real roots of a system of random polynomial equations. In *Foundations of Computational Mathematics: Proceedings of the Smalefest (Hong Kong)*, pages 149–188. World Scientific, 2002.

112. Christoph Koutschan. HolonomicFunctions: A Mathematica package for dealing with multivariate holonomic functions, including closure properties, summation, and integration. *available at www3.risc.jku.at/research/combinat/software/ergosum/RISC/HolonomicFunctions.html*.

113. Khazhgali Kozhasov, Alan Muniz, Yang Qi, and Luca Sodomaco. On the minimal algebraic complexity of the rank-one approximation problem for general inner products. *arXiv:2309.15105*.

114. Kaie Kubjas, Olga Kuznetsova, and Luca Sodomaco. Algebraic degree of optimization over a variety with an application to p-norm distance degree. *arXiv:2105.07785*.

115. Pierre Lairez. Computing periods of rational integrals. *Math. Comp.*, 85(300):1719–1752, 2016.

116. Pierre Lairez, Marc Mezzarobba, and Mohab Safey El Din. Computing the volume of compact semi-algebraic sets. In *Proceedings of the International Symposium on Symbolic and Algebraic Computation*, pages 259–266. ACM, 2019.

117. Jean Bernard Lasserre. *Moments, Positive Polynomials and their Applications*. Imperial College Press Optimization Series. Imperial College Press, 2010.

118. Lieven De Lathauwer. Decompositions of a higher-order tensor in block terms - Part II: Definitions and uniqueness. *SIAM J. Matrix Anal. Appl.*, 30:1033–1066, 2008.

119. Thomas Laurent and James Brecht. Deep linear networks with arbitrary loss: All local minima are global. In *International Conference on Machine Learning*, pages 2902–2907, 2018.

120. John Lee. *Riemannian Manifolds: Introduction to Curvature*. Springer, 1997.

121. John Lee. *Introduction to Smooth Manifolds*. Springer, 2013.

122. Kisun Lee. Certifying approximate solutions to polynomial systems on Macaulay2. *ACM Communications in Computer Algebra*, 53(2):45–48, 2019.

123. Tony Lelièvre, Gabriel Stoltz, and Wei Zhang. Multiple projection MCMC algorithms on submanifolds. *IMA Journal of Numerical Analysis*, 2022.

124. Hao Li, Zheng Xu, Gavin Taylor, Christoph Studer, and Tom Goldstein. Visualizing the loss landscape of neural nets. *Advances in Neural Information Processing Systems*, 31, 2018.

125. Martin Lotz. On the volume of tubular neighborhoods of real algebraic varieties. *Proceedings of the American Mathematical Society*, 143, 2012.

126. Laurent Manivel, Mateusz Michalek, Leonid Monin, Tim Seynnaeve, and Martin Vodička. Complete quadrics: Schubert calculus for Gaussian models and semidefinite programming. *Journal of the European Mathematical Society*, 2023.

127. Maplesoft, a division of Waterloo Maple Inc. Maple.
128. Laurentiu Maxim, Jose Rodriguez, and Botong Wang. Euclidean distance degree of the multiview variety. *SIAM Journal on Applied Algebra and Geometry*, 4:28–48, 2020.
129. Clerk Maxwell. On the description of oval curves, and those having a plurality of foci. *Proceedings of the Royal Society of Edinburgh*, 2, 1846.
130. Dhagash Mehta, Tianran Chen, Tingting Tang, and Jonathan Hauenstein. The loss surface of deep linear networks viewed through the algebraic geometry lens. *IEEE Transactions on Pattern Analysis and Machine Intelligence*, 44(9):5664–5680, 2021.
131. Sean Meyn and Richard Tweedie. *Markov Chains and Stochastic Stability*. Springer, 1993.
132. Mateusz Michalek, Leonid Monin, and Jaroslaw Wiśniewski. Maximum likelihood degree, complete quadrics, and $\mathbb{C}^*$-action. *SIAM J. Appl. Algebra Geom.*, 5:60–85, 2021.
133. Mateusz Michalek and Bernd Sturmfels. *Invitation to Nonlinear Algebra*, volume 211 of *Graduate Studies in Mathematics*. American Mathematical Society, 2021.
134. Guido Montúfar, Yue Ren, and Leon Zhang. Sharp bounds for the number of regions of maxout networks and vertices of Minkowski sums. *SIAM Journal on Applied Algebra and Geometry*, 6(4):618–649, 2022.
135. Alexander Morgan and Andrew Sommese. Coefficient-parameter polynomial continuation. *Applied Mathematics and Computation*, 29(2):123–160, 1989.
136. Gabin Nguegnang, Holger Rauhut, and Ulrich Terstiege. Convergence of gradient descent for learning linear neural networks. *arXiv:2108.02040*.
137. Jiawang Nie, Pablo Parrilo, and Bernd Sturmfels. Semidefinite representation of the k-ellipse. In *Algorithms in Algebraic Geometry*, volume 146 of *I.M.A. Volumes in Mathematics and its Applications*, pages 117–132. Springer, 2008.
138. Partha Niyogi, Steve Smale, and Shmuel Weinberger. Finding the homology of submanifolds with high confidence from random samples. *Discrete Comput. Geometry*, (39):419–441, 2008.
139. Luke Oeding. The quadrifocal variety. *Linear Algebra and its Applications*, 512:306–330, 2017.
140. Barrett O'Neill. *Elementary Differential Geometry*. Elsevier, 2001.
141. Giorgio Ottaviani and Luca Sodomaco. The distance function from a real algebraic variety. *Comput. Aided Geom. Design*, 82, 2020.
142. Giorgio Ottaviani, Pierre-Jean Spaenlehauer, and Bernd Sturmfels. Exact solutions in structured low-rank approximation. *SIAM J. Matrix Anal. Appl.*, 35(4):1521–1542, 2014.
143. Pablo A. Parrilo and Rekha R. Thomas. *Sum of Squares: Theory and Applications*. Proceedings of Symposia in Applied Mathematics. American Mathematical Society, 2020.
144. Dylan Peifer, Michael Stillman, and Daniel Halpern-Leistner. Learning selection strategies in Buchberger's algorithm. In *International Conference on Machine Learning*, pages 7575–7585, 2020.
145. Ragni Piene. Polar classes of singular varieties. In *Annales scientifiques de l'École Normale Supérieure*, volume 11, pages 247–276, 1978.
146. Ragni Piene, Cordian Riener, and Boris Shapiro. Return of the plane evolute. *Annales de l'Institut Fourier*, 2023.
147. Julius Plücker. *Gesammelte wissenschaftliche Abhandlungen*, volume 2. BG Teubner, 1895.
148. Sebastian Pokutta and Elias Wirth. Conditional gradients for the approximately vanishing ideal. In *International Conference on Artificial Intelligence and Statistics*, 2002.
149. Yang Qi, Pierre Comon, and Lek-Heng Lim. Semialgebraic geometry of nonnegative tensor rank. *SIAM J. Matrix Anal. Appl.*, 37:1556–1580, 2016.
150. Adityanarayanan Radhakrishnan, Mikhail Belkin, and Caroline Uhler. Overparameterized neural networks implement associative memory. *Proceedings of the National Academy of Sciences*, 117(44):27162–27170, 2020.
151. Christophe Raffalli. Distance to the discriminant. *arXiv:1404.7253*.
152. John Rice. A theory of condition. *SIAM J. Numer. Anal*, 3:287–310, 1966.
153. Gareth Roberts and Jeffrey Rosenthal. General state space Markov chains and MCMC algorithms. *Probability Surveys*, pages 20–71, 2004.
154. Jose Rodriguez and Botong Wang. The maximum likelihood degree of mixtures of independence models. *SIAM J. Appl. Algebra Geom.*, 1:484–506, 2017.
155. Mutsumi Saito, Bernd Sturmfels, and Nobuki Takayama. *Gröbner Deformations of Hypergeometric Differential Equations*, volume 6 of *Algorithms and Computation in Mathematics*. Springer, 2000.

156. George Salmon. *A Treatise on the Higher Plane Curves*. Hodges and Smith, 1852.
157. George Salmon. *A Treatise on the Analytic Geometry of Three Dimensions*. Hodges, Smith, and Company, 1865.
158. Muhamad Saputra, Andrew Markham, and Niki Trigoni. Visual SLAM and structure from motion in dynamic environments: A survey. *ACM Computing Surveys (CSUR)*, 51(2):1–36, 2018.
159. Anna-Laura Sattelberger and Bernd Sturmfels. D-modules and holonomic functions. *to appear in: Varieties, Polyhedra and Computation, EMS Series of Congress Reports*, arXiv:1910.01395.
160. Luca Sodomaco. *The distance function from the variety of partially symmetric rank-one tensors*. PhD thesis, Universitá degli Studi di Firenze, 2020.
161. Andrew Sommese and Charles Wampler. *The Numerical Solution of Systems of Polynomials Arising in Engineering and Science*. World Scientific, 2005.
162. Henrik Stewénius, Frederik Schaffalitzky, and David Nistér. How hard is 3-view triangulation really? In *IEEE International Conference on Computer Vision*, pages 686–693, 2005.
163. Rud Sturm. Das Problem der Projectivität und seine Anwendung auf die Flächen zweiten Grades. *Mathematische Annalen*, 1(4):533–574, 1869.
164. Bernd Sturmfels. *Solving Systems of Polynomial Equations*. Number 97 in CBMS Regional Conferences Series. American Mathematical Society, 2002.
165. Bernd Sturmfels and Simon Telen. Likelihood equations and scattering amplitudes. *Algebr. Stat.*, 12(2):167–186, 2021.
166. Bernd Sturmfels, Sascha Timme, and Piotr Zwiernik. Estimating linear covariance models with numerical nonlinear algebra. *Algebr. Stat.*, 11:31–52, 2020.
167. Seth Sullivant. *Algebraic Statistics*, volume 194 of *Graduate Studies in Mathematics*. American Mathematical Society, 2018.
168. Gyula Sz.-Nagy. Tschirnhaus'sche Eiflächen und Eikurven. *Acta Mathematica Hungarica*, pages 36–45, 1950.
169. Matteo Tacchi, Jean Bernard Lasserre, and Didier Henrion. Stokes, Gibbs, and volume computation of semi-algebraic sets. *Discrete Comput. Geom.*, 69:260–283, 2023.
170. Matteo Tacchi, Tillmann Weisser, Jean Bernard Lasserre, and Didier Henrion. Exploiting sparsity for semi-algebraic set volume computation. *Found. Comput. Math.*, 22:161–209, 2022.
171. The Sage Developers. *SageMath, the Sage Mathematics Software System*. https://www.sagemath.org.
172. Gerald Toomer. *Apollonius: Conics Books V to VII: The Arabic Translation of the Lost Greek Original in the Version of the Banū Mūsā*. Springer, 1990.
173. Lloyd Trefethen and David Bau. *Numerical Linear Algebra*. SIAM, 1997.
174. Alan Turing. Rounding-off errors in matrix processes. *The Quarterly Journal of Mechanics and Applied Mathematics*, 1948.
175. Hermann Weyl. On the volume of tubes. *Amer. J. Math.*, 61(2):461–472, 1939.
176. James Wilkinson. Note on matrices with a very ill-conditioned eigenproblem. *Numer. Math.*, 19:176–178, 1972.
177. Francis Williams, Matthew Trager, Daniele Panozzo, Claudio Silva, Denis Zorin, and Joan Bruna. Gradient dynamics of shallow univariate ReLU networks. *Advances in Neural Information Processing Systems*, 32, 2019.
178. Lior Wolf and Amnon Shashua. On projection matrices $\mathbb{P}^k \to \mathbb{P}^2$, $k = 3, \ldots, 6$, and their applications in computer vision. *International Journal of Computer Vision*, 48:53–67, 2002.
179. Doron Zeilberger. A holonomic systems approach to special functions identities. *J. Comput. Appl. Math.*, 32(3):321–368, 1990.
180. Liwen Zhang, Gregory Naitzat, and Lek-Heng Lim. Tropical geometry of deep neural networks. In *International Conference on Machine Learning*, pages 5824–5832, 2018.
181. Qunjie Zhou, Torsten Sattler, Marc Pollefeys, and Laura Leal-Taixe. To learn or not to learn: Visual localization from essential matrices. In *IEEE International Conference on Robotics and Automation*, pages 3319–3326, 2020.
182. Günter Ziegler. *Lectures on Polytopes*. Graduate Texts in Math. Springer New York, 1995.